W9-AOB-579

ADVANCES IN URBAN PEST MANAGEMENT

Advances in Urban Pest Management

Edited by

Gary W. Bennett
Purdue University
West Lafayette, Indiana

John M. Owens
S. C. Johnson and Son, Inc.
Racine, Wisconsin

VNR VAN NOSTRAND REINHOLD COMPANY
New York

Library of Congress Catalog Card Number: 85-20192
ISBN: 0-442-20960-6

Manufactured in the United States of America.

Published by Van Nostrand Reinhold Company Inc.
115 Fifth Avenue
New York, New York 10003

Van Nostrand Reinhold Company Limited
Molly Millars Lane
Wokingham, Berkshire RG11 2PY, England

Van Nostrand Reinhold
480 Latrobe Street
Melbourne, Victoria 3000, Australia

Macmillan of Canada
Division of Gage Publishing Limited
164 Commander Boulevard
Agincourt, Ontario MIS 3C7, Canada

15 14 13 12 11 10 9 8 7 6 5 4 3 2 1

Library of Congress Cataloging-in-Publication Data
Main entry under title:
Advances in urban pest management.
Includes index.
1. Urban pests—Integrated control. I. Bennett, Gary W. II. Owens, John M.
SB950.8.A38 1986 628.9′6 85-20192
ISBN 0-442-20960-6

Contents

Preface

The control of pests in the urban environment has historically meant using a chemical pesticide to reduce pest numbers once an outbreak has occurred. This reliance upon pesticides has fueled major concerns and controversies. Thus, there has been a major shift, philosophically and practically, to the concept of pest management—the balanced use of cultural, biological, and chemical measures that are environmentally compatible and economically feasible to reduce pest populations to tolerable levels.

The war between man and pests in his environment never ends, but man has made many advances in the strategies and practices for managing pests. This text is designed as a book that will present the latest technical developments and the most recent thoughts and principles of urban pest management, and take a look into the future from this base. As editors, we have asked chapter authors to formulate a healthy, rigorous, and up-to-date coverage of their subject matter so that the reading audience will have adequate information on which to build a sound, practical foundation for implementing the concepts of pest management for the urban environment.

The book is not intended to answer all questions on urban pest management, but will direct researchers, teachers, practitioners and other pest management professionals into appropriate areas of the scientific literature. Pest control operators, public health officials, nurserymen, grounds maintenance personnel, sanitarians, garden shop personnel, city foresters, golf course superintendents, park superintendents, and other professionals will be afforded a good reference point for the incorporation of alternate strategies into their control programs. Even individuals who live and work in urban environments should be able to use this volume to gain access to readable, up-to-date information on how to handle their pest problems.

The editors express sincere appreciation to the following persons for their contributions and advice: Raymond Beal, U.S. Forest Service; Philip Colbaugh,

Texas A & M University; Robert Corrigan, Purdue University; Lester Ehler, University of California, Davis; Philip Harein, University of Minnesota; Robert Hollingworth, Purdue University; Noel Jackson, University of Rhode Island; William Jackson, Bowling Green State University; George Kennedy, North Carolina State University; M. Keith Kennedy, S. C. Johnson and Son, Inc.; John Neal, U.S.D.A.; John Osmun, Purdue University; William Robinson, Virginia Polytechnic Institute; Erik Runstrom, Purdue University; James Sargent, Great Lakes Chemical Corp.; Brian Schneider, Purdue University; Mark Shour, Purdue University; Robert Smith, University of Arizona; Walter Stevenson, University of Wisconsin; Robert Timm, University of Nebraska; Jeffrey Tucker, Consulting Entomologist, Houston; Vern Walter, Abash Pest Control Co.; and James Yonker, Purdue University.

GARY W. BENNETT
JOHN M. OWENS

Contributors

Judy K. Bertholf
Dow Chemical Co., 13355 Noll Road, Suite 1025, LB 18, Dallas, Texas 75240

David N. Byrne
Department of Entomology, University of Arizona, Tucson, Arizona 85721

Edwin H. Carpenter
Department of Agricultural Economics, University of Arizona, Tucson, Arizona 85721

Robert Davis
U.S. Department of Agriculture-Agricultural Research Service, Stored Products Insects Laboratory, P.O. Box 22909, Savannah, Georgia 31403

Robert V. Flanders
Department of Entomology, Purdue University, West Lafayette, Indiana 47907

Gordon W. Frankie
Department of Entomological Sciences, University of California, Berkeley, California 94720

Jack B. Fraser
Department of Entomological Sciences, University of California, Berkeley, California 94720

J. Kenneth Grace
Department of Entomological Sciences, University of California, Berkeley, California 94720

James I. Grieshop
Department of Applied Behavioral Sciences, University of California, Davis, California 95616

D. James Madder
Culice, Inc., 380 Woolwich Street, Guelph, Ontario, Canada N1H 3W7

Boyd T. Marsh
Environmental Health, Summit County General Health District, 1100 Graham Court, Cuyahoga Falls, Ohio 44224

Rex E. Marsh
Department of Wildlife and Fisheries Biology, University of California, Davis, California 95616

Freeman L. McEwen
Department of Environmental Biology, University of Guelph, Guelph, Ontario, Canada 1NG 2W1

Harry B. Moore
Department of Entomology, North Carolina State University, Raleigh, North Carolina 27650

John M. Owens
S. C. Johnson & Son, Inc., 1525 Howe Street, Racine, Wisconsin 53403

Daniel A. Potter
Department of Entomology, University of Kentucky, Lexington, Kentucky 40546

Michael K. Rust
Department of Entomology, University of California, Riverside, California 92521

Michael J. Sinsko
Indiana State Board of Health, 1330 West Michigan Street, Indianapolis, Indiana 46206

Keith O. Story
Winchester Consultants, 6 Black Horse Terrace, Winchester, Massachusetts 01890

F. Eugene Wood
Department of Entomology, University of Maryland, College Park, Maryland 20742

Alan C. York
Department of Entomology, Purdue University, West Lafayette, Indiana 47907

ADVANCES IN URBAN PEST MANAGEMENT

1

Urban Pest Management: Concept and Context

John M. Owens

S. C. Johnson & Son, Inc.
Racine, Wisconsin

DEFINING URBAN PEST MANAGEMENT

In recent history the control of pests in the urban environment has meant using a chemical to reduce pest numbers once an outbreak has occurred. Pesticides have resulted in many remarkable successes in controlling a wide variety of pests worldwide. However, this reliance on pesticides has fueled major concerns and controversies. Technical and scientific people in certain disciplines have experienced increasing problems with pest resistance to chemicals, while public and regulatory pressures have often sought to restrict pesticide use largely because of environmental concerns. As a consequence, there has been impetus for a major shift, philosophically and practically, to the concept of pest management. To date, pest management programs are much more highly evolved and widely implemented in agriculture than in the urban setting (Bottrell, 1979; Frankie and Koehler, 1978, 1983).

The pest management concept has principally evolved over the past 20-25 years, and authors have proposed various definitions for this process (Bottrell, 1979; Flint and van den Bosch, 1981; Metcalf and Luckman, 1975; Olkowski et al., 1978). These definitions sometimes reflect the different scientific disciplines of their authors or their apparent motives for promoting alternatives to pesticide use. We define pest management as follows: a process consisting of the balanced use of cultural, biological, and chemical procedures that are environmentally compatible and economically feasible to reduce pest populations to tolerable levels. This definition follows that of Rabb (1972) for integrated pest management (IPM) programs in general.

As such, pest management differs from traditional pest control (relying solely on pesticide use) in three fundamental ways. First, pest management is a process and not just a reaction (e.g., spraying an insecticide) to pest presence. This process typically consists of at least five steps:

1. *Sampling* and monitoring of pest population levels.
2. Proper *identification* of the pest(s) present and understanding of their basic biology and behavior.
3. *Comparing pest population levels* to a predetermined level of pest presence deemed acceptable (the concept of a threshold is used in decision making).
4. *Application of pest management procedures,* among which pesticides are frequently considered important tools, as appropriate. The decision to take no action is occasionally made, despite the presence of some pests.
5. Follow-through after application of pest management procedures by continued sampling (*monitoring*) of pest population levels and reapplying or revising any pest management actions as appropriate.

Again, pest management is an ongoing process and is more complex than that typically surrounding pesticide use. As part of this process, the second major difference between pest management and traditional pest control with pesticides is the use of threshold levels of pest presence against which those levels experienced in any particular instance can be compared when deciding what action to take. In agriculture these thresholds are frequently called economic or damage thresholds (which are not equivalent), as they constitute those pest population levels at which subsequent crop damage and yield loss will exceed the cost of applying appropriate pest management procedures to prevent that damage. By contrast, in urban pest management situations these thresholds (where they have been studied and exist at all) are generally related to aesthetic rather than economic considerations. Also, where public health concerns or individual sensitivities dictate, the acceptable level of pest presence may be zero, so that no decision making threshold exists and simple detection forces action to eliminate pest presence.

Another important difference between management and control of pests is that pest management stresses integration of two or more pest management procedures to reduce or limit pest population levels, compared to unilateral use of pesticides. This is significant because pesticide use alone generally allows only temporary reductions in pest population levels and, because of a phenomenon termed "pest resurgence," may occasionally result in increased problems after a limited period of relief. Where research or other experience has identified other nonchemical procedures to replace or augment pesticide use, a more satisfying and long-lasting result is often achieved. This concept of using pest management procedures to augment each other and act in concert with those natural environmental factors that also serve to regulate pest population numbers was an important early contribution of ecological theory (e.g., Clark et al., 1967) to the subsequent evolution of integrated pest management. Much applied research and development was then necessary to demonstrate how these procedures (chemical and nonchemical) can be appropriately applied within the context of the pests' biology and ecological relationships within the environment of the specific situations where the IPM programs are to be implemented.

A fourth difference between pest management and pest control is highlighted by the term "integrated." This term is actually dropped by most authors discussing urban pest management. Sawyer and Casagrande (1983) stress that true integrated pest management (IPM) involves a multidisciplinary approach where all types of pests are routinely considered in the pest management process. Further, this pest management program should be appropriately designed and implemented within (i.e., integrated into) the overall management system for the urban ecosystem. Thus, as a special form of IPM, urban pest management can involve integration of activities at three or more levels in an ecologically and economically acceptable framework (Sawyer and Casagrande, 1983).

THE IMPORTANCE OF MANAGING URBAN PESTS

The term "pest" is applied, based on subjective opinion or objective analysis, to organisms considered damaging to human interests. Control or management of organisms so described is therefore important. In this book we have included discussion of pest management from all the traditional subdisciplines dealing with pests that occur in and around homes, commercial or industrial operations, and towns or cities as constituting urban pest management. Thus, the close proximity and frequent contact of each of us with urban pests in our daily lives reinforces the importance we should attach to this subject.

Many urban pests have substantial medical and public health importance, thereby warranting aggressive pest management action (Busvine, 1980; Gorham, 1979; National Academy of Sciences, 1976). The reader is referred to any of the many available reference texts on medical and veterinary entomology for detailed discussion of the various medically important anthropods (e.g., Harwood and James, 1979). Domiciliary cockroaches are generally considered to be medically important as carriers of various pathogenic bacteria and protozoa, many of which cause food poisoning or other forms of gastroenteritis (Cornwell, 1968). In recent years medical researchers have come to recognize the importance of cockroaches (especially their molted cuticles) as a source of allergens in homes (Wirtz, 1984).

Mosquitoes and biting flies are especially important as disease vectors. Again,there are many reference books that review the historical and present-day medical importance of mosquitoes (Clark et al., 1967; Faust, Beaver, and Jung, 1975; Gillett, 1971). Mosquitoes are important vectors of encephalitis and dog heartworm (*Dirofilaria immitis*) in the United States. The worldwide disease list includes malaria, yellow fever, dengue, and filiariasis (Neilsen, 1979). It is interesting that the medical importance and nuisance aspects (especially in recreational areas) of mosquitoes has spawned the development and implementation of extensive abatement programs in many municipalities (Newson, 1977). Many of these programs are examples of the most sophisticated, true urban pest management currently practiced. Similar to the cockroaches, nonbiting flies (the house fly, *Musca domestica*, and other species of filth flies) are well-known carriers of disease organisms (Greenberg, 1973; Oldroyd, 1964). Recent advances in fly pest management for the poultry, swine, and beef cattle production industries may some day yield procedures or technologies useful toward managing flies in urban areas. An example, albeit for a rather specialized urban situation, is the very successful fly pest management program established for Mackinac Island (Kennedy and Merritt, 1980).

Fleas, especially the increasingly common and important cat flea (*Ctenocephalides felis*), are intermediate hosts for helminth parasites of dogs and cats. Most important among these parasites is the double-pored dog tapeworm (*Dipylidium caninum*), which infests dogs, cats, young children, and some animals. Important tick-borne diseases in the United States are Rocky Mountain spotted fever, which is now widespread east of the Rocky Mountains, and Lyme disease in certain areas of the northeast.

Ranking along with some of the more advanced mosquito abatement programs as examples of true pest management in urban areas are some vertebrate pest management programs. It has frequently been observed that proper vertebrate pest control actually constituted the practice of IPM long before this concept was articulated. Like mosquitoes, vertebrate pests such as rats, mice, starlings, pigeons, sparrows, and bats have significant public health importance (Weber, 1979, 1982). Raccoons and skunks are becoming increasingly important in the urban and suburban neighborhoods of many areas in the United States because of their role as rabies carriers. These vertebrates and others also require management effort around many food industry facilities because of the economic damage they can cause in these operations. There is a wide variety of nonchemical management procedures available for vertebrate pest management, and many of these programs integrate two or more of these procedures for a better and more lasting effect than could be expected from use of chemicals alone.

Home vegetable gardening is one of the ten top activities for homeowners in terms of time spent in this activity (The Gallup Organization, 1981). High populations of plant pests in vegetable gardens can reduce the quantity and quality of the produce yield. Because low populations may do either little or no damage (compared to the cost of applying pest management procedures) or may be held in check by environmental conditions and biological control agents, the concept of economic thresholds is directly applicable to these situations. Similarly, pests of shade trees, ornamental plants, and turf in landscapes can do economically significant damage to valuable landscape plants. However, low populations can often be tolerated.

The objective of urban pest management is not to replace or eliminate pesticide use. However, it seems appropriate to gauge the potential importance of urban pest management as a process that includes pesticide use and offers promise of more satisfying results than have been experienced by past pest control practices by assessing the economics of pesticide use in urban areas (broadly defined). In 1979 nonfarm (approximately, urban) pesticide usage in the United States was estimated at 309.5 million lb of active ingredient (National Research Council, 1980; U.S. Environmental Protection Agency, 1979). This amount represented about 27% of all pesticides used. Other surveys have shown that the intensity of pesticide use (amount applied per unit land area) is greater in urban areas than in most agricultural situations (Von Rumker et al., 1972). These figures amply demonstrate the importance of urban pest problems and consequently justify our interest in urban pest management as a philosophical and practical framework to deal with these problems effectively and efficiently.

THE URBAN ECOSYSTEM

Since urban pest management relies heavily on the consideration of pest situations in the broader context of their ecosystems and stresses use of pest management procedures that augment available natural or environmental pest population regulatory factors, it is worthwhile to compare ecosystems where various types of pest management programs are implemented. Diversity and stability are two important related characteristics of ecosystems that influence the extent to which we expect difficult pest problems to arise. One of the distinguishing features of any pest organism is the ability of its populations to reach high levels or densities (numbers of individuals per unit area) in the absence of sufficient limiting forces. Such pest population fluctuations are sometimes termed outbreaks. Ecosystems where population densities of the various organisms fluctuate relatively little over time are considered to be relatively stable. Classic examples of stable (and productive) ecosystems are tropical rain forests, tidal and freshwater marshes, and coral reef communities. Year in and year out, relative population levels of the different organisms in these ecosystems remain fairly constant. An important contributing factor to this stability is the large number of different species present in these ecosystems and the complex nature of their interaction. Ecosystems that contain relatively high numbers of species, which are often further categorized into complex heirarchies of producers (typically plants) and consumers (typically animals), are termed diverse.

Modern agricultural ecosystems have distinctly and artificially low diversity. An agricultural field or even much larger areas in many regions have cropping systems consisting of one or a very few crop plants (plus what few weeds escape cultivation or herbicide treatment). A consequence of this alteration of the more ecologically diverse, pre-existing native forest or prairie ecosystem is that the overall stability of the agro-ecosystem is low. Certain consumer species, now labeled pests because of human interest in the crop plants, then typically become dominant as they consume the amply available resources (crop plants). Typically, modern agricultural demands substantial energy input to maintain productivity in the face of competition from pests.

Sawyer and Casagrande (1983) state that an urban ecosystem, in its entirety, is so large and complex that no functional enterprise could claim to consider so broad a subject. In attempting to provide a conceptual framework for urban pest management, they avoided consideration and use of the term "urban ecosystem." They prefer to use the term "urban pest ecosystem" to describe all those portions of the urban ecosystem that are of immediate importance to the ecology of these pests. Important components of the urban pest ecosystem are the pest populations, natural regulatory factors that serve to limit pest population size, and the managed resources exploited by the pests (see Fig. 1 in Sawyer and Casagrande, 1983).

Urban pest ecosystems are largely shaped by man's activities. Our structures are artificially contrived spaces designed for use or habitation primarily by one species plus a selection of pets and houseplants. However, these designs are imperfect, as pest species of cockroaches, ants, rodents, wood decay fungi, and other organisms are notably successful at exploiting the resources our structures and lifestyles offer. These few pest species frequently develop high population levels in such simplified ecosystems. Because people are mostly unwilling to cohabit with other organisms besides select types of pets and interior plants, household pest management programs that seek to incorporate biological control organisms (to impose a slightly greater, managed level of ecological diversity) would seem to have limited potential. Rather, certain researchers have studied the effects of removing pest resources such as food, water, and harborage (as pest management procedures) from German cockroaches infesting urban apartments (Bertholf, Owens, and Bennett, in prep., Farmer and Robinson, 1984; Owens, 1980). To date, these pest management procedures have not been particularly effective when used unilaterally. There is a substantial need for architects, design engineers, and sanitarians to understand the weaknesses of buildings and the facilities or equipment they contain with regard to encouraging pest problems.

Our landscapes, parks, vacant lots, and other areas in the urban ecosystem support rather diverse plant and animal (especially insect) communities (Frankie and Ehler, 1978). Studies of the plant flora of some U.S. and European cities have shown them to contain hundreds of species of landscape and garden plants, representing dozens of plant families. Similarly, studies of vacant lots and home lawns have found a rich diversity of insect and other animal fauna. This level of diversity often results in a fair degree of stability against pest outbreaks in our landscapes. Exceptions occur with nonnative or introduced pests, which usually have few if any natural enemies. We also see increased incidence of pest problems in areas (e.g., turf) that receive intense horticultural management in the forms of excessive fertilizer applications and irrigation. Pesticide applications can also disrupt pest population regulatory factors resulting in secondary pest outbreaks. An example is the effect mosquito adulticiding programs can have on natural enemies of certain shade tree pests such as scale insects (Merritt, Kennedy, and Gersabeck, 1983). Compared to the distinctly uniform nature of agricultural ecosystems over wide areas, the species diversity of urban areas is closely related to environmental heterogenity. That is, human activities have altered existing habitats or established new ones in a very patchy manner. Therefore, pest management programs that incorporate biological control agents often require continued monitoring and augmentative releases. Again, there is a need for architects and urban planners to understand the urban pest management issues involved when landscapes, parks, and other urban areas are planned (Roberts and Dill, 1983).

The important presence and influence of humans as part of the fabric of this ecosystem and not just as managers of it is a distinguishing characteristic of urban pest ecosystem. We shape this ecosystem by providing the pests' resources (host plants, food, water, harborage, temperature, humidity) and directly or indirectly affecting the natural regulatory factors (biological control agents) that may act on pest populations. Our attitudes and expectations about the quality of our environment, our style of life, pests, and pest control or management are important areas of consideration lying outside the realm of applied biology or ecology, in the social sciences. Several authors have stressed the importance of social science research to study this human factor in urban pest management and to develop appropriate and effective educational or other programs to change attitudes and practices that contribute to pest problems (Frankie and Levenson, 1978; Levenson and Frankie, 1983; National Research Council, 1980; Sawyer and Casagrande, 1983). The knowledge and attitudes of urbanites regarding pests and pesticides is reviewed in chapter 2 of this book; chapter 8 discusses the needs for education of homeowners and professionals in order to advance the application of urban pest management technology.

PRESENT STATUS OF URBAN PEST MANAGEMENT

As stated earlier, pest management systems are more highly evolved for many agricultural situations than is currently the case for most urban pest situations. Extensive agricultural research has led to sophisticated sampling procedures, identification of economic thresholds for pest presence, computerized models for predicting pest population trends based on various environmental variables, and availability of a variety of pest management procedures. These procedures include pest resistant crop varieties, biological control agents, specialized cultivation techniques, and improved pesticide technology. Such advances have not diminished pesticide use rates for most cropping systems. In the case of fruit and vegetable production (especially for fresh consumption), this pesticide use is largely dictated by the very stringent aesthetic requirements consumers place on the produce they purchase. In most cases even superficial pest damage is unacceptable to consumers and results in severe economic loss to the grower. This forces reliance on multiple pesticide applications to prevent pest presence or build-up in these high cash value crops. Education, leading to even relatively minor attitude changes on the part of consumers and others in the food industry, could allow substantial changes in the viability of true IPM in much of agriculture. Similarly, most homeowners have very low tolerance for pest presence in their homes, yards, or gardens. Thus there is substantial reliance on pesticide use for pest control in urban areas. Consequently, the future prospects for urban pest management seem limited unless educational efforts can change attitudes and practices relative to urban pests. (See also chaps. 2 and 8 of this book.)

True pest management systems, in the sense of the definition in this volume, are either distinctly lacking or infrequently implemented for most urban pest situations. While the development and implementation of urban pest management faces many of the same obstacles as agricultural pest management, it is also severely impeded by an inadequate research base for most situations. In some cases where true pest management programs have been developed and implemented, often under special or limited circumstances and on a limited scale, these programs have not been widely adopted (Davidson, Hellman, and Holmes, 1981; Kennedy and Merritt, 1980; Olkowski et al., 1978). Pest management programs, in urban situations or elsewhere, can be expected to become widely adopted only when they offer superior overall results compared to conventional pest control or where pesticides become ineffective (e.g., because of overriding environmental concerns). Again, widespread adoption of urban pest management will be facilitated by intensive and effective educational efforts for homeowners and professional practitioners. Advances in pesticide technology that have allowed more target-specific and less environmentally disruptive pesticide applications do not constitute true pest management. Neither does the use of pheromone traps, biological or phenological models, and other techniques aimed merely at more precise and appropriate timing of pesticide applications, even though these are important advances.

A notable exception to this pattern of underdevelopment for urban pest management programs is the advanced and complex mosquito abatement programs carried out in many areas to manage these nuisance and public health pests. Successful pilot programs in turf and ornamentals pest management, where some level of pest presence can be accepted, suggest that greater future progress toward true pest management can be expected in these areas. Similarly, practical and reliable pest management programs may evolve relatively rapidly for home vegetable gardening, augmented by technology transfer from agricultural research.

OBJECTIVES

This book is not intended to be a complete, definitive review of the scientific literature relating to urban pest management in all of the subdisciplines we have identified (chap. 9-15). Rather, it is intended to serve as a basis for understanding the essential components and concepts of urban pest management and an entry point into the literature for interested practitioners and researchers. While it is recognized that true pest management programs have not been developed and widely implemented for most urban situations, we have sought to present urban pest management as a viable, practical concept and to review the most recent advances in the development of its technology. In this way, practitioners can begin to view urban pest problems and their subsequent resolution from an ecosystem management perspective. As researchers contribute the understanding and technology to advance development of urban pest management in the future, practitioners will have an important base upon which to implement urban pest management and build their understanding of its complexities. It is hoped that researchers will recognize the deficiencies in the research base needed for development and implementation of urban pest management programs in their subdisciplines and will in turn be stimulated to focus their efforts toward resolving these needs.

REFERENCES

Bertholf, J. K., J. M. Owens, and G. W. Bennett (in prep.). Influence of sanitation on field populations of German cockroaches. (In preparation for *Environmental Entomology.*)

Bottrell, D. R., 1979. *Integrated Pest Management.* Council on Environmental Quality, Washington, D.C., 120p.

Busvine, J. R. 1980. *Insects and Hygiene,* 3rd ed. Chapman and Hall, New York, 568p.

Clark, L. R., P. W. Geier, R. D. Hughes, and R. F. Morris. 1967. *The Ecology of Insect Populations in Theory and Practice.* Methuen, London. 232p.

Cornwell, P. B. 1968. *The Cockroach,* vol. 1. Hutchinson, London. 400p.

Davidson, J., J. L. Hellman, and J. Holmes. 1981. Urban ornamentals and turf IPM. In *Proceedings of Integrated Pest Management Workshop*, G. L. Worf (ed.). National Cooperative Extension Service, Dallas, Texas, pp. 68-72.

Farmer, B. R., and W. H. Robinson. 1984. Harborage limitation as a component of a German cockroach pest management program. *Proc. Entomol. Soc. Wash.* **86**(2):269-273.

Faust, E. C., P. C. Beaver, and R. C. Jung. 1975. *Animal Agents and Vectors of Human Disease,* 4th ed. Lea & Febinger, Philadelphia, 479p.

Flint, M. L., and R. van den Bosch. 1981. *Introduction to Integrated Pest Management.* Plenum, New York, 240p.

Frankie, G. W., and L. E. Ehler. 1978. Ecology of insects in urban environments. *Annu. Rev. Entomol.* **23:**367-387.

Frankie, G. W., and C. S. Koehler, eds. 1978. *Perspectives in Urban Entomology.* Academic Press, New York, 416p.

Frankie, G. W., and C. S. Koehler, eds. 1983. *Urban Entomology: Interdisciplinary Perspectives.* Praeger, New York, 496p.

Frankie, G. W., and H. Levenson. 1978. Insect problems and insecticide use: Public opinion, information, and behavior. In *Perspectives in Urban Entomology,* G. W. Frankie and C. S. Koehler (eds.). Academic Press, New York, pp. 359-399.

The Gallup Organization. 1981. *1981-1982 National Gardening Survey.* Gardens for All (National Association for Gardening), Burlington, Vermont, 181p.

Gillett, J. D. 1971. *Mosquitos.* Weidenfeld and Nicholson, London, 272p.

Gorham, J. R. 1979. The significance for human health of insects in food. *Annu. Rev. Entomol.* **24:**209-224.

Greenberg, B. 1973. *Flies and Disease,* vol. 2. Princeton University Press, Princeton, N. J., 448p.

Harwood, R. F., and M. T. James. 1979. *Entomology in Human and Animal Health,* 7th ed. Macmillan, New York, 548p.
Kennedy, M. K., and R. W. Merritt. 1980. Horse and buggy island. *Nat. Hist.* **89:**34-41.
Levenson, H., and G. W. Frankie. 1983. A study of homeowner attitudes and practices toward arthropod pests and pesticides in three U.S. metropolitan areas. In *Urban Entomology: Interdisciplinary Perspectives,* G. W. Frankie and C. S. Koehler (eds.). Praeger, New York, pp. 67-106.
Merritt, R. W., M. K. Kennedy, and E. F. Gersabeck. 1983. Integrated pest management of nuisance and biting flies in a Michigan resort: Dealing with secondary pest outbreaks. In *Urban Entomology: Interdisciplinary Perspectives,* G. W. Frankie and C. S. Koehler (eds.). Praeger, New York, pp. 277-299.
Metcalf, R. L., and W. H. Luckman. 1975. *Introduction to Insect Pest Management.* Wiley, New York. 592p.
National Academy of Sciences. 1976. *Pest Control: An assessment of Present and Alternative Technologies,* vol. 5, *Pest Control and Public Health.* National Academy Press, Washington, D.C. 282p.
National Research Council. 1980. *Urban Pest Management.* Report prepared by the Committee on Urban Pest Management, Environmental Studies Board, Commission on Natural Resources. National Academy Press, Washington, D.C., 275p.
Neilsen, L. T. 1979. Mosquitoes, the mighty killers. *Natl. Geogr.* **156**(3):426-440.
Newson, H. D. 1977. Arthropod problems in recreation areas. *Annu. Rev. Entomol.* **22:**333-353.
Oldroyd, H. 1964. *The National History of Flies.* Norton, New York, 324p.
Olkowski, W., H. Olkowski, R. van den Bosch, and R. Hom. 1976. Ecosystem management: A framework for urban pest control. *Bioscience* **26:**384-389.
Olkowski, W., H. Olkowski, A. I. Kaplan, and R. van den Bosch. 1978. The potential for biological control in urban areas: Shade tree insect pests. In *Perspectives in Urban Entomology* G. W. Frankie and C. S. Koehler (eds.). Academic Press, New York, pp. 311-347.
Owens, J. M. 1980. Some aspects of German cockroach population ecology in urban apartments. Ph.D. diss., Purdue University, 116p.
Rabb, R. L. 1972. Principles and concepts of pest management. In *Implementing Practical Pest Management Strategies,* proceedings of a U.S.D.A. Cooperative Extension Service workshop. Purdue University, West Lafayette, Indiana, pp. 6-29.
Roberts, F. C., and C. H. Dill. 1983. Urban planning and insect pest management. In *Urban Entomology: Interdisciplinary Perspectives,* G. W. Frankie and C. S. Koehler (eds.). Praeger, New York, pp. 41-66.
Sawyer, A. J., and R. A. Casagrande. 1983. Urban pest management: A conceptual framework. *Urban Ecol.* **7:**145-147.
U.S. Environmental Protection Agency. 1979. *Pesticide Industry Sales and Usage: 1979 Market Estimates.* Office of Pesticide Programs, Economic Analysis Branch, Washington, D.C.
Von Rumker, R., R. M. Matter, D. P. Clement, and F. K. Erickson. 1972. *The Use of Pesticides in Suburban Homes and Gardens and Their Impact on the Aquatic*

Environment, Pesticide Studies Series 2. Office of Water Programs, Washington, D.C.

Weber, W. J. 1979. *Health Hazards from Pigeons, Starlings and English Sparrows.* Thomson, Fresno, Calif., 138p.

Weber, W. J. 1982. *Diseases Transmitted by Rats and Mice.* Thomson, Fresno, Calif., 182p.

Wirtz, R. A. 1984. Allergic and toxic reactions to non-stinging arthropods. *Annu. Rev. Entomol.* **29:**47-69.

2

Attitudes and Actions of Urbanites in Managing Household Arthropods

David N. Byrne

University of Arizona, Tucson

Edwin H. Carpenter

University of Arizona, Tucson

As human populations have become more and more concentrated, people have increasingly provided food and shelter for various arthropods, thus establishing a close, though not often cordial, relationship. The need to control these animals which share a common habitat with humans, has long existed.

Early in this association, we can assume that people, out of necessity, were more willing to accept the presence of arthropod populations than today and shared a different perception of the amount of control needed. Apparently, as the technology was developed to reduce encounters with these animals, the tolerable population levels were also lowered.

The purpose of this chapter is to provide a brief description of the urban environment and current household arthropod management practices, particularly as they relate to the use of pesticides. The need for change in these practices is also discussed, along with the ways in which present antipathy toward arthropods and humanity's elevated expectations for an arthropod-free environment may impact on the potential for such change.

Studies concerning urban ecology have been comprehensively reviewed by Frankie and Ehler (1978). These studies are not as prevalent in the literature as research concerning reports of other areas, such as agricultural ecological systems. Most of the principles developed in examining agricultural arthropods are also applicable in urban situations. The major factor making the urban system unique is, of course, the presence of large human populations. People have widely expressed need to manipulate their environment, far beyond the levels necessary to satisfy their creature comforts, and are willing to expend a great deal of time, effort, and money to do so. The satisfaction of biological needs cannot explain the presence of grass lawns in the arid southwestern United States or exotic fish in New York apartment buildings. These exist as expressions of people's desire to improve their quality of life, which they will protect at great cost.

Referring to earlier work by Boyden and Miller (1978), Sawyer and Casagrande (1982/1983) state that each person's individual response to arthropods is conditioned in part by cultural and educational background, the information available, and personal sensitivities, preferences, and values. Since Western society's intolerance of urban arthropods seems so entrenched (Byrne et al., 1984), we must believe that people see the presence of these animals as impinging upon their perceived quality of life.

CURRENT CONTROL PRACTICES

What steps are people currently taking to minimize their association with arthropods and what are the consequences? Surveys of control practices commonly employed against urban arthropods indicate that both householders and professionals largely depend on pesticides (Bennett, Runstrom, and Wieland, 1983; Byrne and Carpenter, 1983). Although efforts are being made to develop and implement programs that minimize the use of chemicals through encouraging alternative control methods and educating the urban public to tolerate some levels of arthropods, we expect that the use of pesticides will continue to be the first and in most cases the only method of control because of efficacy and low cost.

Levenson and Frankie (1981) see indications that the public is more cautious regarding the use of pesticides. Members of a National Academy of Sciences (NAS) committee expressed the sentiment that one reason for a reduced enthusiasm for chemical pest control stemmed from a concern about the intrinsic wisdom of the purposeful pollution of the environment (National Research Council, 1980). Although there is some evidence of general public concern over the use of pesticides, we have not found that in the minds of the public the risks associated with the use of pesticides in urban settings exceed the benefits. There is little indication that the average homeowner weighs the risks against the benefits of pesticide use when confronted with a pest problem. Consequently, we see it unlikely that urban pest management programs (those not largely dependent on pesticides) for household arthropods will quickly supplant conventional pest control practices based on pesticides.

We believe there is a greater likelihood that urban pest management programs will be implemented by public agencies, such as programs aimed at controlling arthropods that are a threat to public health (although these are not the subject of this chapter). In those situations, the individuals who select the appropriate pest control strategies are most often professionals who have greater access to information and thus are aware of alternatives to chemicals. More important, they are more inclined to deal with arthropods objectively and to have more reasonable expectations relative to control levels.

This chapter is concerned with the need to minimize the use of pesticides in urban ecosystems by householders and the impact of the attitudes of urban dwellers on attaining that objective.

RISK IN THE HOME

While pesticides are undeniably useful, it must be acknowledged that certain substances put the public at risk in terms of short-term and potential long-term health effects (Whorton et al., 1977). We are only now beginning to address this situation. The public is accustomed to taking discretionary risks, so we would not propose that our society has the right to expect a risk-free environment. The risks from pesticides, however, are not calculable as exact probabilities. This uncertainty should lead the public to minimize their exposure to pesticides and to search for alternatives to chemical control.

It appears, however, that the uncertainty surrounding the subject has led the public into a situation in which the untoward effects of pesticides are not taken into serious consideration. In a 1983 survey in which Levenson and Frankie asked about awareness of possible harm from these chemicals, only 65 of 372 people who used pesticides responded positively (Levenson and Frankie, 1983).

It may be that the public's apparent lack of concern is a logical response to mixed signals. Authorities from the scientific community can only report the results of objective research, and defining risks from pesticides requires some speculation. When scientists take stands on opposing sides of an issue, the public has difficulty deciding whom to believe. An example of this confusion is the case for Agent Orange, the U.S. military code name for a 50/50% mixture of the herbicides 2, 4, 5-T and 2, 4-D, used during the Vietnam war from 1965-1970 to defoliate jungle vegetation and protect U.S. troops from ambush. These defoliants contained traces of dioxins (notably 2, 3, 7, 8-tetrachlorodibenzo-para-dioxin) that have been identified in laboratory studies as potential carcinogens (Dow Chemical, 1984). Yet, in spite of recent out-of-court settlements, epidemiological studies of Vietnam veterans exposed to these materials do not so far indicate that these veterans are more at risk than other veterans who were not exposed (Fox, 1983).

More recently, the Environmental Protection Agency (EPA) took action against ethylene dibromide (Federal Register, 1983), commonly used as a fumigant in soil or in warehouses to control nematodes and insects. The EPA took steps to reduce exposure to this material because research has indicated that it was mutagenic and carcinogenic, as well as being capable of causing reproductive disorders in laboratory animals. The result was that we found ourselves in a situation in which prominent toxicologists like B. N. Ames (1984) were publicly dismissing the risks posed by ethylene dibromide, as federal and state regulators were simultaneously purging grocery store shelves of bakery products containing this identified carcinogen.

In another case familiar to the pest control industry, the signals are particularly unclear. The EPA declared in 1976 (Federal Register, 1976) that, while the termiticide chlordane may be the causal agent for the carcinoma found in mouse liver, the evidence that this substance was a human carcinogen was insufficient.

Velsicol Chemical Corporation, the pesticide's principal U.S. manufacturer, in supporting the safety of chlordane, referred to a National Academy of Sciences (NAS) study that found no similar cancerous response in rats and concluded that the mouse strain used in the study examined by the EPA was particularly susceptible to cancer and that, even then, carcinomas occurred only at the highest dosage rate of chlordane (Gibson, 1983). On April 10, 1983, however, officials of the U. S. Occupational Safety and Health Administration appeared on the Columbia Broadcasting System's program "Sixty Minutes" and stated that there was enough evidence to assume that chlordane may cause cancer in human populations. What is the reasoning public to conclude from such conflicting reports? If the health risk is one of the most important influences for the public to minimize their use of pesticides, as stated by NAS (1980), a good case can be made for the fact that this risk is poorly defined.

The situation is not likely to acquire focus in the near future. Apparently, the conflicting reports are expressions of the fact that chemicals have varied behavior and effects in the environment. Our ability to measure and predict these phenomena, while growing, is still fragmented and uneven. This uncertainty cannot be avoided when toxicologists must draw conclusions about human populations based on information gathered from animals of a different species in a laboratory or must extrapolate to events far outside the range of the observed data.

In this milieu there are indications that some segments of the pest control industry have used the uncertainty as an excuse to ignore, or at least minimize, the potential risk associated with pesticides and to further encourage their use. Industry trade journals, which spend much time explaining the benefits of pesticides, allot precious little space to examining the hazards objectively. Articles appear advising pest control operators on how to deal with the "ecofreaks," characterized as having an exaggerated fear of cancer risks, and dismissing reports of Love Canal as fear-producing chemical stories (Ware, 1982). Armed with facts such as that table salt is more toxic than the herbicide picloram or that, with the exception of lung cancer, the death rate from cancer has decreased over the last 30 years (Rapson, 1983), industry members appear to conclude that the risks from pesticides exist at extremely low levels and pass this message along to the public.

The public, for their part, seem more willing than ever to use pesticides, and we feel that their attitudes and behavior are unlikely to change in the immediate future. The assumption that the public is ready to accept urban integrated pest management programs being developed by entomologists and other scientists needs to be examined more carefully. If the available data are examined objectively, the signals indicating incipient change simply are not there. In fact, the public behavior and attitudes indicate just the opposite. The public has almost no affinity for most arthropods and therefore pest control or eradication and not urban pest management, which allows some low level of pest population, will likely dominate in the foreseeable future.

PUBLIC BEHAVIOR

Recently reports in a series of information publications have detailed the behavior of homeowners in this country relative to their use of pesticides. Information collected by von Rumker et al. (1972) from 525 respondents indicated that 92.5% had used pesticides. Frankie and Levenson (1978) found that most people interviewed in Texas cities reported using pesticides personally, although they did conclude that the frequency of personal use did vary from year to year over the course of their three-year study. Savage, Keefe, and Wheeler (1979) report in a nationwide survey that 90.7% of their 8,254 respondents had used pesticides in one form or another. According to their data, 83.7% of the householders applied pesticides in their homes themselves, 21.4% to their garden, and 38.7% in the yard. The study also indicated that the percentage of respondents using pesticides varies in different parts of the country, ranging from 97.1% of the households in the north central region to 83.3% in the southeastern region. In a recent study, Bennett, Runstrom and Wieland (1983) found that 78% of 952 interviewees had used pesticides; the figure did not include those employing professional pest control operators. They went on to express the opinion that there was a dramatic lack of knowledge relative to the selection and application of pesticides. Very few homeowners had sought help in diagnosing their pest problems or in selecting appropriate pest control.

We (Byrne and Carpenter, 1983) found in a telephone survey that 87.1% of the householders questioned used pesticides. Finally, Levenson and Frankie (1983) found that, while the majority of their interviewees used pesticides, only 17% knew of any associated externalities; and many of these were concerned about odor only.

We think it reasonable to conclude from these surveys that current urban pest control programs for household arthropods are largely, even excessively, dependent on the use of pesticides and that decisions to apply pesticides are made with little regard for potential untoward effects.

PUBLIC ATTITUDES

Why, in the face of evidence that the use of pesticides carries some intrinsic danger, is the public so willing to continue using them? As more has been learned about the untoward effects associated with pesticides, why do we at least not see a trend of diminishing use since 1972 instead of continued high levels? As alluded to, the lack of concern about pesticides is greater than widely realized and the abhorrence of arthropods is more deeply ingrained in the public psyche and acted upon more vehemently than past surveyors of public attitudes toward arthropods have recognized.

Attempts have been made to characterize public attitude toward urban arthropods and pesticides, but these reports seen at odds with the wide-scale use of pesticides reported above. We feel this is because public attitudes have not been properly characterized.

Von Rumker et al. (1972) stated that the overwhelming majority of the public was keenly aware of the problems associated with the use of pesticides and expressed a belief that they would welcome information on alternative methods for control. Levenson and Frankie (1981) thought it seemed clear that homeowners are able to tolerate certain levels of insects and suggested that urban pest management programs could be more acceptable if we aimed for a level of control commensurate with those tolerated insect populations. In their survey Levenson and Frankie (1983), as part of a lengthy questionnaire, asked respondents if there were any insects they liked; when approximately half said yes, they used that as a basis for statements such as "Cognitively, those who felt positively toward insects were more aware of harm from chemicals. . . ." Others, like Bennett, Runstrom, and Wieland (1983), in another important survey aimed primarily at determining behavior relative to pesticide use, asked only whether or not people were aware of beneficial insects.

From the results of a recent survey (Byrne et al., 1984), we draw different conclusions from those who suggest we are in a state of transition to a situation in which householders are willing to accept a greater level of interaction with arthropods—this assumption is of linchpin importance to any urban pest management program (Olkowski, 1974). After making a large effort to characterize urban attitudes, we find that householders are almost uniformly intolerant of most arthropods.

In this survey, when asked specifically to describe how they felt about arthropods found inside their home, less than 11% (Table 2-1) said they were even willing to tolerate these animals, and less than 1% described their feelings in terms of enjoyment. The remainder described their feelings as dislike (83.5%) or fear (4.6%). Although not as drastically, people also express a distaste for insects found outdoors since only 6% said they enjoyed the presence of those arthropods.

Conclusions about potential behavior drawn from answers to these types of direct questions shown in Table 2-1, rather than nonspecific questions such as "Are there any insects you like?" (Levenson and Frankie, 1983), are more likely to be productive. We are concerned that past characterizations of public attitude toward arthropods have been incorrect because their questions did not probe deeply enough. The public generally does not like arthropods to the extent that they will not allow even minimal numbers in their homes.

Table 2-1. Results of a cross-tabulation of pesticide use by response to the questions concerning attitudes toward insects found in yards and homes ($n = 1{,}133$).

Question	*Percent of 1,133 respondents* [a]	*Pesticides are employed* [b]	
		Yes	*No*
Which of the following statements best describes how you feel about insects you might find inside your home?[c]			
I am afraid of them	4.4	94.0	6.0
I dislike them	80.2	90.1	9.9
I tolerate them	10.8	72.1	27.9
I enjoy their presence	0.7	75.0	25.0
Which of the following statements best describes how you feel about insects you might find in your yard?[c]			
I am afraid of them	4.1	84.8	15.2
I dislike them	34.1	94.0	6.0
I tolerate them	47.0	84.8	15.2
I enjoy their presence	5.5	74.2	25.8

[a] Percentages do not total to 100 because of missing data on one or more of the variables employed in these cross-tabulations.

[b] Pesticides are employed if the respondent applied them, used Raid® or a similar product, or employed a professional exterminator.

[c] Significant X^2 at $p < .05$.

It should be stated, however, that we did find the public appreciates some arthropods. Byrne et al. (1984) report that when homeowners are allowed the option of expressing affinity for different arthropods, they did rate certain of them relatively high when compared to a deer. For example, scorpions are held in low regard while butterflies are at the top of the list. A homeowner may have an appreciation for the aesthetics of a butterfly's appearance or be aware of beneficial insects (Bennett, Runstrom, and Wieland, 1983), but these factors may have little influence on the decision to use pesticides against ants or cockroaches.

Furthermore, when the attitudes of people about arthropods outside or inside their homes is cross-tabulated with pesticide use (Table 2-1), it can be seen that about nine out of ten people who dislike or fear arthropods also employ pesticides. Seven or eight out of ten people who describe their feelings toward arthropods in terms of toleration or enjoyment also employ pesticides. Stated differently, while feelings toward arthropods definitely do impact upon pesticide use, the degree of difference does not provide much hope that the more affinity people have for arthropods the less likely they are to employ pesticides.

APPLICATION OF URBAN INTEGRATED PEST MANAGEMENT

In addition to our concern over the ability or willingness of the householder to change pest control strategy, we need to address another area. Although there is an apparent need for change, are alternative programs that do not rely so heavily on pesticides readily available?

Most scientists make claims that techniques developed, for the most part in agriculture, and termed "integrated pest management" (IPM) have applicability to the urban sector (National Academy of Sciences, 1980). By way of definition, this simply involves the intelligent selection and use of pest control options in an ecologically, economically, and sociologically sound manner (Sawyer and Casagrande, 1983).

Complaints have been expressed in the past that urban pest problems have only been examined on an ad hoc basis without a workable definition or consensus on what constitutes urban pest management. Other people have stated that, by following the lead of agriculture and adopting a more holistic global view, we could avoid chemical dependency. Our contention is that the two systems exhibit fundamental differences that prevent wholesale changes in the way household pests are managed.

Luckmann and Metcalf (1982) tell us an agricultural community cannot be expected to move from a historical dependency on preventive pest control practices straight into as complex an endeavor as IPM and that for a period of 5-25 years professional educational extension philosophies must be applied, new tools readied, and economic and sociological benefits tallied. In agriculture, at least, existing attitudes and mechanisms would provide reasonable expectations of success. This, however, is not the case with household pest control.

At the core of the problem are the practitioners. Charges are made repeatedly that the key to altering pest management programs, in either the urban setting or in agriculture, is through educational efforts (Koehler, 1983). The sign of the professional is the ability to receive, interpret, and translate into appropriate action information on pest control. For the most part, in agriculture the individuals responsible for selecting pest control strategies deal with those matters on a daily basis and are aware of the need constantly to evaluate the impact and efficacy of control regimen. While this may occasionally be the case in urban settings, more probably the person involved is at best a paraprofessional and more often a layman with no experience at all. In a survey conducted in Arizona (Boyden and Miller, 1978), an assessment was made of information available to householders before they applied a pesticide. Of the interviewees, 88.2% said they knew what they were trying to control before using the pesticides. A much smaller portion of the sample professed to know what these arthropods used as a food source (44.3%), and fewer still (16.3%) told us they knew its life cycle in advance of using pesticides. Not knowing an insect's food source or life cycle makes it difficult to determine accurately the potential costs resulting from its presence. From this, one can conclude that the use of insecticides by householders is in most cases a simple response to the presence of these animals and not the result of a deliberate consideration of the expected benefits. In the absence of these deliberations, the likelihood for adopting urban pest management programs is seemingly reduced.

Luckmann and Metcalf (1982) also say that pest management concepts are not new to applied agricultural entomology, and it can be reasonably assumed that this familiarity would facilitate change. In urban entomology the same tradition does not appear to exist. Relatively few researchers are active in this area, and there is a paucity of extension workers enthusiastically involved in extending information to householders on alternatives to pesticides. In agriculture, the U.S. Department of Agricultural Research Service and university agriculture experiment stations and cooperative extension services are in place to generate new programs and disseminate information on implementation, but no mechanisms of the same magnitude exist in the urban sector. Even if they did, it seems apparent the public is making little effort to look for and obtain information which might convince them of the need to change.

Some evidence also exists that the urban public is overlooking the universities as a principal source of information on the biology of the urban arthropods with which they are trying to deal. Levenson and Frankie (1983) found that only 14% of their respondents discussed their pest problems with university personnel. Byrne and Carpenter (1983) found that only 6.8% did so. Part of the problem is that universities are geared primarily to producing information on arthropods associated with agriculture and have been slow to accept the challenge to conduct research on urban pests; thus the base of information needed to develop urban integrated pest management programs does not exist. In a directory for the Entomological Society of America, Hall and Odland (1979) asked members to list their areas of specialty. Even though each member was allowed to list several areas, urban/industrial entomology was listed by only 150 members, whereas the other specialties were named over 5,000 times.

Another important factor is that the motivation to adopt urban pest management practices does not exist at the same level as in agriculture. The claim has been made that the development of resistance to insecticides strikes at the fabric of pest control. In our view few pest control operators and even fewer householders are aware of this resistance phenomenon. In fact, such a small percentage understand insecticide resistance that the idea plays an insignificant role in the selection of pest control strategy.

Another hurdle for urban IPM is the lack of a clearly defined action level. In agriculture, growers are commonly felt to be receptive to new pest control strategies because reducing the number of applications is economically in their best interests. Little such motivation exists in the urban sector, and what does exist is far outrun by the desire to protect the quality of environment from animals generally abhorred.

These attitudes toward urban arthropods are too firmly entrenched, and if our survey is correct, they are too strongly negative to indicate much hope for immediate change. Attitudes described as dislike are adhered to with more clarity and firmness than those described as indifference, suggesting that attempts to convert public attitudes will undoubtedly encounter such strong resistance that the widespread adoption of urban pest management programs seems a remote possibility. This seems especially true in light of the fact that 75% of the people who say they enjoy insects also use pesticides.

In the absence of an event such as an environmental calamity that might heighten public awareness of the untoward effects associated with pesticides, change is unlikely. As scientists, we have a responsibility to continue our efforts to develop urban pest management schemes, many of which will undoubtedly be successful; overall, however, we will probably have to be content with small victories, since we will not be able to effect a wholesale reduction in the use of pesticides.

REFERENCES

Ames, B. N. 1984. Peanut butter, parsley, pepper and other carcinogens. *Wall Street Journal,* Feb. 14, p. 30

Bennett, G. W., E. S. Runstrom, and J. A. Wieland, 1983. Pesticide use in homes. *Bull. Entomol. Soc. Am.* **29**(1):31-38.

Boyden, S., and S. Miller. 1978. Human ecology and the quality of life. *Urban Ecol.* **3**:263-287.

Byrne, D. N., and E. H. Carpenter. 1983. Behavior of metropolitan and nonmetropolitan residents relative to urban pest control strategies. *Southwestern Entomol.* **8**(3):198-204.

Byrne, D. N., E. H. Carpenter, E. M. Thoms, and S. T. Cotty. 1984. Public attitudes toward urban arthropods. *Bull. Entomol. Soc. Am.* **30**(2):40-44.

Dow Chemical. 1984. *Dioxin, Agent Orange, and Human Health.* Dow Chemical, Midland, Mich., 71p.

Federal Register. 1976. Consolidated heptachlor/chlordane hearings. (February 19) **41**:7552-7558.

Federal Register. 1983. Occupation exposure to ethylene dibromide. (October 7) **48**:45956-46003.

Fox, J. L. 1983. Dioxin's health effects remain puzzling. *Science* **221**:1161-1162.

Frankie, G. W., and L. E. Ehler. 1978. Ecology of insects in urban environments. *Annu. Rev. Entomol.* **23**:367-387.

Frankie, G. W., and H. Levenson. 1978. Insect problems and insecticide use: Public opinion, information, and behavior. In *Perspectives in Urban Entomology,* G. W. Frankie and C. S. Koehler (eds.). Academic Press, New York, pp. 359-399.

Gibson, S. 1983. Is chlordane safe? *Pest Control Technol.* **11**(5):82, 84, 85, 91.

Hall, W. P., and G. C. Odland. 1979. *Directory of North American Entomologists and Acarologists.* Entomological Society of America, Hyattsville, Md., 189p.

Koehler, C. S. 1983. Information transfer in urban pest management. In *Urban Entomology: Interdisciplinary Perspectives,* G. W. Frankie and C. S. Koehler (eds.). Praeger, New York, pp. 151-160.

Levenson, H., and G. W. Frankie. 1981. Pest control in the urban environment. In *Progress in Resource Management and Environmental Planning,* vol. 3, T. O. Riordan and R. K. Turner (eds.). Wiley, New York, pp. 251-272.

Levenson, H., and G. W. Frankie. 1983. A study of homeowner attitudes and practices toward arthropod pests and pesticides in three U.S. metropolitan areas. In *Urban Entomology: Interdisciplinary Perspectives,* G. W. Frankie and C. S. Koehler (eds.). Praeger, New York, pp. 67-106.

Luckmann, W. H., and R. L. Metcalf. 1982. The pest-management concept. In *Introduction to Insect Pest Management,* 2nd ed., R. L. Metcalf and W. H. Luckmann (eds.). Wiley, New York, pp. 1-32.

National Research Council. 1980. *Urban Pest Management.* Report prepared by the Committee on Urban Pest Management, Environmental Studies Board, Commission on Natural Resources. National Academy Press, Washington, D.C., 275p.

Olkowski, W. 1974. A model ecosystem management program. *Proc. Tall Timbers Conf. Ecol., Anim. Control Hab. Manage.* **5**:103-117.

Rapson, W. H. 1983. Benefits and risks of chemicals in the environment. *Pest Manage.* **2**(9):13-16, 18, 20, 22.

Savage, E. P., T. J. Keefe, and H. W. Wheeler. 1979. *National Household Pesticide Usage Study, 1976-1977.* U. S. Environmental Protection Agency, Washington, D.C., 126p.

Sawyer, A. J., and R. A. Casagrande. Urban pest management: A conceptual framework. *Urban Ecol.* **7**:145-157.

Von Rumker, R., R. M. Matter, D. P. Clement, and F. K. Erickson. 1972. *The Use of Pesticides in Suburban Homes and Gardens and Their Impact on the Aquatic Environment.* Pesticide Studies Series 2. Office of Water Programs, U.S. Environmental Protection Agency, Washington, D.C.

Ware, G. W. 1982. Coping with the ecofreak: Let reason prevail. *Pest Manage.* **1**(7):18-22.

Whorton, D., R. M. Krauss, S. Marshall, and T. H. Milby. 1977. Infertility in male pesticide workers. *Lancet*, pp. 1259-1261.

3

Environmental Impact of Pesticide Use

Freeman L. McEwen and D. James Madder
Ontario Agricultural College, Guelph

By its nature, the urban environment has many distinctive features. It is people-dense, space-limited, and those who live there go to great lengths to beautify their properties with ornamental plantings, many of which are subject to pest attack. The close proximity of buildings and the active movement of people makes it possible for many insect and rodent pests to gain access to buildings and become generally distributed within the community. Household pests are often brought home with groceries and go unnoticed until they have become established in the kitchen. Municipal garbage collection and disposal systems provide ideal habitats for a wide variety of insects and rodents, which may radiate widely from dumps and related facilities. Modern building design with multi-story service chases provide easy access for insects and rodents to move freely form one apartment to another and create unique problems for pest control. In many of our communities, wood destroying insects such as termites and carpenter ants become well established and are transported readily when infested timbers are moved from one location to another. Pigeons and other birds may deface buildings or contaminate foods, and mosquitoes and other outdoor biting insects are frequent problems as nuisance pests or disease vectors.

The urbanite may be either fastidious or casual (Byrne et al., 1984), but in most communities urban dwellers take pride in their homes and wish to keep them free of pests that may cause structural problems or reduce aesthetic qualities. The maintenance of attractive lawns and shrubbery is often a high priority. These urbanites try to protect themselves, their homes, and surroundings from pests within their immediate environment. Whether it be cockroaches or dermistids in the pantry, clothes moths or carpet beetles in their fabrics, termites or carpenter ants in timbers, mildew in the basement, cluster flies in the attic, mosquitoes in the back yard, black spot on their roses, leaf miners in their birch trees, or dandelions in the lawn, homeowners want to remove these problems and rely heavily on pesticides to accomplish this task (Bennett, Runstrom, and Weiland, 1983). A survey in the mid-1970s in the United States indicated that 90% of householders used pesticides in the home, garden, or lawn. More than 500 different formulations of pesticides were found in these homes (U.S. Environmental Protection Agency, 1979). Thus pest control in the urban environment involves the use of a wide spectrum of pesticides, and while a significant part of this is done by professional pest control operators, much is done by homeowners. Many homeowners have little knowledge of the nature of the pesticides they use (U.S. Environmental Protection Agency, 1979). It is not surprising, therefore, that neighbors are often disturbed when they see the person next door using pesticides rather indiscriminately, under conditions that permit some of the spray to drift into the neighboring yard. Surveys of homeowners have shown that less than 50% read the pesticide label, and many ignore common safety procedures, for example, storage of pesticides in uncontrolled situations (Bennett, Runstrom, and Weiland, 1983). This chapter will attempt to address some of the concerns about pesticides in the urban environment and to put in perspective the significance of the problem.

WHY PESTICIDES ARE USED

Quantification of losses due to pests in and around the home is difficult. Some surveys done by pest control companies indicate that it is very extensive, with damage from termites alone estimated in the United States to be $1.17 billion annually (Granovsky, 1983). The importance of urban pest control is also indicated by commercial activity in this area. In the United States, an estimated 15,000 pest control companies with 50,000 technicians used $130-200 million worth of chemicals and had billings of $2.2 billion in 1980. Of this amount $1.5 billion was for general pest control and $0.7 billion for the control of termites (Pinto, 1981). In Canada 336 companies spend an estimated $2.5 million for chemicals and had billings in 1982 of approximately $35 million (Gardner, 1984, personal communication). The nature of the pests contributing to the business of pest control companies varies widely from region to region (National Research Council, 1980; Pinto, 1981), however, and in the United States termites constitute the largest single dollar item. Major pests as identified by homeowners in Indiana, in decreasing order of complaint were: ants, flies, fleas, mosquitoes, mice, hymenoptera, spiders, cockroaches, termites, rats, silverfish, and centipedes (Bennett, Runstrom, and Weiland, 1983). Surveys of pest control companies indicate that the most common complaints from homeowners concern: cockroaches, ants, termites, fleas, stored product pests, silverfish, firebrats, ticks, and fabric pests. The mention of weeds is noticeably lacking in these lists. This may result from the perception that weeds are not thought of as pests by homeowners and pest control companies often do not perform lawn maintenance. However, weed control constitutes a major use of pesticides in the urban environment. In excess of 2.5 million kg active ingredient (ai) of herbicides are used annually in the United States in the home and garden (U.S. Environmental Protection Agency, 1979). Surveys in the United States indicate that in 1979, 140 million kg of pesticides were used for the control of pests in urban environments when industrial, commercial, and government use is added to that for home and garden (National Research Council, 1980).

NATURE OF ENVIRONMENTAL EFFECTS

Pesticides may be introduced into the urban environment either directly through efforts to control pests in that environment or indirectly through drift from agricultural and silvicultural applications, pesticide spills, industrial wastes, and improper disposal of pesticides. This discussion will address the effects of the pesticides used directly in the urban environment in relation to their primary uses, that is, pest control in lawns, turfgrass, ornamentals, and rights-of-way; structural pest control; control of household pests; and biting fly control. The nature of the effects of these pesticides on the urban environment is dependent on the specific component of the environment treated; the size of the treated area; the nature of the pesticide used; and the rate of application. The specific characteristics of the pesticides that influence their fate and effects upon the environment include rates and modes of metabolic activation or degradation, hydro- or lipo-philicity, partition coefficients, adsorption or chemisorption to abiotic and biotic components, vapor pressure, and volatility. Extremely complex interactions are involved, few of which have been studied thoroughly in any ecosystem.

SIGNIFICANCE AND ENVIRONMENTAL CONCERNS

Lawns, Turfgrass, Ornamentals, Rights-of-way

Of primary concern in the control of pests on lawns, turfgrass, and rights-of-way is the control of weeds and brush. It has been estimated that in the United States there are more than 2 million ha of home lawns and an additional 4 million ha of other types of turf, including golf courses. Phenoxy herbicides have been and remain extensively used for weed control in these areas (Bovey, 1980*a*).

Few groups of synthetic organic compounds have caused as much controversy and concern as the phenoxy herbicides. Public concern has been aroused by the use of Agent Orange (a formulation of 2, 4, 5-T) in Vietnam from 1962 to 1969 (Bryan, 1983; Lathrop, 1984; Long and Hanson, 1983; Young, Kang, and Shepard, 1983); by the accidental exposure of humans or the environment to significant concentrations of phenoxy herbicides, their contaminants, or products produced as a result of explosions and fires at petrochemical production facilities, for example as Seveso, Italy, in July, 1976 (Hay 1976*a*, 1976*b*; Layman, 1983; Rice, 1981); by the leaching of toxic wastes from industrial dump sites, for example at Love Canal, New York (Boddington et al., 1983; Heida, 1983; Severo, 1984); and by the spraying of oil contaminated with dioxin near Times Beach, Missouri (Long and Hanson, 1983). The primary concerns have not been with the herbicides themselves but with secondary products produced during their synthesis or combustion, that is, polychlorinated dibenzo-p-dioxins (PCDDs) or other dioxins (National Research Council of Canada, 1978*a*, 1981; Poiger and Schlatter, 1983; Reggiani, 1983; Espito, Tiernan, and Dryden, 1980). By far the most toxic dioxin is 2, 3, 7, 8-tetrachlorodibenzo-p-dioxin (TCDD). TCDD has been shown to cause chloracne, liver enzyme stimulation, thymic atrophy, immunosuppressive effects, and teratogenesis at exposure rates of less than 0.1 ug/kg (single dose) in a variety of laboratory animals and some of these effects in man (Bleinburg et al., 1964; Poland et al., 1971). TCDD has the extremely low mammalian LD_{50} of 1 to 100 ug/kg (single dose) (Schwetz et al., 1973). 2, 4, 5-T (2, 4, 5-trichlorophenoxyacetic acid) does contain trace amounts of TCDD, however, 2, 4-D (2, 4-dichlorophenoxycetic acid), by the nature of a different synthetic method (Bovey, 1980), does not contain TCDD (National Research Council of Canada, 1983*b*) but may contain other dioxins. The controversy concerning 2, 4, 5-T and TCDD has resulted in all of the phenoxy herbicides undergoing extensive reviews of their toxicology (National Research Council of Canada, 1978*a*, 1983*b*; U.S. Environmental Protection Agency, 1979*c*, 1979*d*). The result has been the removal of most of the registrations of 2, 4, 5-T in the United States (U.S. Environmental Protection Agency, 1979*a*, 1979*b*) and Canada. Since 2, 4, 5-T is no longer commonly used, most of this discussion will concern 2, 4-D, the most commonly used herbicide in the urban environment.

In the United States, the annual use of 2, 4-D for all purposes is on the order of 27 million kg. Between 10 and 15% (2.7 to 4.1 million kg) of its use is on domestic lawns and turf and rights-of-way (Bovey, 1980*a*). Typical application rates for 2, 4-D range from 4 to 6 kg/ha on rights-of-way and less than 1 kg/ha when used on lawns and turfgrass (*Weed Control Manual*, 1982).

When applied to grasses at normal rates the majority, as much as 99%, of that applied is adsorbed onto the thatch and grass, with less than 1% reaching the soil (Thompson, Stephenson, and Sears, 1984). 2, 4-D on grass and in soil has shown half-lives of several days up to three weeks (Thompson, Stephenson, and Sears, 1984; National Research Council of Canada, 1978*a*). Degradation in grass thatch and soils occurs primarily through the actions of microorganisms. Degradation may show a lag phase associated with the growth of microbial populations (Kaufmann and Kearney, 1976; Loos, 1975; Young, 1980). Phenoxy herbicides are generally water soluble, and while they adsorb to soil colloids and humic substances, they may also desorb and leach in sandy or mineral soils. Field studies have shown that some leaching does occur in soils, but the relatively fast degradation rate of 2, 4-D in this environment minimizes the effects of this movement. The vast majority of recoverable residues in field situations are within 5 cm of the soil surface (Thompson, Stephenson, and Sears, 1984; Young, 1980). Some movement of the phenoxy herbicides may occur due to storm runoff and the resultant movement of soil particles (Barnett et al., 1967).

The intensive long-term use of 2, 4-D might be expected to cause significant deleterious effects to the microbial fauna and flora in soils. When exposed to high concentrations of 2, 4-D, bacteria and fungi have shown either stimulatory or inhibitory responses in growth and reproduction. While some changes have been noted in the field, no significant long-term effects have been reported (Mullison, 1981; Pimentel, 1971).

The direct toxicity of phenoxy herbicides to birds and wildlife is moderately low. Single dose LD_{50} values for mammals and birds are in the range of 100-1,000 mg/kg with no effect levels as low as 10-20 mg/kg/day (Mullison, 1981; Pimentel, 1971; Thomson, 1981; Turner, 1977). Maximum residues on grasses resulting from 2, 4-D are approximately 10 mg/kg wet weight with residues declining rapidly after application. Thus, the direct toxicity to vertebrates is unlikely to be as significant as the consequences of the change in compositional structure of the plant community resulting from the selective nature of these herbicides. Changes in the plant community in rights-of-way precipitated by the use of 2, 4-D or 2, 4, 5-T have resulted in either the enhancement or decline of vertebrate species (National Research Council of Canada, 1978*a*). For example, populations of rodents have declined because of a reduction of broadleafed forage, while populations of deer and elk have increased (Johnson and Hansen, 1969; Krefting and Hansen, 1969; Wilbert, 1963). Similarly, populations of birds preferring to nest in brush and small trees have declined, while ground-nesting species have increased in abundance (Dwernychuk and Boag, 1973). These changes would not be expected to occur in the urban environment except where extensive rights-of-way are adjacent to or pass through urban areas.

Significant reductions in earthworm populations have been reported when granular formulations of 2, 4-D have been used (National Research Council of Canada, 1978*a*). However, no effects have been reported at normal rates to populations of wireworms, springtails, mites, and other soil microarthropods (Rapoport and Cangioli, 1963). The most dramatic effects on invertebrates have been attributed to shifts in community structure (National Research Council of Canada, 1978*a*). Reported LD_{50} values for bees range from 11 to 104 ug/bee, and mortality of bees in treated areas has been noted (Pimentel, 1971). However, this toxicity has not been reported consistently and has been attributed to the carriers or surfactants in some formulations (Moffett and Morton, 1975). Reductions in flowers produced by the shift in species composition is probably more critical to pollinator populations than direct toxicity.

Phenoxy herbicides may be introduced into aquatic ecosystems by drift from spray operations, runoff from treated areas, or by direct application for the control of the aquatic weeds. Resultant water concentrations range to a maximum of 3 mg/L from these operations and dissipate within several weeks (National Research Council of Canada, 1978*a*). Some ester formulations of 2, 4-D and 2, 4, 5-T are toxic to fish at 1 mg/L, and therefore a direct toxicity hazard is present (Kenaga, 1974). However, ester formulations are rarely used in urban areas because of their high degree of volatility and the potential hazard to fish. Studies concerning the direct toxicity of 2, 4-D to aquatic invertebrates have produced variable data with some studies reporting no effects, while others observing 50% reductions in populations of almost all benthic organisms. These findings do not seem to be correlated to application rate or formulation but may reflect differences in species selectivity or assessment methods used by the researchers (Mullison, 1970; Pimentel, 1971). Secondary effects of the phenoxy herbicides on aquatic systems may be significant. The decomposition of herbicide-killed macrophytes results in a sudden release of nutrients that could stimulate algal blooms. This release would result in the production of anaerobic conditions resulting in faunal kills (Hurlbert, 1975; National Research Council of Canada, 1978*a*). However, present evidence indicates that, when used according to labeled directions, any effects of 2, 4-D on aquatic systems are transient (Mullison, 1970, 1981).

Another source of concern of the effects of phenoxy herbicides in the urban environment has been atmospheric contamination and drift associated with agricultural weed control operations. The extensive use of ester formulations of phenoxy herbicides in grain crops has resulted in concentrations of 2, 4-D at 0.01 to 1 ug/m^3 with occasional reports of 10 to 20 ug/m^3 in urban air in western North America (Adams, Jackson, and Bamesberger, 1974; Grover et al., 1976). These residues have resulted in deleterious effects to plants (e.g., shelter belt trees and tomatoes) (National Research Council of Canada, 1978*a*). Significant effects on animals at these concentrations are unlikely, and concentrations in the urban environment have been reduced substantially because of the increased use of low-volatility ester formulations and the substantially less volatile amine formulations.

In general, the 2, 4-D related phenoxy herbicides are unlikely to directly cause detrimental effects in the urban environment. If deleterious effects were to be noted these would probably be a result of a change in community structure and the resulting secondary shifts in faunal composition.

Insects are the second major group of pests attacking turfgrass and lawns in the urban environment. During the past 30 years the chlorinated cyclodiene insecticides aldrin, deildrin, heptachlor, and chlordane have been used successfully to control soil-inhabiting insects in lawns and turfgrass. Usually a single application of one of these insecticides provided control for several years (Fleming and Maines, 1954; Nash and Woolson, 1967). However, this long residual life contributed to cancellation of the registration of most of these insecticides except where there was no viable alternative (e.g., termite control). As a result, the primary insecticides currently used on turfgrass in North America are: diazinon, chlorpyrifos, trichlorofon, fonofos, carbaryl, bendiocarb, chlordane, and the bacteria *Baccillus popilliae*. Because of the large number of insecticides used on turfgrass this discussion will focus on those most commonly used, i.e., diazinon, chlorpyrifos, and historically, chlordane. These insecticides are used to control a wide variety of insects, primarily white grubs, Japanese beetles, chinch bugs, the european chafer, and sod webworms.

In general, the primary means of degradation of pesticides in soils is through biological rather than chemical processes. Thus the metabolic activity of microorganisms is usually the most significant aspect influencing persistence, with pH and moisture playing secondary roles (Chapman, 1982). The highest insecticide residues resulting from the application of insecticides to lawns, turfgrass, and rights-of-way would be expected in sparsely vegetated target areas with most of the residues being at or close to the soil surface. The maximum persistence of these insecticides would be expected in soils with high organic content under cool dry conditions. The formulation used also modifies residues and the degradation rate, i.e., granular formulations produce lower initial residues but are more persistent than emulsifiable concentrates (Harris and Svec, 1968).

A study concerning the degradation of diazinon in sod and penetration through the thatch layer showed that when diazinon (50EC) was applied at 4.0 kg ai/ha, 98% of the initial application residues were in the grass and thatch layer, less than 2% in the root layer, and less than 1% in the underlying soil (Sears and Chapman, 1979, 1982). By 14 days after the application, less than 2% remained in the grass and thatch layer, with less than 1% in the root zone or the underlying soil. This study was conducted under conditions of relatively low thatch and high soil moisture and are typical of those reported for many organophosphate insecticides in most grass/soil systems. Typical half-lives for diazinon in soil range from 2 to 8 weeks (Laskowski et al., 1983; Pimentel, 1971), with little movement down the soil column. The failure of many insecticides to control turf insects has been attributed to their lack of movement through the thatch layer (Niemczyk, Kruger, and Lawrence, 1977).

Diazinon has LD_{50} values for mammals in the 76–267 mg/kg range and for birds of 3.5–4.3 mg/kg (single oral dose) (Pimentel, 1971; Turner, 1977). While these latter values are low, no direct toxic effects would be expected or have been reported to mammals or birds at field application rates.

The toxicity of diazinon to terrestrial invertebrates in turf/soil systems is well documented. Deleterious effects are reported against a variety of collembola, coleoptera, and annelids. Recovery to pretreatment levels in field studies was reported to occur within a year of treatment (Malone, Winnett, and Helrich, 1967). Because of the short half-life of diazinon, long-term effects would not be expected except where repeated applications were made.

In most instances, diazinon would not be expected to cause phytotoxic effects when used in the urban environment. However, russeting has been reported on some apple varieties, and some varieties of lettuce show phytotoxic effects when exposed to diazinon at normal rates (Thomson, 1982).

Diazinon is not registered for use in aquatic systems and as such must enter these systems either through runoff from treated areas or as drift from nearby treatments. Diazinon degrades relatively rapidly in aquatic systems with a half-life of less than 5 days, however, diazinon is very toxic to both vertebrate and invertebrate aquatic organisms. Twenty-four hour LC_{50} values for fish range from 50 to 380 ug/L. Insects, molluscs, and amphipods had LC_{50} values (3–96 h) of less than 500 ug/L with some values as low as 1 ug/L. Deleterious sublethal effects were observed at concentrations much lower than those mentioned above and "safe" levels of diazinon would be on the order of less than one-fiftieth of the LC_{50} values (Miller, Zuckerman and Charig, 1966; Morgan, 1983; Thomson, 1982). The application of diazinon to a small stream at 3 ug/L resulted in increased drift of benthic organisms and changes in the composition of benthic fauna. These changes were not persistent however, as the communities were restored to "normal" within 4 weeks of treatment (Miller, Zuckerman, and Charig, 1966).

The application of chlorpyrifos (4E) at 4.0 kg ai/ha to sod resulted in 97% of the initial residues in the grass and thatch zone, 3% in the root zone, and less than 1% in the underlying soil. By 14 days after the application, 34% of the initial residue remained in the grass thatch, 2% in the root zone, and less than 1% in the underlying soil (Sears and Chapman, 1979, 1982). As would be expected, chlorpyrifos had a slower degradation rate than that of diazinon when applied under the same conditions. As seen with diazinon, very little chlorpyrifos moved down the soil column. Typical half-lives for chlorpyrifos in soils range up to 30 weeks (Laskowski et al., 1983; National Research Council of Canada, 1978*b*).

Direct toxicity of chlorpyrifos to terrestrial vertebrates would not be expected because of its relatively high LD_{50} values, i.e., in excess of 100 mg/kg, although LD_{50} values for some species of birds are less than 10 mg/kg (National Research Council of Canada, 1978*b;* Pimentel, 1971). In field studies using normal application rates, deleterious effects of chlorpyrifos on mammals, birds, amphibians, and reptiles have not been reported, with one exception. The exception is that of Hurlbert et al. (1970), who noted significant mortality of mallard ducklings at treatment rates of 0.01-1.1 kg ai/ha (EC). In this instance, mortality probably resulted from the ingestion of large quantities of moribund chlorpyrifos-contaminated insects.

Most arthropods in turf treated with chlorpyrifos at 1.12 kg ai/ha were killed immediately after treatment (Cockfield and Potter, 1983). Three and one-half months after treatment, collembolan, predatory mesostigma mite, and sacrophagus orbatid mite populations recovered to levels that exceeded those of control plots. No deleterious effects have been reported to populations of earthworms (Thomson and Sans, 1974), although few detailed experiments have been done on these subjects. More significant long-term effects might be expected to occur following the application of chlorpyrifos in comparison to diazinon because of the slower degradation of chlorpyrifos in soil.

Chlorpyrifos has been shown to cause significant honey bee mortality at rates as low as 0.012 kg ai/ha. However, turf-spraying operations should rarely expose significant numbers of pollinators to chlorpyrifos (Atkins, 1972).

Chlorpyrifos has rarely been shown to cause significant damage to plant populations, including grasses when used at registered rates. At application rates above 4.5 kg ai/ha leaf tips may yellow; however, growth was not affected (National Research Council of Canada, 1978*b*). At registered rates, phytotoxic effects may be apparent on roses, azaleas, camelias, and variegated ivy (Thomson, 1982).

Chlorpyrifos may be introduced into aquatic systems in urban environments, either by drift from spraying operations, runoff from treated areas, or by direct application for mosquito control. Chlorpyrifos does cause dramatic effects on bentic systems, resulting in the mortality of most aquatic insects and many other invertebrates (Hurlbert et al., 1970, 1972; National Research Council of Canada, 1978*b*). It is also highly toxic to fish with 24 and 96 hr LC_{50} values of 4-550 ug/L (Macek, Hutchinson, and Cope, 1969; National Research Council of Canada, 1978*b*). While chlorpyrifos has been shown to degrade relatively rapidly in water, i.e., half-life of less than one week, its ability to adsorb onto organic matter results in toxic effects to susceptible invertebrates for an extended period (Hurlbert et al., 1970; Rawn, 1977).

Historically, chlordane has been widely used for the control of soil insects, including turf insects. Recent restrictions on its use in North America have significantly reduced its use on lawns and turfgrass. Chlordane is a mixture of products, some of which degrade to the insecticidal heptachlor epoxide. Thus, this discussion considers the effects of chlordane being that of the technical mixture and heptachlor epoxide (National Research Council of Canada, 1974).

The application of chlordane (8E) at 9.0 kg ai/ha resulted in 92% of the initial residues in the grass and thatch layer, 8% in the root zone, and 1% in the underlying soil. After 56 days residues in the grass and thatch decreased to 60%, while residues in the root zone remained almost constant at 8%, and less than 1% in the underlying soil (Sears and Chapman, 1979, 1982). The degradation rate of chlordane in soils is very slow in relation to other insecticides registered for use on turfgrass. Half-lives have been reported to vary from 4 months to 11 years in soil (National Research Council of Canada, 1974) depending on the soil type, temperature, pH, and microbial populations. Chlordane does not show significant mobility in a variety of soil types even when exposed to extensive rainfall. Thus movement of chlordane would be more likely associated with the movement of soil particles and not leaching (Harris, 1972; Nash and Woolson, 1967; Wilson, 1973).

Acute toxic effects of chlordane on terrestrial vertebrates resulting from turf application would not be expected, as mammalian LD_{50} values range from 100 to 1,000 mg ai/kg, and from 120 to 1,200 mg ai/kg for birds (National Research Council of Canada, 1974; Pimentel, 1971). However, no-effect levels are much lower, i.e., 3-20 mg ai/kg for mammals and 0.3-200 mg ai/kg for birds. When applied to a marsh at 1.12 kg ai/ha, reproductive activity of birds in the marsh was dramatically reduced. This was attributed to reduction in the food supply (chlordane decimated the aquatic insect population) of the birds rather than direct toxic effects (Hurlbert et al., 1970).

An normal application rates used on turfgrass chlordane did not significantly reduce populations of soil mites. Chlordane at 2.2 kg/ha has little effect on symphylids but caused significant reductions of pauropods. Earthworms are extremely susceptible to chlordane (Pimentel, 1971), with reductions following treatments lasting as long as three years at high application rates. With chlordane applied at 2.2 kg ai/ha, earthworm populations were reduced for more than a year. As would be expected, populations of predacious and phytophagous insects were reduced for up to three years following the application of chlordane at 8.9-11.2 kg ai/ha (National Research Council of Canada, 1974).

Chlordane is not registered for use in aquatic systems. Its current uses, primarily as a termiticide and to a reduced extent on turf, substantially limit the contamination of aquatic systems. Because of the extremely low solubility of chlordane in water (6-9 ug/L) and its adsorptive nature, chlordane is unlikely to be found in aquatic systems at high concentrations. Reported residues in water rarely exceed 250 ng/L, with most reports below 20 ng/L (Kutz, 1983; National Research Council of Canada, 1974). However, relatively high concentrations have been reported in aquatic sediments, i.e., 2.9 mg/kg (Kutz, 1983). LC_{50} values for fish are as low as 71 ng/L, although most 96 hr LC_{50} values are in the 10-200 ug/L range. Thus, direct fish toxicity would not be expected, unless chlordane was accidentally introduced into aquatic systems at high concentration. Most LC_{50} values (48 or 96 hr) for aquatic invertebrates are in the range of 1-100 ug/L. However, these concentrations are rarely encountered in natural waters (Kutz, 1983). Secondary deleterious effects may occur, as chlordane has been shown to biologically concentrate in some aquatic systems (National Research Council of Canada, 1974; Pimentel, 1971).

The relatively long residual life of chlordane in soils and its demonstrated deleterious effects on both terrestrial and aquatic invertebrates would suggest that it would cause more significant effects than diazinon or chlorpyrifos when used for insect control in lawns, turfgrass, or on rights-of-way.

Structural Pest Control

Control of structural pests requires the use of pesticides with long persistence. Creosote and pentachlorophenol have been used extensively for the control of fungi that cause wood decay and therefore are perhaps the most widely used pesticides in the world. In recent years pentachlorophenol has been the primary wood preservative used. In the United States, it is estimated that 18-20 million kg of pentachlorophenol is used as a pesticide on an annual basis (U.S. Environmental Protection Agency, 1983). It is a broad spectrum pesticide and is effective for the control of fungi, algae, and bacteria as well as insects. Used extensively in wood preservation, it is also an ingredient in a number of paints and may be incorporated into shingles, concrete materials, and other building materials.

Pentachlorophenol is produced through successive stages of the chlorination of phenol at high temperatures (Plimmer, 1973). The end product is not pure and contains a variety of incompletely chlorinated phenols as well as chlorinated dioxins, chlorinated diphenyl ethers, and chlorinated dibenzofurans. Concern over the use of pentachlorophenol has been generated primarily by the presence of these secondary products. Some of these, e.g., two isomers of hexachlorodibenzo-p-dioxin, have been shown to cause benign and neoplastic liver tumors in mice of both sexes and female rats

(Johnson, et al., 1973; Langer, Brady, and Briggs, 1973; U.S. Environmental Protection Agency, 1983).

Studies concerning the direct toxicity of pentachlorophenol to vertebrates produced the following data: mammalian LD_{50} values of 27-175 mg ai/kg, birds LD_{50} values of 4,000-5,000 mg/kg, and fish actively avoided concentrations as low as 0.2 mg ai/L (Pimentel, 1971; U.S. Environmental Protection Agency, 1983). Direct toxicity of pentachlorophenol to vertebrates is unlikely, primarily because of the nature of its use (impregnated in building materials) and its highly adsorptive nature. Because of its very broad spectrum of activity and its persistence, it could be anticipated that pentachlorophenol could become a significant environmental contaminant. This has not occurred, probably because the compound is only slightly soluble in water, is highly adsorptive to organic material, and thus is confined largely to the materials treated.

One of the concerns that has been raised regarding pentachlorophenol-treated wood is the problem that might be present when treated woods are burned or sawdust burned from mills in which pentachlorophenol-treated timber has been processed. It could be expected that in these situations additional chlorinated dioxins and furans could be produced and present some environmental hazard. It is known that the use of shavings from sawmills where pentachlorophenol has been used for lumber treatments have resulted in adverse reproductive effects in poultry when these shavings have been used for bedding.

The control of termites is the largest market for professional pest control services in the United States. Some areas in the southeastern United States have more than 50% of the structures in the region treated for the control of termites (U.S. Environmental Protection Agency, 1979*e*). The pesticides used for termite control in North America are chlordane, heptachlor, aldrin, dieldrin, lindane, pentachlorophenol, and chlorpyrifos. Pentachlorophenol, as discussed above, is used almost exclusively in building materials while dieldrin and lindane are rarely used for termite control. Chlorpyrifos, which is more costly than the other registered products, has been registered in the United States for the shortest period (since 1980) and thus has not been used extensively (U.S. Environmental Protection Agency, 1983). The most commonly used termiticides are chlordane, heptachlor, and aldrin. Each of these insecticides has undergone extensive reviews concerning their toxicological effects on vertebrates and their effects on the environment (National Research Council of Canada, 1974, 1978*b;* U. S. Environmental Protection Agency, 1972, 1976*a*, 1976*b*, 1976*c*, 1976*d*, 1978, 1982, 1983). As a result of these reviews, the registered uses of the chlorinated hydrocarbon insecticides have been restricted. These restrictions have in part resulted from the qualities of these pesticides that make them excellent termiticides, i.e., their very slow degradation rates in soil.

Treatments with these pesticides are usually highly effective and may involve any of the use of pesticide impregnated building materials, the injection of a concentrated preparation of one of these materials into the soil around foundations when new structures are being erected, or injection of the pesticide in the soil around older structures to control existing infestations. The primary targets for these treatments in North America are in the genera *Reticulitermes* and *Coptotermes* and to a lesser extent termites in genus *Zootermopsis* (Pinto, 1981; U.S. Environmental Protection Agency, 1983).

Chlordane has shown residual efficacy for as long as 34 years when injected into the soils around foundations and other cyclodienes may be effective for more than 20 years. Chlorpyrifos used under similar conditions has been shown to be effective for at least 15 years with the tests continuing. This extremely long residual life is necessary for the control of termites; however, it also allows the pesticides to be present in the urban environment for long periods of time. Studies concerning the resulting soil residues have indicated that these insecticides move only a few centimeters during 17-21 years of exposure to the climate (U.S. Environmental Protection Agency, 1983). As these pesticides are usually impregnated in building materials or applied in or under the foundations of buildings, exposure to weathering influences is minimal, and their movement through the soil usually insignificant. All treated areas must be covered by an impervious barrier or a layer of untreated soil. In addition, these insecticides are not applied to areas that would be frequented by vertebrates and the vast majority of invertebrate organisms. However, as discussed with chlordane, the contamination of aquatic ecosystems may result in deleterious consequences. While chlorpyrifos is much less persistent than the cyclodiene insecticides under most circumstances, it too is toxic to aquatic organisms at very low concentrations. Improper application procedures may result in contamination of shallow wells and groundwater especially in areas with sandy soils (Montgomery, 1982).

When used according to recommended practices, these insecticides should provide no significant threat to the urban environment. However, each of these insecticides has the potential for significant environmental contamination if the treatments are not performed properly or if treated soil is moved to dumps or handled in such a fashion that the insecticides could enter surface or ground waters.

Public concern has recently been aroused concerning the residues of insecticide in the air of buildings following their treatment for the control of termites and possible effects on human health. As a result the risks and benefits of the pesticides used for termite control in the United States have been reviewed by the National Academy of Sciences (1982) and by the Environmental Protection Agency (1983). The final conclusions of these reviews are that the benefits from the use of the termiticide products outweigh the potential risks, although additional toxicological information is required to fully assess the potential health effects. The termiticides were placed in a restricted classification, and the cyclodiene insecticides cannot be used in plenum houses.

Household Pests

More than 83% of households in the United States use pesticides in the house. Most of the use of pesticides in the home is associated with the control of insects, especially those that infest foods or inhabit food preparation areas (U.S. Environmental Protection Agency, 1979*e*). Those insecticides registered for this use have relatively low mammalian toxicity and low residual activity. The most commonly used insecticides in the house include natural and synthetic pyrethrins, propoxur, dichlorvos, carbaryl, diazinon, and malathion. These insecticides have oral LD_{50} values for mammals ranging from 76 to 1000 mg ai/kg (Pimentel, 1971). The more toxic compounds are formulated and packaged in such a fashion that even if accidental ingestion occurred, exposure to lethal doses would be unlikely.

The use of these insecticides in the home restricts their distribution to the structure itself and thus limits the contamination of the urban environment. It is therefore unlikely that any significant effects on the urban environment would be expected to occur from this use. The improper use or disposal of these insecticides resulting in the contamination of terrestrial urban ecosystems would probably result in localized reductions of insect populations but few, if any, other short-term effects. Long-term effects would be unlikely, as these compounds degrade rapidly when exposed to environmental factors and microbial populations, and they do not biologically concentrate. If aquatic ecosystems were contaminated, short-term localized effects might occur, as in general these insecticides are toxic to aquatic organisms at low concentrations.

The rodenticide warfarin is also used to a significant extent in the home (U.S. Environmental Protection Agency, 1983). If it is handled carelessly the detrimental effects to mammals are obvious. The possibility of rodenticides entering the urban ecosystem in significant quantities is unlikely; however, their effects on urban ecosystems are poorly documented.

In general, the use of pesticides in the home should not result in any significant contamination of or detrimental effects to urban ecosystems if they are used and disposed of properly.

Biting Fly Control

A primary source of annoyance to humans in the urban environment is the blood-seeking habits of biting flies (National Research Council of Canada,

1982*b;* Robinson and Atkins, 1983). These insects may not only cause psychological stress but also transmit several significant diseases of man and household pets in North America. The primary group of diseases which still cause significant illness are encephalitides, i.e., Saint Louis encephalitis, Western Equine encephalitis, Venezualian Equine encephalitis, and the California encephalitis group (Harwood and James, 1979; National Research Council of Canada, 1982*b*). To reduce the biting activity from mosquitoes or black flies a variety of control measures, including the use of insecticides, are used against their larval and adult stages.

The primary insecticides used for the control of mosquito larvae in North America are temephos, chlorpyrifos, malathion, fenthion, methoxychlor, pyrethroids, methoprene, diflubenzuron, and the bacteria *Bacillus thuringiensis* var. *israeliensis*, H-14. These insecticides range from some of the most selective pesticides registered, i.e., methoprene and *Bti*, to those which would result in significant effects on these aquatic systems (e.g., chlorpyrifos). The areas treated are often in or around the urban environment and include snow-melt and floodwater pools in woodlots and ditches, treeholes, tires and other containers, sewage lagoons, catch-basins, and marshlands (Harwood and James, 1979; National Research Council of Canada, 1982*b*). The effects of this variety of insecticides, used in a variety of formulations on these different habitats, are simply too diverse to discuss in any detail. In most instances these insecticides degrade rapidly in aquatic systems, although adsorption and desorption from organic materials and the use of granular or slow-release formulations may lengthen their residual activity. The shortest half-life is that of methoprene, several hours, and the longest is chlorpyrifos, with a half-life of up to one week. The more selective and faster degrading larvicides are preferable for the environment; however, the less selective longer-lasting larvicides provide better control, especially against species with asynchronous larval populations (*Culex, Culiseta, Anopheles*). The more selective insecticides also tend to be significantly more expensive than the less specific mosquito larvicides. Often those persons involved in mosquito control select the more selective insecticides for use in environmentally sensitive areas (i.e., adjacent to, or contiguous with, lotic systems) and the less selective insecticides in confined habitats where contamination of lotic systems is unlikely. The use of any of these insecticides as mosquito larvicides is unlikely to have permanent deleterious effects on these aquatic systems, although population fluctuations of invertebrates following treatments with chlorpyrifos have required weeks to months to return to control levels (Hurlbert, 1975; Hurlbert et al., 1970; National Research Council of Canada, 1982*b*).

Blackfly larvae, unlike mosquito larvae, are characteristically found in lotic ecosystems (Harwood and James, 1979; National Research Council of Canada, 1982*b*). Treatments of large rivers for blackfly control is typically done by point source injection of the insecticide for a given period of time, producing a pulse of insecticide that travels downriver. The primary insecticides used are methoxychlor and temephos (National Research Council of Canada, 1982*b*, 1983*a*).

The treatment of lotic systems with methoxychlor or temephos for blackfly control does not result in direct toxic effects to fish (National Research Council of Canada, 1983*a*). It does result in catastrophic drift of aquatic invertebrates and a reduction in the numbers of benthic organisms in the treated streams and rivers (Helson and West, 1978; National Research Council of Canada, 1983*a*; Wallace et al., 1973). Single treatments may alter the benthic community several hundred kilometers downstream from the point of application. The benthic communities recover from treatment; however, populations of predacious insects with long life cycles require several years to recover, especially if the entire river is treated. Populations of phytophagous species with short life cycles and high reproductive rates recover much faster (Helson and West, 1978; National Research Council of Canada, 1983*a*; Wallace et al., 1973). While temephos is more selective than methoxychlor, temephos is not as effective as methoxychlor when used at the low stream temperatures often encountered in temperate zones. Recent research has focused on the development of particulate formulations of these pesticides that would increase their selectivity to blackfly larvae.

Control programs directed against adult mosquitoes are common in many North American cities. These programs involve the application of insecticides by ground fogging equipment or ground or air ultra low-volume equipment. The application of insecticides by these methods may result in the exposure of virtually all of the abiotic and biotic components of the urban environment to the insecticide. The primary insecticides used for mosquito adulticiding in North America are propoxur, malathion, pyrethrins, methoxychlor, fenthion, dichlorvos, and chlropyrifos. These insecticides would not be expected to cause long-term damage to the urban environment; however, short-term effects have been noted especially when a large urban area is treated in a short period of time (Bowen, Gavin, and Brust, 1976; Manitoba Ministry of Health, 1982). The most commonly used insecticides for large-scale mosquito adulticiding are propoxur and malathion.

Propoxur can be applied by ground or aerial application methods for mosquito control. The ground application methods include thermal fogging, cold fogging, and ultra low-volume application, while aerial applications usually use ultra low-volume methods. Rates of application range from 17.5 to 27.5 g ai/ha for ground application and from 32.5 to 125 g ai/ha for aerial application (National Research Council of Canada, 1982*a*).

Direct toxicity of propoxur to mammals and birds at rates used for mosquito adulticiding would not be expected. Acute oral LD_{50} values for birds range from 3.6 to 60.4 mg ai/kg with the majority of values falling within 5 and 15 mg ai/kg. Long-term dietary and multiple-dose studies indicate that birds can ingest much more propoxur than acute studies would indicate with no effect if the birds are exposed to lower rates. Acute oral LD_{50} values for mammals range from 37 to 191 mg ai/kg, with dermal values in excess of 1,000 mg ai/kg (Chemagro, 1976; National Research Council of Canada, 1982*a;* World Health Organization, 1974). Thus direct exposure of the ingestion of foodstuffs exposed to propoxur at the rates used for mosquito control would not normally cause significant deleterious effects in mammals. Propoxur, like a large number of carbamate pesticides, may under go nitrosation and produce a potential carcinogen (Crosby, 1976). In the case of propoxur, the product nitrosopropoxur has been shown to be mutagenic and tumorigenic in some studies. However, it is not certain whether the contribution of nitrosopropoxur or any nitrosated pesticide would significantly influence the carcinogenic risk already associated with the nitrosation of natural products (Crosby, 1976; National Research Council of Canada, 1982*a*).

When propoxur is applied to soils, half-lives range from one to four months with the primary means of degradation being microbial decomposition (Tu and Miles, 1976). The degradation rate is dependant upon moisture, soil organic content, and microbial populations. The application of propoxur at rates of 138-1,100 g ai/ha did not cause severe reductions in terrestrial invertebrate populations, and reductions that did occur recovered to or exceeded control populations within 3-4 months of treatment. Extremely high rates, i.e., 20 Kg ai/ha, can cause deleterious effects on a variety of species including earthworms; however, these rates are unrealistic. Deleterious effects of propoxur on microbial populations have been reported; however, concentrations used have been relatively high, i.e., usually in excess of 25 mg ai/kg soil (National Research Council of Canada, 1982*a*).

Most studies concerning the phytotoxicity of propoxur have used application rates much higher than those encountered in mosquito control operations. At these high rates phytotoxicity may result, but at the lower rates used in field operations phytotoxicity has not been reported. As the carrier used with propoxur is usually a light oil, phytotoxicity may result from this carrier, although this has not been reported during mosquito control operations (National Research Council of Canada, 1982*a*).

Propoxur is highly toxic to honeybees with an LD_{50} of 1.35 ug/bee (Atkins et al., 1977). This rate would be encountered by bees in areas treated for mosquito control (National Research Council of Canada, 1982*a*). Thus, when possible, applications are made when bee flight activity is at a minimum, i.e., dusk or dawn. Significant mortality of bees has been reported following large-scale mosquito adulticiding (Manitoba Ministry of Health, 1982).

Propoxur does not degrade rapidly in aquatic systems with half-lives under natural conditions of less than 7 days and most report less than 3 days. Degradation in aquatic systems may be by hydrolysis or the action of microorganisms (Eichelberger and Lichtenberg, 1971; Flint and Shaw, 1971). Acute LC_{50} values for fish exposed to the propoxur range from 1.3-50 mg/L (National Research Council of Canada, 1982*a*). These values are considerably higher than those resulting from application for adult mosquito control, i.e., maximum 36.4 ug/L and usually less than 2 ug/L (Bowen, Gavin, and Brust, 1976). Thus, direct toxicity to fish is unlikely unless extremely high concentrations are encountered. Toxicity to aquatic invertebrates, especially insects, shows substantial variability with LC_{50} values ranging from 2 ug/L to in excess of 10 mg/L. Generally, insects are more susceptible than other invertebrates. Toxic effects to a variety of insects including chironomids, mayfly naiads, and blackfly larvae, may result from the rates used for mosquito adulticiding (National Research Council of Canada, 1982*a*).

Large-scale mosquito adulticiding has been performed in many cities in North America, usually in response to an outbreak of mosquito-vectored encephalitides. Western Canada has had outbreaks of Western Equine Encephalitis that have resulted in large numbers of horse fatalities, as well as illness and death of humans (National Research Council of Canada, 1982*b*). As a result, monitoring programs are maintained to provide warning as to the incidence of the disease in birds, the appearance of the disease in horses, and the population levels of the vector *Culex tarsalis*. In 1981 it became apparent that a threat to man was imminent, and a decision was made to apply propoxur to the city of Winnipeg and other urban areas in southern Manitoba. In this operation, 490,000 ha were treated with 2.04 million L of a 12% solution of propoxur applied by large aircraft with ultra low-volume techniques. Intensive environmental monitoring followed the treatments. Water analysis for propoxur in streams, rivers, and water treatment facilities indicated no residues in excess of 1.9 ug/L and residues disappeared rapidly. These levels were considered insignificant and posed no significant threat to aquatic life. Studies concerning toxicity to honeybees indicated direct mortality to bees in treated areas. Mortality was in excess of 75% of the bees returning to colonies from treated areas. Aside from these observations, no short-term or long-term deleterious effects were observed (Manitoba Ministry of Health, 1982). This is similar to the observations made following a treatment of Winnipeg with propoxur in 1975 (Bowen, Gavin, and Brust, 1976).

The second insecticide commonly used in mosquito adulticiding is malathion. Its high margin of safety to mammals and birds in combination with its amenability to ultra low-volume application has made it widely used as a mosquito adulticide (Matsumura, 1975; McEwen and Stephenson, 1979). Application rates used for ground applications range from 50 to 100 ml ai/ha and for aerial application from 219 to 438 ml ai/ha.

Direct toxicity of malathion to birds would not be expected during mosquito control operations as LD_{50} values for birds range from 400 to 1,500 mg ai/kg (Cyanamid International, 1973). Similarly, mammalian LD_{50} values are in excess of 1,000 mg ai/kg (Pimentel, 1971). Chronic effects are also unlikely, as doses required to induce deleterious effects in both birds and mammals are relatively high and malathion degrades rapidly in the environment.

When applied to soil, malathion degrades quickly with a half-life of less than 8 days. The primary route of degradation in soil is through microbial decomposition and to a lesser extent hydrolysis (Mulla, Mian, and Kawecki, 1981). Malathion does not cause significant deleterious effects to microorganisms at registered rates. In general, malathion has a low toxicity to many terrestrial invertebrates. Deleterious effects would not be expected to populations of many invertebrates including earthworms, millipedes, centipedes, predacious mites, collembola, and carabids. However, some parasitic hymenoptera and predacious coleoptera are susceptible at rates used for mosquito control (Mulla, Mian, and Kawecki, 1981).

Because of the wide use of malathion on food crops it has been tested extensively for phytotoxicity. Malathion has been shown to cause damage to a wide variety of plants, especially when formulated as an emulsifiable concentrate. However, at the rates used for mosquito control, phytotoxicity would not be expected. The half-life of malathion on or in plants ranges from 2 to 168 days (Mulla, Mian, and Kawecki, 1981).

Malathion is very toxic to honeybees with an LD_{50} value of 0.079 ug/bee (Anderson and Atkins, 1968). Significant bee mortality (31-42% of caged bees in the treated area) has been reported following the application of malathion for mosquito control. Mortality is especially high if applications are made during the normal flight period of the bees.

The application of malathion as a mosquito adulticide would result in its direct application to water as well as the possibility of drift and runoff from adjacent treated areas into aquatic systems. Malathion has a wide range of toxicity to fish with LC_{50} values (48 or 96 hr) ranging from 0.09-79 mg/L (Mulla, Mian, and Kawecki, 1981). Concentrations resulting from mosquito adulticiding are a maximum of 0.03 mg/L and usually less than 2.0 mg/L (Bowen, Gavin, and Brust, 1976). Thus the possibility exists that some fish mortality may occur if unusually high concentrations are present. Studies concerning the acute toxicity of malathion to amphibians have produced LC_{50} values of 200-420 mg/L, which are well above normal expected concentrations. The toxicity of malathion to aquatic invertebrates shows a wide range. LC_{50} values (96 hr) for snails exceed 26 mg/L, and oysters exceed 1 mg/L, while *Daphnia* are very susceptible, i.e., LC_{50} 0.2 ug/L. Shrimp and crabs are also susceptible at concentrations expected during mosquito adulticiding and field studies have shown 16-80% mortality of two species of shrimp in treated areas. Significant mortality would also be expected to occur in a variety of aquatic insects. Thus it would be expected that significant short-term effects would result from the treatment of water with malathion during mosquito adulticiding operations. Long-term effects would not be expected because of the short half-life of malathion in water.

Malathion has also been used for large-scale mosquito adulticiding in Winnipeg. In 1983 an outbreak occurred like that of 1981, and the material used for control was malathion. In 1983, 1.0 million ha were treated using 157,000 L of a 95% solution of malathion. Environmental monitoring found no significant damage to vegetation, birds, or most other natural biota, although honeybees did sustain significant mortality and some changes were noted in enzyme activity in pickerel in treated areas. From the above discussion it is evident that some short-term deleterious effects may result from mosquito adulticiding operations; however, these effects would not persist.

THE PESTICIDE CONTROVERSY

The introduction of modern pesticides in the 1940s brought with it not only a very effective means of controlling pests but also sustained controversy concerning the potential negative aspects of pesticide use. While this debate has primarily concerned agricultural and forestry uses (perhaps as these are the most extensive uses), controversy has also surrounded their use in urban environments. Most of this concern has focused on human toxicity and long-term chronic effects on humans, while relatively little concern has been focused on the urban environment. This environment is already highly modified and has been seriously affected by almost all of man's activities. The identification of those effects, if any, due to the use of pesticides is almost impossible.

It should be noted that the use of any pesticide must be made with discretion as the improper use of these compounds can result in serious deleterious effects on the environment. As discussed above, the pesticides used in the urban environment can and do cause some deleterious effects in that environment. The use of the pesticide must always result in benefits that far outweigh any deleterious effects that may occur to the ecosystem. The pesticide controversy stems largely from a lack of understanding held by the public with respect to pesticides and their real toxicity. It also results from differences between people with respect to their tolerance of pests, the degree to which aesthetics is important to them, and their knowledge about and concern for the environment in which they live. While few would question the need for a pesticide application to eliminate a cockroach infestation in a restaurant or the control of mosquitoes that carry a health-threatening disease, there are many who see no rationale in killing weeds in lawns that are used only for aesthetic purposes. This situation will not change and those who see no harm in weeds in lawns and parks, some defoliation of ornamental trees, or a few insects running around the home are not likely to conclude that any risk that might be associated with pesticides is worth taking.

The evidence presented here would indicate that when used in accordance with proper application and safety techniques, most pesticides do not pose a significant threat to the urban environment.

REFERENCES

Adams, D. F., C. M. Jackson, and W. L. Bamesberger. 1974. Quantitative studies of 2, 4-D esters in the air. *Weeds* **12:**280.

Anderson, L. D., and E. L. Atkins. 1968. Pesticide usage in relation to bee-keeping. *Annu. Rev. Entomol.* **13:**213.

Atkins, E. L. 1972. Rice field mosquito control studies with low volume Durban sprays in Coulsa County: V. Effects on honey bees. *Mosq. News* **32:**538.

Atkins, E. L., L. D. Anderson, D. Kellum, and K. W. Neuman. 1977. *Protecting Honey Bees from Pesticides,* Leaflet 2883. University of California, Division of Agricultural Sciences.

Barnett, A. P., E. W. Hauser, A. W. White, and H. H. Halliday. 1967. Loss of 2, 4-D in washoff from cultivated fallow land. *Weeds* **15:**133.

Bennett, G. W., E. S. Runstrom, and J. A. Weiland. 1983. Pesticide use in homes. *Bull. Entomol. Soc. Am.* **29**(1):31.

Bleinburg, J., M. Wallen, R. Brodkin, and I. L. Applebaum. 1964. Industrially acquired prophyria. *Arch. Dermatol.* **89:**793.

Boddington, M. J., V. M. Douglas, C. E. Duncan, M. Gilbertson, D. L. Grant, D. Hallett, L. McCellana, J. R. Roberts, J. Singh, and M. Whittle. 1983. Dioxins in Canada. *Chemosphere* **12**(4-5):477.

Bovey, R. W. 1980*a*. Alternatives to the phenoxy herbicides. In *The Science of 2, 4, 5-T and Associated Phenoxy Herbicides,* R. W. Bovey and A. L. Young (eds.). Wiley, New York, p. 462.

Bovey, R. W. 1980*b*. Chemistry and production of 2,4,5-T and other phenoxy herbicides. In *The Science of 2,4,5-T and Associated Phenoxy Herbicides,* R. W. Bovey and A. L. Young (eds.). Wiley, New York, p. 485.

Bowen, W. G., H. C. R. Gavin, and R. A. Brust. 1976. Environmental monitoring of insecticide applications during the emergency mosquito control operation in Manitoba, 1975. *Can. J. Public Health* **67**(suppl.):63.

Bryan, C. D. 1983. The veterans' ordeal. *Net Republic* **188**(25):26.

Byrne, D. N., E. H. Carpenter, E. M. Thoms, and S. T. Cotty. 1984. Public attitudes toward urban arthropods. *Bull. Entomol. Soc. Am.* **30**(2):40.

Chapman, R. A. 1982. Persistence of insecticides in water and soil with special reference to pH. In *Advances in Turfgrass Entomology,* H. D. Niemczyk and B. G. Joyner (eds.). Hammer Graphics, Piqua, Ohio, p. 79.

Chemagro. 1976. *Technical Information: Baygon Insecticide.* Chemagro Corp., 4p.

Cockfield, S. D., and D. A. Potter. 1983. Short-term effects of insecticidal applications on predaceous arthropods and oribatid mites in Kentucky Bluegrass turf. *Environ. Entomol.* **12:**1260.

Crosby, N. J. 1976. Nitrosamines in foodstuffs. *Residue Rev.* **64:**77.

Cyanamid International. 1973. *Malathion Insecticide for Adult Mosquito Control,* 35p.

Cyanamid International. 1977. *Modern Mosquito Control,* 32p.

Dwernychuk, L. W., and D. A. Boag. 1973. Effect of herbicide induced changes in vegetation on nesting ducks. *Can. Field Nat.* **87:**155.

Eichelberger, J. W., and J. J. Lichtenberg. 1971. Persistence of pesticides in river water. *Environ. Sci. Technol.* **5:**541.

Esposito, M. P., T. O. Tiernan, and F. E. Dryden. 1980. *Dioxins,* Report No. EPA/600/2-80-197. U.S. Environmental Protection Agency, Washington, D.C., 351p.

Fleming, W. E., and W. W. Maines. 1954. Persistence of chlordane in soils of the area infested by Japanese Beetle. *J. Econ. Entomol.* **47:**165.

Flint, D. R., and H. R. Shaw. 1971. *The Mobility and Persistence of Baygon in Soil and Water,* Report 30589. Chemagro Corp.

Granovsky, T. A. 1983. Economic impact of termiticides. *Pest Control* **51:**14.

Grover, R., L. A. Kerr, K. Wallace, K. Yoshida, and J. Maybank. 1976. Residues of 2,4-D in air samples from Saskatchewan, 1966-1975. *J. Environ. Sci. Health* **B11:**331-347.

Hanson, W. R. 1952. Effects of some herbicides and insecticides on biota of North Dakota marshes. *J. Wildl. Manage.* **16:**299.

Harris, C. R. 1972. Factors influencing the biological activity of technical chlordane and some related components in soil. *J. Econ. Entomol.* **65:**341.

Harris, C. R., and H. J. Svec. 1968. Toxicological studies on cutworms: III. Laboratory investigations on the toxicity of insecticides to the black cutworm, with special reference to the influence of soil type, soil moisture, method of application and formulation on insecticide activity. *J. Econ. Entomol.* **61:**965.

Harwood, R. F., and M. T. James. 1979. *Entomology in Human and Animal Health,* 7th ed. Macmillan, New York, 548p.

Hay, A. 1976*a*. Toxic cloud over Seveso. *Nature* **262:**636.

Hay, A. 1976*b*. Seveso: The aftermath. *Nature* **263:**538.

Heida, H. 1983. TCDD in bottom sediments and eel around refuse dump near Amsterdam, Holland. *Chemosphere* **12:**503.

Helson, B. V., and A. S. West. 1978. Particulate formulations of Abate and methoxychlor larvicides: Their selective effects on stream fauna. *Can. Entomol.* **110:**591.

Hurlbert, S. H., M. S. Mulla, and H. R. Wilson. 1972. Effects of an organophosphorus insecticide on the phytoplankton, zooplankton and insect populations of freshwater ponds. *Ecol. Monogr.* **42:**269.

Hurlbert, S. H., M. S. Mulla, J. O. Keith, W. E. Westlake, and M. E. Dusch. 1970. Biological effects and persistence of Dursban in freshwater ponds. *J. Econ. Entomol.* **63:**43.

Johnson, D. R., and R. M. Hansen. 1969. Effects of range treatment with 2,4-D on rodent populations. *J. Wildl. Manage.* **33:**125.

Johnson, R. L., P. J. Gehring, R. J. Kociba, and B. A. Schwetz. 1973. Chlorinated dibenzodioxins and pentachlorophenol. *Environ. Health Perspect.* **5**(exp. issue):171.

Kaufmann, D. D., and P. C. Kearney. 1976. Microbial transformations in the soil. In *Herbicides: Physiology, Biochemistry, Ecology,* vol. 2, L. J. Audus (eds.). Academic Press, New York, p. 163.

Kenaga, E. E. 1974. 2,4,5-T and derivatives: Toxicity and stability in the aquatic environment. *Down Earth* **29:**137.

Krefting, L. W., and H. L. Hansen. 1969. Increasing browse for deer by aerial applications of 2,4-D. *J. Wildl. Manage.* **33:**784.

Kutz, F. W. 1983. Chemical exposure monitoring. *Residue Rev.* **85:**277.

Langer, H. G., T. P. Brady, and P. R. Briggs. 1973. Formation of dibenzodioxins and other condensation products from chlorinated phenols and derivatives. *Environ. Health Perspect.* **5**(exp. issue):3.

Laskowski, D. A., R. L. Swann, P. J. McAll, and H. D. Bidlack. 1983. Soil degradation studies. *Residue Rev.* **85:**139.

Lathrop, G. D. 1984. *An Epidemiological Investigation of Health Effects in Air Force Personnel following Exposure to Herbicides Baseline Morbidity Study Results,* USAF Report, Feb. 24, 1984, Federal Government Report 459.

Layman, P. L. 1983. Europe provides test case of human exposure to dioxin. *Chem. Eng. News* **61:**61.

Long, J. R., and D. J. Hanson. 1983. Dioxin issue focus on three major controversies in U.S. *Chem. Eng. News.* **61:**23.

Loos, M. A. 1975. Phenoxyalkanoic acids. In *Herbicides: Chemistry, Degradation and Mode of Action,* P. C. Kearney and D. D. Kaufman (eds.). Marcel Dekker, New York, p. 202.

Macek, K. J., C. Hutchinson, and O. B. Cope. 1969. The effect of temperature on the susceptibility of bluegills and rainbow trout to selected pesticides. *Bull. Environ. Contam. Toxicol.* **3:**174.

McEwen F. L., and G. R. Stephenson. 1979. *The Use and Significance of Pesticides in the Environment.* Wiley, New York, 538p.

Malone, C. R., A. G. Winnett, and K. Helrich. 1967. Insecticide-induced responses in an old field ecosystem: Persistence of diazinon in the soil. *Bull. Environ. Contam. Toxicol.* **2:**83.

Manitoba Ministry of Health. 1982. *Western Equine Encephalitis in Manitoba.* Manitoba Government Publications, Winnipeg, 296p.

Matsumura, F. 1975. *Toxicology of Insecticides.* Plenum Press, New York, 512p.

Miller, C. W., B. M. Zuckerman, and A. J. Charig. 1966. Water translocation of diazinon-C14 and parathion-S35 off a model cranberry bog and subsequent occurrence in fish and mussels.*Trans. Am. Fish. Soc.* **95:**345.

Moffett, J. O., and H. L. Morton. 1975. How herbicides affect honey bees. *Am. Bee J.* **115:**178.

Montgomery, J. J. 1982. Water wells. *Pest Control* **50:**48.

Morgan, H. G. 1983. Sublethal effects of diazinon on stream invertebrates. Ph.D. diss., University of Guelph, Guelph, Ontario.

Mulla, M. S., L. S. Mian, and J. A. Kawecki. 1981. Distribution, transport and fate of the insecticides malathion and parathion in the environment. *Residue Rev.* **81:**2.

Mullison, W. R. 1970. Effects of herbicides on water and its inhabitants. *Weed Sci.* **18:**738.

Mullison, W. R. 1981. *Public Concern about the Herbicide 2,4-D* Dow Chemical, 33p.

Nash, R. G., and E. A. Woolson. 1967. Persistence of chlorinated hydrocarbon insecticides in soils. *Science* **157:**924.

National Academy of Sciences. 1982. *An Assessment of the Health Risks of Seven Pesticides Used for Termite Control.* Committee on Toxicology, Board on Toxicology and Environmental Health Hazards, Commission on Life Sciences, Washington, D.C.

National Research Council. 1980. *Urban Pest Management.* Report prepared by the Committee on Urban Pest Management, Environmental Studies Board, Commission on Natural Resources. National Academy Press, Washington, D.C., 272p.

National Research Council of Canada. 1974. *Chlordane: Its Effects on Canadian Ecosystems and Its Chemistry,* NRCC 14094. Ottawa, Ontario, 189p.

National Research Council of Canada. 1978*a*. *Phenoxy Herbicides: Their Effects on Environmental Quality with Accompanying Scientific Criteria for 2,3,7,8-tetrachlorodibenzo-p-dioxin (TCDD),* NRCC 16079. Ottawa, Ontario, 314p.

National Research Council of Canada. 1978*b*. *Ecotoxicology of Chlorpyrifos.* NRCC 16079. Ottawa, Ontario, 314p.

National Research Council of Canada. 1981. *Polychlorinated dibenzo-p-dioxins: Criteria for Their Effects on Man and His Environment,* NRCC 18575. Ottawa, Ontario, 248p.

National Research Council of Canada. 1982*a*. *Effects of Propoxur on Environmental Quality with Particular Reference to Its Use for the Control of Biting Flies,* NRCC 18572. Ottawa, Ontario, 238p.

National Research Council of Canada. 1982*b*. *Biting Flies in Canada: Health Effects and Economic Consequences,* NRCC 19248. Ottawa, Ontario, 157p.

National Research Council of Canada. 1983*a*. *Impact Assessments in Lotic Environments: methoxychlor,* NRCC 20645. Ottawa, Ontario, 130p.

National Research Council of Canada. 1983*b*. *2,4-D: Some Current Issues,* NRCC 20647. Ottawa, Ontario, 99p.

Niemczyk, H. D., H. R. Krueger, and K. O. Lawrence. 1977. Thatch influences movement of soil insecticides. *Ohio Rep.* **62:**26.

Pimentel, D. 1971. *Ecological Effects of Pesticides on Nontarget Species.* Office of Science and Technology, Washington, D.C. 220p.

Pinto, L. J. 1981. *The Structural Pest Control Industry: Description and Impact on the Nation.* National Pest Control Association, Vienna, Va., 26p.

Plimmer, J. R. 1973. Technical pentachlorophenol: Origin and analysis of base insoluable contaminants. *Environ. Health Perspect.* **5**(exp. issue):41.

Poiger, H., and C. Schlatter. 1983. Animal toxicology of chlorinated dibenzo-p-dioxins. *Chemosphere* **12:**453.

Poland, A. P., D. Smith, D. Metter, and P. Possick. 1971. A health survey of workers in a 2,4-D and 2,4,5-T plant. *Arch. Environ. Health* **22:**316.

Rapoport, E. H., and G. Cangioli. 1983. Herbicides and the soil fauna. *Pedobiologial* **2**:235.

Rawn, G .P. 1977. Disappearance and bioactivity of Dursban insecticide in temporary pools. M. S. thesis, University of Manitoba, Winnipeg.

Reggiani, G. 1983. Toxicology of TCDD and related compounds: Observations in man. *Chemosphere* **12**:463.

Rice, D. 1981. Colorado court involvement in chemical spill cleanup activities. *Am. J. Public Health* **71**:1044.

Robinson, W. H., and R. L. Atkins. 1983. Attitudes and knowledge of urban home-owners towards mosquitoes. *Mosq. News* **43**:38.

Schwetz, B. A., J. M. Norris, G. L. Sparschu, V. K. Rowe, P. J. Gehring, J. L. Emerson, and C. G. Gerbig. 1973. Toxicology of the chlorinated dibenzo-p-dioxins. *Environ. Health Perspect* **5**(exp. issue):87.

Sears, M. K., and R. A. Chapman. 1979. Persistence and movement of four insecticides applied to turfgrass. *J. Econ. Entomol.* **72**:272.

Sears, M. K., and R. A. Chapman. 1982. Persistence and movement of four insecticides applied to turfgrass. In *Advances in Turfgrass Entomology,* H. D. Niemczyk and B. G. Joyner (eds.). Hammer Graphics, Piqua, Ohio, p. 78.

Severo, R. 1984. Accord reached on cleaning up Niagara wastes. *New York Times* January 11, p. A1.

Thompson, A. R., and W. W. Sans. 1974. Effects of soil insecticides in southwestern Ontario on nontarget invertebrates: Earthworms in pastures. *Environ. Entomol.* **3**:305.

Thompson, D. G., G. R. Stephenson, and M. K. Sears. 1984. Persistence, distribution and dislodgable residues of 2,4-D following its application to turfgrass. *Pestic. Sci.* **15**:353.

Thomson, W. T. 1981. *Agricultural Chemicals,* vol. 2, *Herbicides.* Thomson, Fresno, Calif., 274p.

Tu, C. M., and J. R. W. Miles. 1976. Interactions between insecticides and soil microbes. *Residue Rev.* **64**:17.

Turner, D. J. 1977. *The safety of Herbicides 2,4-D and 2,4,5-T,* Forestry Commission Bulletin, Report 57. Her Majesty's Stationery Office, London, 56p.

U.S. Environmental Protection Agency. 1972. *Report on Aldrin/Dieldrin,* Report No. EPA/540/5-72/001, 106p.

U.S. Environmental Protection Agency. 1976*a*. *Heptachlor in Relation to Man and the Environment,* Report No. EPA/540/4-76/007, 73p.

U.S. Environmental Protection Agency. 1976*b*. *Notice of Intent to Cancel Registered Uses of Products Containing Chlordane and Heptachlor,* Report No. EPA/540/4-76/003, 64p.

U.S. Environmental Protection Agency. 1976*c*. *Pesticidal Aspects of Chlordane in Relation to Man and the Environment,* Report No. EPA/540/4-76/006, 88p.

U.S. Environmental Protection Agency. 1976*d*. *Pesticidal Aspects of Chlordane and Heptachlor in Relation to Man and the Environment: A Further Review, 1972-1975,* Report No. EPA/540/4-76/005, 93p.

U.S. Environmental Protection Agency. 1978. *Pentachlorophenol: Position Document 1,* Report No. EPA/SPRD-80/85, 70p.

U.S. Environmental Protection Agency. 1979*a*. *Decision and Emergency Order Suspending Registrations for Certain Uses of 2-(2,4,5-trichlorophenoxy) Propinic Acid (Silvex),* Report No. EPA/SPRD-80/102, 113p.

U.S. Environmental Protection Agency. 1979*b*. *Decision and Emergency Order Suspending Registrations for the Forest, Rights-of-Way, and Pasture Uses of 2,4,5-trichlorophenoxyacetic Acid,* Report No. EPA/SPRD-80/103, 113p.

U.S. Environmental Protection Agency. 1979*c*. *Silvex: Position Document 1/2/3,* Report No. EPA/SPRD-80/52, 114p.

U.S. Environmental Protection Agency. 1979*d*. *2,4,5-T: Position Document 1.* Report No. EPA/SPRD-80/76. 150p.

U.S. Environmental Protection Agency. 1979*e*. *National Household Pesticide Usage Study, 1976–1977,* Report No. EPA/540/9-80-002, 126p.

U.S. Environmental Protection Agency. 1982. *Wood Preservative Pesticides Creosote, Pentachlorophenol and the Inorganic Arsenicals (Wood Uses): Position Document 2/3,* Report No. EPA/540/9-82-004, 906p.

U.S. Environmental Protection Agency. 1983. *Analysis of the Risks and Benefits of Seven Chemicals used for Subterranean Termite Control,* Report No. EPA/540/9-83-005, 70p.

Wallace, R R., A. S. West, A. E. R. Downe, and H. B. N. Hynes. 1973. The effects of experimental blackfly (Diptera:Simuliidae) larviciding with Abate, Dursban and methoxychlor on stream invertebrates. *Can. Entomol.* **105:**817.

Weed Control Manual. 1982. Meister (*Ag Consultant and Fieldman*), Willoughby, Ohio.

Wilbert, D. E. 1963. Some effects of chemical sagebrush control on elk distribution. *J. Range Manage.* **16:**74.

Wilson, D. M., and P. C. Oloffs. 1973. Persistence and movement of α and γ chlordane in soils following treatment with high purity chlordane. *Can. J. Soil Sci.* **53:**465.

World Health Organization. 1974. Proxpur. In *1973 Evaluations of Some Pesticide Residues in Food,* Pesticide Residue Series, No. 3. Geneva, Switzerland, 330p.

Young, A. L., H. K. Kang, and B. M. Shepard. 1983. Chlorinated dioxins as herbicide contaminants. *Environ. Sci. and Technol.* **17:**530.

Young, A. Y. 1980. Phenoxy herbicides and micro organisms. In *The Science of 2,4,5-T and Associated Phenoxy Herbicides,* R. W. Bovey and A. L. Young (eds.). Wiley, New York, p. 139.

4

Importance of Sanitation

Boyd T. Marsh

Summit County General Health District
Cuyahoga Falls, Ohio

Judy K. Bertholf

Dow Chemical
Dallas, Texas

Sanitation is an important concept relative to the control of many pest species in the urban environment. From a pest management standpoint, sanitation is a reflection of the available food, water, and harborage in an environment. These three components of sanitation are identified as factors that regulate population numbers (Clark et al., 1967). From a public health standpoint, sanitation can be defined as a modification of the environment in such a manner that a maximum of health, comfort, safety, and well-being occurs to mankind (Johnson, 1960).

There is a wide range of problems that can result from unsanitary environmental conditions. Those conditions that provide food, water, harborage, or concealed routes of movement all influence whether pest species will be present, can become established, and can sustain themselves. By improving sanitary conditions we attempt to control environmental factors that will impact upon pest populations. By reducing any or all of these factors the population of many pest species can be correspondingly reduced (Jackson and Marsh, 1978).

The conditions that constitute an acceptable sanitation level will vary, depending on the specific pest and environment involved. Obviously, sanitary practices and standards will differ between health care facilities, food handling establishments, animal facilities, or multi-family dwellings. Practices and regulations will also differ between public institutions and private dwellings. Therefore, each situation is unique and should be evaluated and treated as such.

General sanitation evaluation checklists and surveys have been used in the past (National Pest Control Association, 1972; Gupta, 1973, 1975; Bennett, 1978). These surveys have had limited success because they invariably lack specificity for each particular situation and because it is difficult to quantify the level of sanitation. Evaluation systems are often arbitrarily established without information regarding the relative importance of various sanitation deficiencies. Furthermore, these systems are subject to the evaluator's own interpretation.

The impact that improved sanitation ultimately has on a pest population will depend on the situation. In some cases sanitation alone may be an effective control strategy, in other situations it may have little direct effect toward reducing pest numbers. However, sanitation is important when it is incorporated with other components of an integrated pest management system. Sanitation may, in fact, enhance other control strategies in an indirect but measurable way.

Unfortunately, in many instances, controlling pests by improving an unsanitary environment is more costly in the short term and takes a planned and sustained effort. This improvement is often perceived by the public as being more complicated than a chemical control program. As a result only token efforts are often made to alleviate unsanitary conditions.

PROBLEMS CREATED BY UNSANITARY ENVIRONMENTS

Available food, water, and harborage for many pests species is often generated as a by-product of human activity (Frankie and Ehler, 1978). At one time the accumulation of garbage and trash from backyard burning as well as the unburned residual, provided a ready food supply for many pests. Due in large measure to air pollution restrictions and community-wide waste collection, backyard burning is not as common as it once was. These trends have reduced the available food resources in urban areas, and consequently certain types of pests such as larder beetles have become less common.

Many communities have encouraged the use of plastic and heavy paper refuse sacks. This method of household trash containment is convenient for the individual resident, but in areas with rodent or stray dog populations, bags are frequently torn open, and contents strewn about (Marsh, 1971). It is not clear if the dogs smell the food in the plastic bags or learn the food is there from previous encounters with open bags (Beck, 1973). This activity makes the garbage available to other pests and also increases the unsightliness of the urban environment. It has been reported that the presence of free-ranging dogs in poor neighborhoods is probably correlated with the excessive garbage available (Beck, 1973).

The situation becomes more complex when industrial buildings and multiunit housing are considered. Common trash-handling methods include outside dumpsters, metal trash chutes, compactors, and incinerators. Each of these methods is subject to pest infestations and should be regularly inspected. It is critical to monitor and control (keep sanitary) the trash-handling areas of such facilities if there is to be successful pest control within the structure (Nixon, 1984).

In many central cities a large number of structures have been demolished without subsequent rebuilding. The resulting vacant lots often become weed covered and serve as depositories for illegally dumped garbage or trash. The weed growth and garbage on these lots add stress to the monetary resources of a community. As a result the garbage may not be frequently collected (Kohuth and Marsh, 1973, 1976; Marsh, 1970).

In communities where refuse collection is controlled by the community (collection by the community or by contract), there is a high probability that these materials will be regularly removed from individual premises. However, if collection is less frequent, the food availability to various pests increases, and the removal of garbage and trash from inside the structure may also become less frequent.

Unsanitary conditions also have a blighting influence on a neighborhood and community. As a neighborhood begins to show signs of decay (unsanitary conditions) more of the same seems to be encouraged. There may be health problems because of overcrowding and poor sanitation. Often a lack of basic facilities and services results from inadequately maintained structures. An obvious outcome of these situations is unsanitary conditions that can influence the numbers and types of pests as well as their proximity to humans (Hinkle and Loring, 1977).

Chronic cockroach infestations in high-density public housing units are perpetuated by a number of factors, including poor sanitation and lack of environmental controls (Todaro, 1984). This observation can be extended to include other pests such as rats and mice.

Federal funding for urban rodent control efforts dramatically called attention, not only to the rat, but to those unsanitary conditions that contribute to the presence of rodents, citizen concerns about rodents, health-related issues, and rodent control activities (Marsh, 1977). Food sources that make rat control more difficult include inadequately stored garbage, animal feed, fruit and vegetables from home gardens, animal feces, food inside the house or food operation, and the backyard burning of garbage (Jackson and Marsh, 1978). Water is usually available to rodents either through garbage with high moisture content, poorly drained sewer and drain systems below ground, or surface water. Harborage is also usually abundant (Brooks, 1973).

The seriousness of pest problems created by unsanitary conditions is magnified if we consider the close association between many human health problems and related pest species. As the population of a pest increases and those pests and their wastes come into closer association with urban dwellers, the potential for human disease problems also increases.

Pigeon populations, with sustained roosting or nesting, may result in an accumulation of droppings on buildings, park benches, and sidewalks. These accumulations provide the potential for the development of fungal diseases such as histoplasmosis. Pigeons have also been shown to carry salmonella, encephalitis, toxoplasmosis, pseudoterberculosis, pigeon ornithosis, New Castle disease, and cryptococcosis (Martin and Martin, 1984). Mites and other bird ectoparasites may also affect humans.

Cockroaches have also been implicated in the transmission of disease to humans by mechanically transmitting pathogens to foods ingested or objects handled by humans. The most obvious association is with food products, and most common pathogens are gastro-intestinal related (Alcamo and Frishman, 1980; Frishman and Alcamo, 1977).

Rat bites are another obvious consequence of a close association with, and exposure to, a pest species. The bites of other mammals, particularly dogs, also constitute a major problem in many urban areas. It has been estimated that free-ranging dogs may well account for a majority of reported bites. Besides trauma, disfigurement, and even death as a result of dog bites, an overlying concern is rabies (Beck, 1973; Sampson, 1984). Vaccination efforts have substantially reduced the rabies incidence in domestic animals. However, many wild animals have become semi-domesticated and live off urban garbage. Whenever domestic animals become associated with wild animals, there is always a possibility of introducing rabies into the domestic animal population (McLean, 1970). It should be emphasized that rats are not involved with rabies transmission.

Food operations attract pests by their very nature. A pest infestation will present potentially serious health hazards through food contamination and add to owner costs, as contaminated foods must be disposed of (National Research Council, 1980). Understanding by the facility owner or manager of proper sanitation measures can reduce the possibility of pests becoming established or of their persisting as a chronic problem. The routine delivery of raw food products provides pests with access to indoor areas. Refuse storage is frequently in close proximity to food operations. Once pests are inside, food preparation practices, food storage, and inadequate sanitation can all contribute to the establishment and continuation of pest problems.

The presence of pests results in damaged product and at the very least damaged goodwill with the customer (Truman, Bennett, and Butts, 1976).

INFLUENCE OF SANITATION ON THE ENVIRONMENT, PESTS, AND PEOPLE

Any environment has a carrying capacity, which is the number of organisms that a particular environment will support (Odum, 1959). Those controllable environmental factors that impact on populations are known as limiting factors. By reducing any one or more of these factors, we tend to increase competition for the remaining resources. Increasing competition within a pest population results in decreased pest numbers through mortality or dispersal. Additionally, if the decrease in one or more of the limiting factors can be maintained, the change (reduction in pest numbers) will become permanent. Subsequent reduction of a limiting factor already in least supply can have an even greater impact in reducing a pest population. This alteration of habitat can result in the long-term reduction of a pest species (National Research Council, 1980).

In some cases a well-implemented sanitation program alone has been shown to be effective in reducing pest populations. This was demonstrated in Baltimore (Davis, 1953) when the effectiveness of rodent baiting alone was compared to improved sanitation (Fig. 4-1). In the latter part of this study,

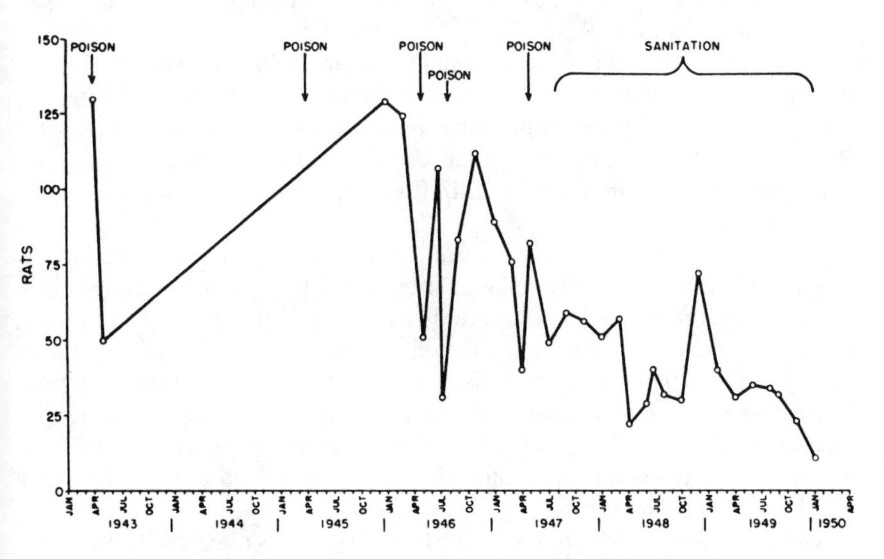

Figure 4-1. Norway rat population subjected to poison and finally to sanitation.

sanitation efforts were implemented and rodenticides were not used. While baiting resulted in temporary population reductions, improved sanitation resulted in a sustained population decrease over time. In fact the carrying capacity of that particular habitat had been reduced.

Mosquito control, prior to the wide availability of insecticides, was based on survey, habitat modification, and breeding site elimination. The elimination of breeding sites by simple drainage played a considerable role in eliminating malaria from the United States (National Academy of Sciences, 1976). Urban yellow fever was eliminated from southern U.S. cities by the destruction or removal of artificial water containers in and around dwellings. Public health agencies recognized that property owners play a major role in reducing the numbers of pest- or disease-causing mosquitoes by eliminating water-holding reservoirs in and around their homes. This form of nonchemical control is mandated by law in some cities of the world where malaria remains an ever present threat (e.g., Singapore). In such areas, this form of sanitation (breeding source removal) is very effective in limiting mosquito populations. (Chapter 15 further addresses this subject.)

Although insecticides have been the mosquito control tool of choice during recent years, environmental improvement efforts have become increasingly popular in large measure because of insecticide resistance and concerns over environmental contamination (National Academy of Sciences, 1976). Sanitation to prevent or reduce insect infestation through removal of breeding and hibernating sites is a sound insect control principle with broad applicability (National Research Council, 1969).

Inadequately stored garbage, animal feces, and infrequently turned compost piles all provide breeding sites for many filth, and some biting, fly species. It is estimated that proper refuse sanitation alone would provide 90% of fly control in many urban areas (Johnson, 1960). Additionally, insecticides will not adequately control filth flies while breeding sites continue to exist (Williams, 1982).

While it is clear that improving sanitary conditions will generally have a positive impact on the presence or numbers of many pest species, this relationship (improved sanitation and reduced pest numbers) is not direct and clear-cut. Good sanitation will reduce high populations of the house mouse but is apparently a less satisfactory method of control than for rats. Even if substantial sanitation improvements are made, the house mouse cannot be eliminated by sanitation alone because of the ability of that animal to survive in small areas on limited amounts of food and water (Marsh and Howard, 1976*a,* 1976*b,* 1976*c*).

The German cockroach has been shown to be less affected by improvements in sanitation than once suspected. Once a population is established, the adaptability of this pest makes it virtually impossible to improve sanitary conditions to a level that will reduce the population (Owens, 1980; Bertholf, 1983). Recent studies have shown that, although German cockroach populations are disrupted by improved sanitation (Bertholf, 1983; Bennett, Runstrom, and Bertholf, 1984), the incorporation of other control methods is necessary in order to effect any significant population reduction (Bennett, 1984; Bennett, Runstrom, and Bertholf, 1984).

Many vertebrate pests are also less affected by sanitation. Most frequently a bird's food source is separated from its roosting or nesting site. With a less restrictive home range, sanitation and environmental control techniques are difficult and often less effective for reducing bird numbers than for many other pest species (Schneider and Jackson, 1968). Spilled grain at railroad tracks and the purposeful feeding of birds (pigeons at public places) provides a reliable food source and assures a continuing nuisance.

In the scientific literature, little data exists relative to the impact of sanitation when it is combined with other strategies in integrated pest management programs. In a series of experiments, Gupta examined the relationship between sanitation and the effectiveness of insecticide treatments for German cockroach control (Gupta et al., 1973, 1975). He concluded that, over time, the insecticide treatments were more effective in homes with good sanitation. Later studies have shown that the effectiveness of caulking (to eliminate or reduce harborage availability) as a German cockroach control strategy may also depend on the level of sanitation (Farmer and Robinson, 1984).

In a general sense, improvement in basic sanitation in the urban environment has a positive influence on the total environment. Improving sanitary conditions does reduce the numbers of many pest organisms and may largely eliminate certain species from an area. (See chapters 9 and 10 for information on sanitation as related to plant pest management.)

SOCIAL AND ECONOMIC IMPLICATIONS ASSOCIATED WITH HOW PEOPLE LIVE

A major deterrent to using environmental improvement practices as preferred methods of pest control is the inappropriate expectation that these efforts should bring about total elimination of a pest species. Public demand for pest-free homes and gardens, weed-free lawns, and unblemished flowers all result in the use of pesticides for the "quick fix." This attitude can be attributed to where we live, how we live, and our expectations as to the quality of life. Since unsanitary conditions are usually multiple and pervasive in nature, no one person or group generally has sufficient control over those factors that will allow a readily demonstrable direct decrease in pest numbers.

The all too common result is a continuing dependence upon chemical control methods with little consideration for the need of environmental improvements.

Recent surveys have been made to assess the attitudes of the public toward urban arthropods and pesticide use (Byrne et al., 1984; Bennett, Runstrom, and Wieland, 1983; Wood et al., 1981). In these studies sanitation was recognized by the public as a factor in urban pest control. While poor sanitation was often given as the major cause of an infestation, good sanitation was listed as a less effective method of control than conventional chemical applications (Wood et al., 1981).

A great number of pest problems occurs in low income areas, where the range of personal and family problems may well assume higher priorities than do the unsanitary conditions that contribute to the pest problem. Under these conditions it is virtually impossible to mobilize the sustained efforts necessary for environmental improvement that will impact on pest numbers.

SANITATION PRACTICES AND TECHNIQUES

Sanitation practices and techniques can be classified either as preventive or corrective. While no structure is totally immune to pest infestation, preventive sanitation can reduce susceptibility. For example, proper site preparation and sanitation during construction will reduce the chance of a subterranean termite infestation (Ebeling, 1978). Ideally, preventive sanitation practices become a way of life (American Institute of Baking, 1979).

Corrective sanitation frequently comes to the attention of the professional pest control operator. The professional can insist the client correct the unsanitary conditions before control methods are implemented or subcontract the job out (Katz, 1979). Some firms have taken the approach of performing the sanitation measures themselves and charging accordingly.

In urban environments, habitat modification can include a number of activities including (National Research Council, 1980) proper storage, collection and disposal of garbage and trash; good housekeeping within structures; proper food storage and handling; elimination of pest harborage indoors and outdoors; demolition or renovation of abandoned or dilapidated structures; maintenance and upkeep of occupied and secondary structures; enforcement of sanitation, health, and housing codes; elimination of pest entry points (proper screening, caulking); and permanent elimination of pest breeding sites and other sources of pest infestation.

In a general sense, these activities apply to the most common urban pests, including rodents, birds, bats, cockroaches, filth flies, mosquitoes, termites, and stored product pests.

When dealing with industrial and public health pest problems, a sanitation program must be tailored to meet the specific needs of the situation. In many instances the use of nonchemical control measures becomes critical because of the presence of nontarget animals, health concerns, government regulations, and high public attendance.

Sanitation has been shown to be the best preventive measure against many pest problems in amusement parks. Routine inspections and reports are necessary follow-ups once a sanitation program has been established (Hofmeister, 1982).

Animal research facilities require similar considerations. Key elements in a preventive sanitation program include exclusion, proper drainage, proper feed storage, regular animal waste disposal and replacement of animal bedding, and inspection of incoming products (Tucker, 1984).

Sanitation practices are particularly important in the food service industry (National Pest Control Association Committee, 1972; Truman, Bennett, and Butts, 1976; American Institute of Baking, 1979; Cohn, 1982). Sanitation strategies include locating the food establishment in as clean and pest-free an environment as possible; maintaining the structural integrity of the building; designing equipment for easy cleaning; and initiating routine management practices designed to reduce or eliminate the likelihood of pest infestations (inspection on incoming supplies, proper storage of perishables, regular and scheduled inspection and cleaning of the premises, proper disposal of garbage and waste water, and regular machinery maintenance). The key to the development of good quality control in sanitation is the establishment of an education program.

ROLE OF EDUCATION IN IMPROVING SANITATION

Education is essential if we are to improve the public's knowledge, attitudes, and actions concerning sanitation issues. Often the public does not recognize the relationship between their actions and a resulting pest problem. Just as frequently, individuals may acknowledge a relationship to their actions but do not care (littering is an example). An inherent reason for any education experience is to alter an individual's behavior so that the target (deleterious) behavior is expunged or changed to something more positive (Johnson, 1960; Kohuth and Marsh, 1973, 1974, 1976, 1977; Marsh, 1970).

Several agencies and organizations are involved with public education on sanitary issues. These include public health and regulatory officials, state extension personnel, and pest control operators. These groups are most often involved in correcting individual problems. To effectively deal with community-wide sanitation issues these groups must work together. Unfortunately, the communication and cooperation between these groups is often not as good as it should be (Beck, 1981; Martarano, 1973).

The effectiveness of cooperation and education in establishing integrated pest management (IPM) programs was demonstrated in Pittsburgh. Health officials, pest control operators, and IPM consultants worked together to establish an integrated program for cockroach control. Active involvement by both public housing tenants and management was also essential. An ongoing education effort was established to ensure tenants were aware of the role they played, both in perpetuating pest problems and in their successful solutions. Pamphlets, newsletters, and training sessions were used to stress the importance of sanitation for successful pest management. Testing later showed that many of the sanitation practices were adopted by the tenants (Todaro, 1984).

Another example was the Federal Urban Rat Control Programs, which consisted of ten program elements encompassing all of the possible areas that contribute to the presence of rodents or their immediate and permanent removal. Two of the program elements, (a) community organization and citizen participation and (b) community education, were designed to inform the public of their role and responsibility in improving those environmental conditions that contribute to the presence of rats and to develop procedures for the involvement of citizens in the process of community improvement (National Pest Control Association, 1968).

The basic objectives of the federal guideline for Urban Rat Control projects were stated as "education and motivation of citizens in the principles of environmental sanitation to reduce and prevent rat infestation, and to improve the livability of the disadvantaged areas" (U.S. Department of Health, Education, and Welfare, 1972, p. 1).

There was a purposeful effort by health agency staff to train community residents (Health Educator Aides) in recognizing environmental influences that related to pests and to use these trained individuals to inform other members of the community (Knittel, 1970; U.S. Depart. Health, Education and Welfare, 1969*a,* 1969*b*). The use of indigenous persons encouraged the development of training techniques and materials appropriate for local needs. Unfortunately, as the federal budget for rat control was reduced, health educator aides were among the first staff to be terminated from the program.

Informational materials focusing on single issues like on-premises garbage storage were designed for the individual homeowner or apartment dweller. Multi-media approaches for identifying pests and environmental problems (Marsh, 1970), ways to organize and keep neighborhood groups together (Kohuth and Marsh, 1976) and educational school programs (Kohuth and Marsh, 1973, 1974, 1977) are a few of the activities used in urban rat control projects.

Yale University, under contract to the then Department of Health, Education and Welfare, developed guidelines for evaluating the content and motivational impact of the educational program. These efforts were helpful in improving the educational materials used by the rodent programs (U.S. Depart. Health, Education, and Welfare, 1972). Many of the principles and materials developed through the Urban Rat Control projects can be applied to other environments and pests.

Considerable efforts have been directed toward the development and implementation of school curricula, the theory being that if young people are taught the effects of environmental abuses, they might be motivated, in both the short and long term, to maintain proper habits (Kohuth and Marsh, 1974).

In Urban Rat Control Programs the goals of community organizations and participation of citizens expanded beyond mere rodent control. By involving citizens in the sanitation improvement process, the chances of informed personal choice and "buying into" sanitation improvement solutions also increases and improves the possibility of better neighborhood environments and fewer pests (U.S. Depart. Health, Education and Welfare, 1972).

This same principle can also be applied to industries where individual food operations or warehouses are subject to pest problems. These problems can be reduced if employees understand the factors that contribute to the situation. Employee incentives to implement control measures based on sound sanitation principles should be incorporated into personnel management programs.Active involvement of both personnel and management is essential for solving such problems. Individual instruction, on-the-job training, and an understanding of each individual cleaning schedule is critical in correcting unsanitary conditions (American Institute of Baking, 1979).

ROLE OF GOVERNMENT IN IMPROVING SANITATION

Government at all levels has a role in improving sanitation. That role may take one of three basic forms: (a) encouraging sanitation improvement, (b) implementing sanitation improvement, and (c) requiring sanitation improvement through regulatory code enforcement activities. Government encouragement of sanitation improvement is frequently accomplished through information and education efforts.

The implementation of sanitation improvement by government involves some of the basic services the public has come to expect. These services include garbage and trash collection, street sweeping, animal control services, weed cutting, and vacant lot sanitation. The implementation of sanitation improvement may also include the formation of districts where environmental management is assumed as a specific activity by a designated public agency. For example, in mosquito control districts those activities relating to control (adulticiding, larvaciding, water management, public information) are assumed by that agency.

Government action requiring sanitation improvement generally takes the form of laws or regulations. Regulation takes place at all levels of government, but the most meaningful and effective level for sanitation improvement occurs at the local level. Health agencies and local government units generally have the power to enact and enforce laws to protect and promote the health, safety, morals, order, peace, comfort, and general welfare of the people. This is known as state police power, which local health agencies and local units of government derive from delegation by the state legislature (Grad, 1981).

It is interesting that the federal government does not have police power in its usual sense because it is a government of limited, delegated powers. Therefore, the federal government exercises control in the public health field through two powers: the power to regulate interstate commerce and the power to tax and spend money.

Most of the local regulations directed toward improving sanitation are classified as nuisances. These are conditions that have been recognized as a nuisance by the courts, without a statutory or other legislative finding or declaration to that effect (Grad, 1981). Nuisances are a miscellaneous group of conditions that may obstruct, cause inconvenience to the public, or adversely affect the property rights of a private person in the vicinity of the condition. Examples of nuisances relating to sanitation in urban areas include garbage that is inappropriately stored, transported or disposed of; uncut weeds; accumulation of animal feces; presence of rodents or other pests inside residential structures; and rodents exterior to buildings.

Other local regulations that influence sanitation and sanitation practices include: food service and handling (regulate cleaning, trash accumulation inside and outside the structure, and pest infestations inside and immediately outside the structure); housing codes (provide parameters for owner and tenant responsibilities in regard to such factors as garbage and trash, pests, and maintenance); solid waste (regulate the proper methods of disposal [U.S. Depart. Health, Education and Welfare, 1975]).

While local responsibility is rather extensive in improving sanitation, there are few state laws dealing specifically with urban pest management. Generally, control of insects, rodents, or other general nuisances is delegated by state law to local government, often to local health agencies (National Research Council, 1980).

Although the federal government has no legal authority to make policy for urban pest management, there are a number of federal agencies and laws that impact upon urban pest management.

1. *U.S. Environmental Protection Agency (EPA).* The Federal Insecticide, Fungicide and Rodenticide Act (FIFRA) is concerned with pesticide control not with pest management. Pest control companies are licensed under its authority, applicators are certified, and chemicals are registered and labeled under its provisions. Although pest management does not appear to have been a major consideration in solid waste programs required by The Resource Conservation and Recovery Act (RCRA), these programs do impact on urban pest problems.
2. *U.S. Department of Housing and Urban Development (HUD).* HUD has an interest in urban pest management because housing deterioration, unsanitary conditions, pests, and human health issues have been recognized in federal legislation on urban renewal. It has been calculated that the potential annual cost for pest control in HUD's 4,000,000 living units could range from $40-80 million (National Research Council, 1980).
3. *U.S. Department of Health and Human Services (HHS).* HHS interest in pest management is primarily related to disease prevention (National Research Council, 1980). The Food and Drug Administration (FDA) is responsible for regulating the handling, processing and transportation of food. Regulations require that food must not be adulterated nor prepared, packaged, or held under conditions whereby it may become contaminated. The Good Manufacturing Practice Regulations (GMPs) consist of guidelines for food processors that have the same effect as law.
4. *U.S. Department of Agriculture (USDA).* USDA has no statutory authority for pest management in urban areas, but its animal and plant health inspection service is charged with preventing the importation and spread of pests. Food Safety and Quality Service is responsible for federal meat and poultry inspection programs.
5. Other federal agencies that have some regulatory involvement include *Department of the Interior.* The National Park Service maintains urban parks and buildings, while the Fish and Wildlife Service conducts research and provides recommendations on vertebrate pests and their control. *Veterans Administration (VA)* operates pest control services for its hospitals and buildings. *Department of Defense* has recently estimated that their pest management programs cost $110 million per year (National Research Council, 1980).

Government does play a role in improving sanitation. Through education, services, and by requiring sanitation improvement, government at all levels is involved. The Urban Rat Control Program provides an example of positive involvement by the government. The goal of the program was to eliminate rats from urban residential areas by correcting environmental conditions that promoted their proliferation (Kennedy, 1981). The effectiveness of the project depended on changes in the residents' attitudes and behavior in relation to garbage storage and associated good housekeeping practices (Kennedy, 1981). The intent was to focus the community resources so as to maximize the coordinated benefits of the program. The following items were proposed for accomplishment primarily by local governments (U.S. Depart. Health, Education and Welfare, 1972): Inspection of premises for rat problems and "code" violations; refuse and garbage collection and disposal; renovation or demolition of structures; clean-up of vacant lots; ratproofing; removal of abandoned automobiles and other materials providing harborage; code enforcement; sewer cleaning and repair; street sweeping; rat baiting; public education, citizen participation; and employee career development. As a result of the program, there was cooperation between the municipal agencies involved. The resulting skills and management techniques have since been used in other programs. For example, a program to control mosquitoes in Puerto Rico was developed using the Urban Rat Control Program as an administrative base.

FUTURE NEEDS

Sanitation has long been recognized as a foundation for sound pest control programs. It is logical to assume that by controlling factors that promote the presence of pests we can limit their numbers. Although this is a simple concept, the difficulties in accomplishing necessary sanitation seem enormous.

Generally, the urban environment is controlled or owned by many entities (homeowners, apartment dwellers, business, government). Each of those owners has some level of responsibility in maintaining an environment free of garbage, trash, high weeds, and so on. When that responsibility is not properly exercised, pests may well become a problem.

Improved sanitation measures have been shown to be effective in controlling certain urban pests (Davis, 1953; Jackson and Marsh, 1978; Johnson, 1960; Marsh and Howard, 1976; National Academy of Sciences, 1976; National Research Council, 1980). However, the task requires the education, interest, and cooperation of everyone involved.

Local laws and ordinances have not proven to be substantially effective in reducing urban pest problems. Individuals, not neighborhoods, industrial complexes, and the like are generally cited with nuisance, housing, food, and related ordinances. Although individuals must assume responsibility for a portion of the urban pest problem, citing each person under present local laws and regulations is an impossible task.

To improve sanitation and subsequently reduce urban pest problems, major efforts must be initiated to educate the public about the importance of improved sanitation for reducing urban pest populations; examine the present local laws and regulations as they relate to urban pest control and broaden their scope beyond the individual property owner; and actively seek public and private sector involvement in promoting citizen education and improved understanding of their responsibilities with regard to pest control activities.

Most important, further research is needed to assess the role of sanitation, both alone and when it is incorporated into IPM systems with other pest management strategies (e.g., chemicals). Specifically, more studies are needed to quantify the changes in pest population due to sanitation programs. Although sanitation has been shown to be essential in the control of many urban pests there is a great deal more we must learn to successfully maintain reduced urban pest populations.

REFERENCES

Alcamo, J. E., and A. M. Frishman. 1980. The microbial flora of field-collected cockroaches and other arthropods. *J. Environ. Health* **42**(5):263-266.

American Institute of Baking. 1979. *Basic Food Plant Sanitation Manual*, 3rd ed. American Institute of Baking, Manhattan, Kansas, 255p.

Beck, A. M. 1973. *The Ecology of Stray Dogs*. York Press, Baltimore, 98p.

Beck, J. E. 1981. The reforging of relationships. *Pest Control* **49**(12):8.

Bennett, G. W. 1978. Evaluating pesticides in urban environments. *Chem. Times and Trends* **2**(1):55-61.

Bennett, G. W. 1984. German cockroaches: Bionomics and control (pt. 3). *Pest Manage.* **3**(5):28-31.

Bennett, G. W., E. S. Runstrom, and J. K. Bertholf. 1984. Examining the where, why and how of cockroach control. *Pest Control* **52**(6):42-43, 46-50.

Bennett, G. W., E. S. Runstrom, and J. A. Wieland. 1983. Pesticide use in homes. *Bull. Entomol. Soc. Am.* **29**(1):31-38.

Bertholf, J. K. 1983. The influence of sanitation on German cockroach populations. Ph.D. diss., Purdue University, 71p.

Brooks, J. E. 1973. A review of commensal rodents and their control. *Critical Reviews in Environ. Control* **3**(4):405-453.

Byrne, D. N., E. H. Carpenter, E. M. Thoms, and S. T. Cotty. 1984. Public attitudes toward urban arthropods. *Bull. Entomol. Soc. Am.* **30**(2):40-44.

Clark, L. R., P. W. Geier, R. D. Hughes, and R. F. Morris. 1967. *The Ecology of Insect Populations in Theory and Practice*. Methuen, London, 232p.

Cohn, D. H. 1982. A good sanitation program. *Pest Control* **50**(5):24-26, 29.

Davis, D. E. 1953. The characteristics of rat populations. *Quant. Rev. Biol.* **28**(4):373-401.

Ebeling, W. 1978. Past, present, and future directions in the management of structure-infesting insects. In *Perspectives in Urban Entomology*, G. W. Frankie and C. S. Koehler (eds.). Academic Press, New York, pp. 221-247.

Farmer, B. R., and W. H. Robinson. 1984. Is caulking beneficial for cockroach control? *Pest Control* **52**(6):28, 30, 32.

Frankie, G. W., and L. E. Ehler. 1978. Ecology of insects in urban environments. *Annu. Rev. Entomol.* **23:**367-387.

Frishman, A. M., and J. E. Alcamo. 1977. Domestic cockroaches and human bacterial disease. *Pest Control* **45**(6):16-20, 46.

Grad, F. P. 1981. *Public Health Law Manual.* American Public Health Association, Washington, D.C., 243p.

Gupta, A. P., Y. T. Das, J. R. Trout, W. R. Gusicora, D. S. Adam, and G. J. Bordash. 1973. Effectiveness of spray-dust-bait combinations and the importance of sanitation in the control of German cockroaches in an inner-city area. *Pest Control* **41**(9):20-26, 58-62.

Gupta, A. P., Y. T. Das, W. R. Guisicora, D. A. Adam, and L. Jargowsky. 1975. Effectiveness of 3 spray-dust combinations and the significance of correction treatment and community education in the control of German cockroaches in an inner-city area. *Pest Control* **43**(7):28, 30-3.

Hinkle, L. E., and W. C. Loring, eds. 1977. *The Effect of the Man-Made Environment on Health and Behavior,* Publ. No. 77-8318. U.S. Department of Health, Education and Welfare, Center for Disease Control, Atlanta, Ga., 315p.

Hofmeister, S. 1982. Amusement parks. *Pest Control* **50**(3):32-36.

Jackson, W. B., and B. T. Marsh. 1978. Environmental control of rats. *Pest Control* **46**(8):12-16.

Johnson, W. H. 1960. *Sanitation in the Control of Insects and Rodents of Public Health Importance.* U.S. Department of Health, Education and Welfare, Public Health Service, Communicable Disease Center, Atlanta, Ga.

Katz, H. 1979. Integrated pest management then and now. *Pest Control.* **46**(11):16-18, 36-40, 81.

Kennedy, F. D. 1981. *The Urban Rat Control Program.* U. S. Department of Health and Human Services, Center for Disease Control, Atlanta, Ga., 30p.

Knittel, R. E., 1970. *Organization of Community Groups in Support of the Planning Process and Code Enforcement Administration,* Publ. No. 1997. U.S. Department of Health, Education and Welfare, Public Health Service, 69p.

Kohuth, B. J., and B. T. Marsh. 1973. Environment in education: A pragmatic look. *Educ. Leadership* **30**(7):656-658.

Kohuth, B. J., and B. T. Marsh. 1974. *An Educational Guide for Planning an Improved Human Environment.* Cleveland, Ohio, 255p.

Kohuth, B. J., and B. T. Marsh. 1976. *Neighborhood Improvement Guide.* Cleveland Department of Public Health, Cleveland, Ohio, 112p.

Kohuth, B. J., and B. T. Marsh. 1977. Urban environmental education: Aiming at results in Cleveland. *Environ. Educ. Rep.* **5**(9):3-4, 13.

McLean, R. G. 1970. Wildlife rabies in the United States: Recent history and current concepts. *J. Wildl. Dis.* **6:**229-235.

Marsh, B. T. 1970. Publicity techniques: Cleveland's rodent control and environmental improvement program. *Ohio Sanitarian* **14**(4):16-19.

Marsh, B. T. 1971. The plastic sack as a premise refuse storage method. *Ohio Sanitarian* **15**(2):16-20.
Marsh, B. T. 1977. Developing and committing other resources for urban rat control. *Ohio J. Environ. Health* **21**(5):3-6.
Marsh, R. E., and W. E. Howard. 1976*a*. House mouse control manual, part I. *Pest Control* **44**(8):23-33, 62-64.
Marsh, R. E., and W. E. Howard. 1976*b*. House mouse control manual, part II. *Pest Control* **44**(9):21-28, 53-54.
Marsh, R. E., and W. E. Howard. 1976*c*. House mouse control manual, part III. *Pest Control* **44**(10):27-38, 43-45.
Martarano, C. 1973. We need to learn more about each other. *Pest Control* **41**(9):40, 42.
Martin, C. M., and L. R. Martin. 1984. Solving those pesky pigeon problems. *Pest Control Technol.* **12**(4):44-48.
National Academy of Sciences. 1976. *Pest Control: An Assessment of Present and Alternative Technologies,* vol. 5, *Pest Control and Public Health.* National Academy Press, Washington, D.C., 282p.
National Pest Control Association. 1968. *Current Guidelines for Federal Rat Control Project Grants,* Service Letter 1209. Elizabeth, N.J., 4p.
National Pest Control Association Sanitation Committee. 1972. *Sanitation and Pest Control Floor-Level Inspection Manual.* Vienna, Va., 22p.
National Research Council. 1969. *Principles of Plant and Animal Pest Control,* vol. 3, *Insect-Pest Management and Control.* Committee on Plants and Animal Pests, Agricultural Board, Washington, D.C.
National Research Council. 1980. *Urban Pest Management.* Report prepared by the Committee on Urban Pest Management, Environmental Studies Board, Commission on Natural Resources. National Academy Press, Washington, D.C., 272p.
Nixon, J. 1984. Cockroaches and rodents make life tough when you're talking trash. *Pest Control* **52**(8):33, 37-38.
Odum, E. P. 1959. *Fundamentals of Ecology,* 2nd ed. W. B. Saunders, Philadelphia, 546p.
Owens, J. M. 1980. Some aspects of German cockroach population ecology in urban apartments. Ph.D. diss., Purdue University, 116p.
Sampson, W. W. 1984. The urban canine. *J. Environ. Health* **46**(6):306-310.
Schneider, D. E., and W. B. Jackson, ed. 1968. *Proceedings, Fourth Bird Control Seminar,* September 16-18. Bowling Green State University.
Todaro, W. 1984. Public housing: The pest control industry's ultimate challenge. *Pest Control Technol.* **12**(8):61, 62, 64, 76.
Truman, L. C., G. W. Bennett, and W. L. Butts. 1976. *Scientific Guide to Pest Control Operations,* 3rd ed. Purdue University/Harvest, Cleveland, Ohio, 276p.
Tucker, J. B. 1984. "Don't come near my research animals with that pesticide!" *Pest Control Technol.* **12**(2):46-53.
U.S. Department of Health, Education and Welfare. 1969*a*. *Health Educator Aide.* Public Health Service, Cincinnati, 139p.
U.S. Department of Health, Education and Welfare. 1969*b*. *Generating Community Action for Environmental Health.* Public Health Service, Cincinnati, 106p.
U.S. Department of Health, Education and Welfare. 1972. *Evaluation Guidelines*

and Measurement of Four Program Activities and a Manual for Their Use. Public Health Service, Atlanta, Ga., 92p.

U.S. Department of Health, Education and Welfare. 1975. *APHA-CDC Recommended Housing Maintenance and Occupancy Ordinance.* Publ. No. 75-8299. Center for Disease Control, Washington, D.C., 48p.

Williams, R. 1982. Fly control: Sanitation shoos this bothersome pest. *Pest Control* **50**(6):70-75.

Wood, F. E., W. H. Robinson, S. K. Kraft, and P. A. Zungoli. 1981. Survey of attitudes and knowledge of public housing residents toward cockroaches. *Bull. Entomol. Soc. Am.* **27**(1):9-13.

5

Inspection, Diagnosis, Pest Population Monitoring, and Consultation in Urban Pest Management

Keith O. Story

Winchester Consultants
Winchester, Massachusetts

Typically when a fly causes annoyance at the dinner table or a mouse is seen in a garage or a carpet beetle larva is seen on a favorite woolen cardigan, action is promptly taken against that particular fly, mouse, or carpet beetle larva. Often this reaction becomes an over-reaction to the individual pest, evoking latent hunter-killer instincts until the object of fury is destroyed. It is rare for people to view these and other pests as merely symptoms of more fundamental problems. This is unfortunate because, unless these other problems are recognized and dealt with, it is probable that the pest incidents will be repeated. Moreover, these other problems, which may include sanitation and building maintenance deficiencies, may be even more serious than problems from the pests themselves.

The general public can largely be excused for not knowing the various relationships between particular pests and conditions conducive to these pests. But people who provide a service to the general public, whether it be a food service, a maintenance service, or some other service, should have some understanding of pests. They should know how their own actions or inactions may influence pests. For instance, a home builder should know that soil-wood contact may lead to wood rot or attack by subterranean termites. Likewise if vents in eaves are too large and inadequately screened they may become access points for squirrels and other pests. Indeed the rationale behind many good building practices relates to pests rather than other engineering criteria.

In the food industry, levels of in-house knowledge regarding pests should be substantial for two reasons: to fulfill a duty to the consumer in providing wholesome food and to comply with federal, state, and local laws. The Federal Food, Drug and Cosmetic Act requires that measures be taken to avoid food adulteration, and provisions of this act are reflected in various additional codes governing sanitary standards for producing and handling food. Section 402 of the act particularly relates to pest control considerations, since it states that food shall be deemed adulterated "if it has been prepared, packed, or held under insanitary conditions whereby it may have become contaminated with filth, or whereby it may have been rendered injurious to health." In other words, food does not have to actually be contaminated in order to violate the act. The presence of pests or conditions conducive to pests would be sufficient to constitute a violation (Truman, Bennett, and Butts, 1976).

Notwithstanding the spirit and letter of these acts and codes it is still quite common to find food industry personnel who are dealing with the symptoms rather than the causes of pest problems. Two examples recently encountered illustrate the worst extremes of treating symptoms. The first involved a world-class restaurant where dinner for two typically cost in excess of $150. This restaurant had a German cockroach problem, which simply reflected a broader problem of poor food management and sanitation problems. In order to avoid offending the sensitivities of their clientele the restaurant management took action to prevent cockroaches appearing on the dining tables. This action consisted of employing a full-time person who stood just inside the kitchen and whose job it was to raise the covers of dishes and remove any cockroaches before the tray of dishes was taken into the dining room. The second example involved a food processor who supplied food ingredients to a food manufacturer. The food manufacturer noted that rodent droppings and hairs were present on many of the pallets received from the food processor and demanded that action be taken to prevent this. In response to this demand the food processor installed an air compressor and established a protocol for removing rodent droppings and hairs by air blast from all pallets immediately before they were transported to the complaining manufacturer.

In both of these examples the food industry personnel truly believed that the symptoms were the problem. Their actions, though stupid and inadequate, were not motivated by indifference or greed. They were the result of ignorance. These people, like the average consumer, were reacting in the best way they knew.

In the case of the restaurateur and the food processor their misguided actions were costing as much or more than programs aimed at removing both the causes of pest problems and the pests themselves. But in other cases a client's ignorance about the importance of pests creates a climate in which there is an unwillingness to spend much money on pest prevention and control. This is most commonly manifest when pest control work is contracted out on a low-bid basis with no requirement or expectation of eliminating pests. Typically such low-bid pest control, whether it is in an apartment block or a hospital, becomes a routine matter of harvesting the most accessible pests. In such cases the client expects little improvement in the pest situation and the contractor places few demands on the client for improvements in sanitation or maintenance.

Other examples of ignorance about pests and pest control include a widespread belief that eradication of some common pests, such as cockroaches, is impossible. In view of the frequent failure of contractors to eradicate pests, sometimes after decades of service work, this belief is perhaps understandable. In addition, there is a widely held belief that the measures necessary to control or eradicate pests are worse than the pests themselves.

All these types of ignorance can and must be addressed if the urban pest situation is to be improved. Many people can play a role in communicating the necessary information. They include university extension entomologists, government agency personnel, suppliers of materials and professional pest control operators. However, a key role is played by independent pest management consultants who can objectively put the pests and pest management options in perspective. Given the right perspective the attitudes of people can change dramatically. In the case of the restaurateur and food processor, once they learned that removing cockroaches from plates of food and rodent droppings from pallets did not meet their legal obligations or eliminate risk to their clients, they quickly agreed to a more comprehensive approach to the problem.

Concerning low-bid contract pest control, once clients recognize the health or economic importance of the pests they are more likely to give higher priority to pest eradication or prevention (Weber, 1979, 1982). For instance, most hospitals put their pest control contract needs out to bid, and typically the winning bid is too low to permit an adequate pest management effort. The result is that most hospitals accommodate many more pests than patients, but there is an assumption or belief by hospital staff that such pests do not adversely impact the patients. However, it is now well established (Beatson, 1972; Burgess and Chetwyn, 1979) that common pests in hospitals, such as cockroaches and Pharaoh ants, do carry pathogenic organisms and are at least theoretically capable of transmitting them directly, or indirectly via food and surgical dressings, to patients. (See Tables 5-1 and 5-2). When these facts are presented to hospital administrators or medical staff, it dispels

Table 5-1. Organisms isolated from cockroaches in five hospitals.

Organism	*Found in cockroach gut*	*Found on cockroach externally*	*Persistent for up to (days)*	*Found in cockroaches dead for 10 days*	*Found in environment*	*Known associated human diseases*
Citrobacter diversus	*	*	10	*	*	Urinary tract infections
Citrobacter freundii	*	*	10	*	*	Urinary tract infections
Enterobacter aerogenes	*	*	-	-	*	Urinary tract infections, sepsis
Enterobacter aglomerans	*	*	7	-	*	Urinary tract infections, sepsis
Enterobacter cloacae	*	*	-	-	*	Urinary tract infections, sepsis
Enterobacter hafniae	*	-	7	-	*	Gastroenteritis
Escherichia blattae	*	-	9	-	*	Pathogenesis unknown
Escherichia coli	*	*	2	-	*	Gastroenteritis; urinary, biliary, and peritoneal infections
Klebsiella pneumoniae	*	*	8	*	*	Pneumonia, urinary tract disease
Proteus mirabilis	*	*	6	-	*	Gastroenteritis, wound infection, urinary tract infection
Proteus morganii	*	*	8	*	*	Gastroenteritis, wound infection, urinary tract infection
Proteus rettgeri	*	*	6	*	*	Gastroenteritis, wound infection, urinary tract infection
Proteus vulgaris	*	*	10	*	*	Gastroenteritis, wound infection, urinary tract infection
Pseudomonas aeruginosa	-	*	-	-	*	Gastrointestinal and urinary tract infections
Serratia marcescens	*	*	-	*	*	Pathogenesis uncertain

Table 5-2. Hospital sites from which bacteria were isolated from Pharaoh ants.

Location	*Species of bacteria*	*Known associated human diseases*
Kitchens	*Staphylococcus aureus*	Pneumonia, bacteremia wound infection, osteomyelitis
	Pseudomonas aeruginosa	Gastrointestinal and urinary infections
	Bacillus spp.	Pathogenesis varies with species
	Escherichia coli	Gastroenteritis and urinary tract infections, etc.
	Staphylococcus dublin	Pathogenesis unknown
	Klebsiella spp.	Urinary tract infections, pneumonia
	Acinotobacter anitratus	Pathogenesis unknown
	Staphylococcus epidermidis	Skin abscesses
	Pseudomonas fluorescens	Pathogenesis unknown
	Bacillus cereus	Pathogenesis unknown
	Bacillus pumilis	Pathogenesis unknown
	Escherichia coli intermediate II	Urinary tract infections, etc.
	Clostridium welchii	Gas gangrene
	Providence subgroup	Sepsis, urinary tract infections
	Sarcina lutea	Pathogenicity unknown
Wards, casualty rooms, and intensive-care units	*Streptomyces pyogenes*	Pyodermatitis, septicemia
	Pseudomonas aeruginosa	Gastrointestinal and urinary tract infections
	Escherichia coli intermediate II	Urinary tract infections, etc.
	Neisseria sicca	Pathogenicity unknown
	Streptomyces faecalis	Pathogenicity unknown
	Pseudomonas fluorescens	Pathogenesis unknown
	Proteus spp.	Gastroenteritis, urinary tract and wound infection
Washrooms and toilets	*Escherichia coli* type I	Urinary tract infections
	Clostridium cochlearium	Pathogenesis unknown
	Providence subgroup	Sepsis, urinary tract infection
	Escherichia coli intermediate II	Urinary tract infections, etc.
	Klebsiella spp.	Urinary tract disease, pneumonia

any thoughts that these pests are mere nuisances and encourages commitment of more in-house and contracted pest management efforts. Indeed, in hospitals and such facilities as food-handling establishments, pharmaceutical plants, and recombinant DNA laboratories, it can be argued that there is no acceptable level of a pest that can carry diseases or compromise sterile conditions. Likewise, once clients understand that a single cockroach or mouse can cause fires that can destroy a mainframe computer or an airplane's flight controls, they will be more serious about pest control. In fact the terms "pest control" or "pest management" are increasingly being replaced in high risk or high value situations by the terms "pest elimination," "pest eradication," or "pest extermination." These terms imply that the goal of the anti-pest measures is total freedom from pests, and such a goal is achievable for most, if not all, indoor pests in individual buildings. Moreover, the cost of achieving pest freedom can be very economical when compared with potential losses resulting from the pests. For primarily outdoor pests in urban areas, such as mosquitoes and rats, eradication is a less common objective, particularly when the potential for reinvasion is high. Whether the acceptable number of pests is zero or something above zero, an integrated pest management (IPM) program requires a planned approach. Typically this plan involves a series of actions that fall into the following five categories: inspection; diagnosis and reporting; planning of the pest management or eradication strategy; implementation of strategy; and evaluation of results.

These five categories are listed in the sequence they would be carried out, but it is important to recognize that in practice it is a cyclical series of steps rather than a linear series. Following the evaluation of results, a new inspection would need to be conducted, a new diagnosis made, and so on.

Inevitably the diagnosis and consequent strategy in future cycles will be different than in past cycles because the pest situation will have changed and some of the environmental factors that affect the pest potential will have changed. For instance, a program aimed at killing a mouse infestation in a building may result in mouse carcasses in wall voids that might serve as future breeding grounds for blowflies or dermestid beetles. Likewise, the elimination of bats or birds roosting in or on a building may create at least a temporary problem of invading bat and bird ectoparasites and related scavengers. And the elimination of Argentine ants from an area may allow the recolonization of other species of ants that were formerly suppressed by the more aggressive Argentine ants.

Clearly the interrelationships between pests and environmental factors, including other pests, can be complex and are dynamic. A good inspection will help reveal these relationships and other factors pertinent to conducting an urban pest management program.

THE INSPECTION

The purpose of the inspection is to identify clues regarding the identity of any pests, the extent of any infestation, the source of the infestation, and any factors favoring survival of these or any potential pests. In addition, the inspection must reveal any circumstances that may restrict pest management options.

How well an inspection is carried out will largely determine the suitability and effectiveness of later efforts and special emphasis will be placed on this aspect.

Equipment

Proper equipment is essential for conducting an inspection, and for inspection of a building in which small insects have been reported the following items are useful.

Flashlight for illuminating dark areas favored by many pests. Note that the flashlight should be adequate for the job and in high-ceilinged areas or areas with deep voids a three-to-six-cell flashlight may be needed. In some situations an explosion-proof flashlight should be used.

Mirror with extension arm for use in conjunction with a flashlight to facilitate inspection of hard-to-see areas such as behind sinks or under work surfaces.

Screw top vials for collecting specimens for later study.

Hand lens for on-the-spot study of specimens.

Screwdriver and other tools for probing and gaining access to equipment where insects may hide.

Pyrethrum aerosol or dust for flushing cryptic insects such as cockroaches from inaccessible voids.

Notebook or inspection form for recording findings.

Map of site and colored dots for marking locations where problems exist.

In addition to such standard equipment, certain situations may require other inspection tools. For instance, white stockings pulled over the shoes and the bottoms of pants are useful for estimating the density of adult fleas. For assessment of populations of mosquito larvae a water dipper may be needed.

While most inspection tools and techniques rely on the senses of the inspector, there are now ways of extending these senses so that inspection accuracy and productivity are increased. Thus stethoscopes are useful for pinpointing ants, wasps, and other noisy insects in wall voids. And ultraviolet lamps are valuable in rodent inspections, since rodent urine fluoresces when exposed to ultraviolet light. Moisture meters are increasingly used for identifying areas that may be conducive to a carpenter ant infestation or to infection by wood-rotting fungi (Mallis, 1982). Moreover, since wood infested by termites has a higher moisture content, moisture meters can be valuable in detecting active infestations when no other evidence is visible.

Even more sophisticated is the use of beagles trained to detect termites. Well-trained beagles can gain access to areas inaccessible to a human and with their keener senses of hearing and smell they can detect active termite infestations where human inspectors cannot. With often more than 50% of a home inaccessible to a human inspector, assistance from a beagle can remove much of the guesswork. However, as with any tool a lot depends on the skill and thoroughness of the user. For instance, unless the handler puts the beagle in an attic, it will not detect an attic infestation.

Inspection Procedures

The equipment used and the inspection procedures will vary with the target pest and the urban situation (Davis et al., 1974; Southwood, 1978). In small premises an inspection may take only a few minutes, but in large building complexes the inspection and related diagnosis and reporting may take many days or even weeks.

The inspection of a restaurant for German cockroaches can be used as an example of how to carry out an inspection. While the details will vary, most of the principles of carrying out an inspection will be the same as in other situations.

It is usually best to begin the restaurant inspection by talking with people who work on the premises. They should be asked where they see pests and when they first started seeing them. This may give clues to their origin as well as their current whereabouts. As quickly as possible the inspector should attempt to confirm the identity of the pest by looking for live specimens, dead specimens, egg cases, or fecal stains. Once the identity of the pest is confirmed, and knowing the habits and habitat preferences of the pest, the inspector can target his inspection accordingly.

Before undertaking a detailed examination of the restaurant the inspector should take an overall look at the premises, both on the outside and inside. He should build up a picture of traffic patterns for both people and merchandise. In particular the food flow into, through, and out of the premises should be known. Perhaps the cockroaches were introduced with food supplies and the inspector should find out where incoming goods are stored and whether storage practices, such as on-floor storage or retaining cardboard cartons, encourage the problem. If introduction with food supplies is confirmed, the inspector should identify the particular items and their supplier for future reference.

In some cases the cockroaches may have first entered from outside or from adjoining buildings and the inspector should check possible entry points such as sewer lines, conduits for electricity or water, or crevices in walls connecting with other infested buildings. In other cases the cockroaches may have first been introduced by kitchen staff on their clothing or with other personal belongings. Questioning of staff and examination of staff lockers may shed light on this possibility.

Identifying the origins of the cockroaches is valuable from the long-term viewpoint of preventing re-infestation. But a key part of the inspection is to note all those factors that enable cockroaches to survive and multiply once they have gotten into the restaurant. Since cockroaches need food, the inspector must note any exposed food storage, such as fresh vegetables. He must also note any food spills and any buildup of food material in or under equipment, in mops and brooms, inside the rims and baskets of floor drains, and around the wheels of mobile items. The inspector should check garbage handling procedures to find out if garbage is placed in sealed containers and regularly removed from the premises. And he should also check less obvious food sources, such as rodent bait stations, for signs of cockroach activity.

Cockroaches need water as well as food and the inspector must check for leaks or condensation that may support an infestation. Procedures for drying floors and work surfaces after the daily cleanup should be investigated. As for food sources, the inspector must also consider less obvious sources of water, such as planters, ice machines, toilet pans, and fish or lobster tanks.

In addition to food and water, cockroaches need hiding places or harborages in which to rest and breed, and these actual or potential harborages must be identified during the inspection. Once again the inspector should use his knowledge of the target pest to focus his efforts. German cockroaches tend to prefer dark crevices close to water. They show a special preference for bare wooden surfaces, perhaps because such surfaces are easier to climb and because wood retains the aggregation pheromone secreted by German cockroaches better than some other surfaces. In any case the inspector must pay particular attention to fixtures made of wood, such as storage racks, wooden tables, and cabinets. But he must also check behind and under machinery, in the corners of rooms at floor or ceiling level, behind notice boards, and around the legs or wheels of equipment as well as in machinery voids.

The inspector should not only use his sense of vision to identify cockroach harborages but also his sense of smell. Often there will be a characteristic odor that will help identify a preferred harborage area. Sometimes cockroaches become trapped and the inspector should look for such trap sites as spiders webs, light fittings, and empty beer bottles. In other words, the inspector should look for every clue and every advantage in identifying the problems.

As a general rule the harder it is to gain access to a potential harborage, the more likely it is to be infested. The inspector must insist on examining locked areas such as liquor stores and staff lockers. Suspended ceilings must be inspected. Indeed, if questioning reveals that previous pest control measures have included the use of highly repellent insecticides that have scattered the infestation, the inspector should spend as much time exploring above head height as below.

In many situations the judicious use of flushing pyrethrum aerosols or dust will help identify harborages where other clues are absent. Of course, for voids in electrical machinery and for power outlets, junction boxes, and circuit breaker boxes, care should be taken to use nonconducting flushing formulations and injection equipment.

In addition to obtaining information directly relating to the pest situation the inspector should obtain information that may affect the implementation or choice of pest management measures. In a restaurant situation, such information would include the cleaning methods and schedules, the times when the kitchen is in use, and the times when the restaurant is open to the general public.

In this example of a restaurant infested with German cockroaches, since cockroaches are most active at night it would be wise to inspect the premises late at night as well as in the daytime. The pest data obtained at night is often a useful supplement to daytime data based on observing kitchen activities and questioning staff.

In the case of other pests the inspector must adjust to the different food preferences and habits of the particular pest. For instance, an inspection for Gypsy moths would focus more on examination of oak or apple trees than pine trees because, although Gypsy moths are known to be capable of eating 500 species of trees and shrubs, they have strong preferences for certain species, particularly hardwoods.

Many stored product pests have food preferences that, if known, can be an aid in identifying sources of entry or continuing reservoirs of infestation (Mallis, 1982). For instance, the confused flour beetle, red flour beetle, and sawtoothed grain beetle thrive on processed or mechanically damaged dry food such as breakfast foods, flour, and whole meal, as well as on grain previously damaged by weevils or grain borers. The drugstore beetle can eat similar foods and others too, but because of its relatively slow breeding rate, an inspection should focus more on foodstuffs that have been undisturbed for long periods. The breeding and feeding of the foreign grain beetle are primarily related to damp and moldy grains, since these beetles feed on molds and dead insects. An inspection for this beetle should therefore focus on out-of-condition grains.

Carpenter ants can be used as another example to illustrate how a knowledge of habits of the pest can focus an inspector's efforts and in many cases help pinpoint a nest site. Carpenter ants require voids for nesting purposes and, since they have a particular preference for damp voids, they frequently excavate nesting cavities in damp wood. Outdoors an inspection should therefore focus on the following: hollow stumps or rot holes in trees close to a building subject to ant invasion; firewood stacked near the building; railroad ties used in landscaping; wooden rain gutters, particularly if they are blocked; porch pillars; shingles abutting flat roofs and other areas where moisture is a problem; wooden decking and exterior sills. Indoors an inspection for carpenter ants would focus on the following sites: behind bathtub panels; under dishwasher machines; window and exterior door frames; around fireplaces, particularly if the chimney flashing is leaky; and firewood stored indoors. Of course, other locations may be used as carpenter ant nest sites, and in addition to excavated voids, the inspector must consider natural voids such as hollow ceiling beams and hollow doors, especially bathroom doors. As for other inspections, it is essential to ask the building occupants relevant questions, such as "where do you see most carpenter ants," "have you an existing moisture problem anywhere," or "have you ever had a moisture problem."

In the example of an inspection for German cockroaches in a restaurant, reference was made to seeking cockroach signs, such as fecal stains and odor, as well as the cockroaches themselves. With other pests greater emphasis may be focused on their signs, perhaps because of difficulty in seeing the pests. Even with large pests, such as rodents, the inspection will largely reveal signs such as droppings, foot tracks, grease smears, and gnawing damage rather than the rodents. With other pests the signs may be equally distinctive. For instance, in grain storage bins, heavy webbing over the surface of the grain is characteristic of a meal moth infestation. And with an infestation of powder post beetles and furniture beetles, the insects may never be seen, but their presence will be indicated by the small round exit holes in the wood and the fine frass that is pushed out by emerging beetles. This frass is quite different from the coarse shredded wood expelled by carpenter ants from their galleries.

When inspecting for such signs and for the pests themselves the inspector is engaged in what might be called "macro-inspection." However, in some situations, particularly in food manufacturing and processing establishments, it may be useful to carry out a "micro-inspection." Such an inspection may involve using cotton swabs moistened with light mineral oil to wipe surfaces over which insects and mites may have crawled. These swabs can be placed in sealed vials for later microscopic examination for arthropod fragments (Harris, 1950; Heuerman and Kurtz, 1955; Kardatzke, Rhoderick, and Nelson, 1981) that may not have been visible to the naked eye. In addition, samples of finished food materials can be examined microscopically as a final check for evidence of infestation at some point in the production pathway.

While such micro-inspection procedures are not yet commonly implemented by commercial pest control operators, they indicate the potential of the inspection process. Indeed, the inspection scope is moving even beyond micro-inspection to chemical and even bacteriological inspection (Frankenfeld, 1948; Harris, 1960). Chemical inspection relates to pre-treatment analysis for pesticide residues to avoid later allegations of pesticide misuse. Thus, chemical inspection may be conducted to analyze for cyclodienes in well water or in the air in a building prior to treatments against termites using cyclodiene insecticides such as chlordane. The data in Tables 5-1 and 5-2 are the result of bacteriological inspection to evaluate the potential dangers from pests.

Clearly micro-inspection, chemical inspection, and bacteriological inspection requires sophisticated equipment and might be considered more scientific than simply looking at the conditions in and around buildings. However, good inspection, like criminal detection, requires both an analytical and intuitive ability, together with a strong desire to defeat the pest, and this cannot be achieved by mere equipment.

DIAGNOSIS AND REPORTING

As mentioned earlier, the purpose of the inspection is to identify clues. These clues form the basis of a diagnosis of the pest situation. In some cases on-site inspection may fail to confirm the identity of the pest, but later examination of the clues may provide a definitive answer. The diagnosis goes beyond simply describing the pest situation and should draw on other information such as past experience and scientific literature to suggest the reasons for the infestation and its importance.

In some cases the inspection of one building may yield clues that suggest a follow-up elsewhere before a good diagnosis can be made. As an example, the inspection of a hospital revealed a widely scattered infestation of Pharaoh ants. Fortunately the inspection was thorough enough to reveal a small colony in a stack of clean linen recently received from an outside laundry. This clue was followed up by a visit to the laundry where an extensive Pharaoh ant infestation was confirmed. So in this case the diagnosis of the hospital pest situation included reference to infested incoming supplies, and the resulting strategy for eliminating the infestation included treatment of the laundry as well as of the hospital.

In many cases the people who live or work in a building will not understand the importance of pests. Reference has already been made to people thinking some pests such as cockroaches are mere nuisances rather than potential disease carriers. Even in high technology environments the pest control expert has a role to play in focusing attention on the importance of the pests and how they might specifically impact on work in the premises. For example, in a medical research laboratory test animals were maintained on highly controlled diets, including low protein diets. Even though there was a high population of cockroaches in the laboratory and the researchers were experiencing inconsistent animal research results, it required a pest control expert to point out the link between the poor results and the cockroaches. The test animals were eating the cockroaches and thus supplementing their inadequate diet with high protein cockroaches.

In another case, medical staff in a hospital were reluctant to allow pest eradication measures to be implemented against Pharaoh ants in an intensive care unit. However, when a pest control expert revealed that Pharaoh ants could invade intravenous drip systems and sterile dressing packs the objections to eradicating the ants were withdrawn. Clearly the ants were more likely to compromise the health of a patient than any reasonable measures taken to eradicate the ants.

So a diagnosis of a pest situation should deal not only with the pests but also with the consequences of an infestation, which may not be obvious to specialists in other fields.

The facts and clues discovered in the inspection and the ultimate diagnosis should be recorded. These records will form a basis for future action by the pest control specialist and in-house personnel. Good records are also valuable in dealing with regulatory personnel and in litigation relating to pest problems.

Some pest control companies attempt to use a standard inspection and diagnosis form for all situations. Such forms typically are made up of a checklist to guide the inspector and reduce the chances of overlooking factors favoring pests (Truman, Bennett, and Butts, 1976). However, no standard form can be suitable for totally different types of premises, and it may be necessary to use different forms for different pests or categories of premises (see Fig. 5-1) or to prepare a specific form for each situation. Certainly, items checked for action on a standard form may have less impact and be less likely to be acted on than if the same items appeared on a more customized form.

Some of the disadvantages of a standard inspection and diagnosis report form can be reduced by leaving space on the form for special observations and comments. Such a space will discourage the inspector from having tunnel vision or from being constrained by the limitations of the form.

All inspection and diagnosis reports should contain the following data: the location of the situation, e.g., street address or map reference; the date of the inspection and the number of the report; the duration and time of day of the inspection; the name of the person conducting the inspection; the name and signature of the person completing the inspection and diagnosis report form if different from the above; the name of site personnel spoken with or accompanying the inspector during the inspection; the scope of the inspection and in particular any areas not inspected; pests observed; pest signs observed; physical conditions or procedures conducive to pest entry; physi-

Results of your consultation with the client.
Where has he seen cockroaches?
How many?
Have sightings occurred repeatedly?
How long has he had an infestation problem?

Your survey of the kitchen:
Did you see any evidence of cockroaches (insects, fecal stains, droppings, egg capsules) in cracks and crevices near food and water?

____ around sinks?
____ around dishwashers?
____ around stoves?
____ around ovens?
____ in bandsaws for cutting meat?
____ near worktables?
____ near cutting boards?
____ in electric mixers?
____ in blenders?
____ in toasters?

Did you find any evidence of cockroaches further from food and water?

____ in drop ceilings?
____ behind wall paneling?
____ in switch boxes?
____ in electric motors?
____ in door hinges?
____ in the rubber seals behind cooler doors?

Did you find weaknesses in sanitation procedures that may contribute to cockroach infestation such as:

food scraps
____ in cracks of cutting boards?
____ in cracks of counter tops?
____ in drains?
____ under carts?

food accumulations
____ beneath stoves?
____ beneath work tops?
wet or grease soaked
____ rugs?
____ mops?

Did you find out whether cockroaches are entering the kitchen:
____ in cardboard boxes?
____ in linen deliveries?
____ fresh vegetable deliveries?

Figure 5-1. Inspection/diagnosis form for cockroaches in commercial kitchens (reproduced with permission of the Nor-Am Chemical Co.).

cal conditions or procedures conducive to pest survival; relevant comments by on-site personnel (e.g., pest sightings); damage or risks from the pests.

Some inspectors find it useful to take photographs to illustrate situations such as poor sanitation. Such photographs can help management understand the situation, but if they illustrate embarrassing or illegal conditions they must be treated as confidential documents. In some situations, particularly food plants, photography by pest management inspectors is forbidden because of fears that incriminating photographs may later be used by other parties to support legal actions. Some pest management contractors consider photographs so important in gaining customer cooperation that they refuse to undertake work unless photography is allowed in support of written reports.

In addition to initial reports, where there is an ongoing program of pest management there will be follow-up inspections and the reports on these inspections will address additional questions such as the following. Have sanitation problems highlighted in the previous inspection report been corrected? Have maintenance problems highlighted in the previous inspection report been corrected? Is rodent control equipment (e.g., multiple catch traps) clean and in working order? Is insect control equipment (e.g., electric fly traps) clean and in working order? Are bait stations, traps, etc., correctly labeled and identified on the floor plan? Have their been any changes at the site (e.g., new additions, remodelling, etc.) that might affect the pest situation?

Inevitably an inspection and diagnosis report will include information that some people will find disagreeable. This is particularly true of sanitation problems because such problems frequently reflect human failings. Nonetheless, if the objective is to improve the situation the truth must be told. While an inspector should note good sanitation practices and any improvements, the job is primarily to find those faults that help perpetuate a pest problem. Of course, whether or not these faults are corrected will largely depend on the attitude and the authority of the person to whom the report is directed.

The importance of making a correct diagnosis of a pest problem cannot be overemphasized, since it may substantially determine the pest management strategy. For instance, if a Norway rat infestation is diagnosed when in reality it is a roof rat infestation, the control measures may be totally inappropriate because of the different food and habitat preferences and greater agility of the roof rat. Correct diagnosis of a larger yellow ant infestation will focus control measures on foundations or outdoor sites, since these ants feed outdoors and rarely cause a nuisance indoors. Sometimes correct diagnosis will influence the timing of treatments. Thus diagnosis of an Indian meal moth infestation may suggest that ultra low volume (ULV) applications of insecticide should be timed for late evening, say 9 PM, since this is the peak flying time of this species. However, applications aimed against adult Mediterranean flour moths might be better timed much later at night, since peak flying is just before dawn.

PEST POPULATION MONITORING

In describing the inspection process emphasis has been placed on examining the situation and detecting pest levels at a particular point in time. However, such an isolated examination does not reveal whether a pest population is increasing or decreasing. Knowing about population changes can be important in order to judge the success or failure of pest management measures and in determining the need for additional or repeated efforts where pest population levels violate previously established damage or action thresholds (Kardatzke, Rhoderick, and Nelson, 1981). Pest population monitoring can provide such information. However, unlike the situation in many agricultural IPM programs, examples of proven pest population monitoring procedures from among the various urban pest management disciplines are in short supply. Further, even if population numbers were monitored, in most cases there are no set action threshold levels against which decisions can be made regarding application of pest management procedures. Readers are referred to some basic IPM reference texts for further information on the concepts of pest population monitoring and management decisionmaking (Apple and Smith, 1976; Flint and van den Bosch, 1981; Metcalf and Luckmann, 1975).

Pest population monitoring is based on carrying out repeated surveys using the same methodology each time so that results can be meaningfully compared. The methodologies vary with the target pest and the situation, but the various monitoring methods fall into the following four categories: manual sampling and counting, trapping and counting, assessing tracking activity (rodents), and census baiting.

Manual sampling and counting is widely used in the agricultural sector by crop scouts. In its simplest form it consists of collecting or examining in situ a representative sample of leaves or fruits and noting the number of insects, mites, fungal lesions, etc. For certain soil insects, plants may be uprooted and the degree of injury to the root zone recorded. In ornamental plantings and vegetable gardens, comparing monitoring data against established damage or action thresholds is a practical concept as the host plants can withstand some level of pest presence (Davidson and Gill, 1980). In other urban situations manual sampling is particularly suitable for pests that are readily accessible and easily counted. For example, bodies of water can be regularly sampled by dipper in order to monitor populations of mosquito larvae (Breeland and Mulrennan, 1983). Stored grain and flour can be similarly sampled and the samples examined for stored product pests.

Another example of manual sampling is the recording of mosquito bites on a selected portion of the sampler, usually the forearm, during a standard period of time. Since different species of mosquito bite at different times of the day, the methodology must take this into account (Pratt and Barnes, 1959).

Trapping is perhaps the most widely employed method of monitoring urban pests, and traps may be based on mechanical systems, sticky surfaces, attractant light, food attractants, or pheromones (Barak and Harein, 1982; Donohoes and Barnes, 1934; Lanier, 1981; Loschiavo, 1974, 1975; Loschiavo and Atkinson, 1967, 1973). Mechanical trapping systems, such as pitfall traps and perforated cylinder grain-probe traps have the benefit of simplicity and low maintenance, but they are not especially attractive to pests and are therefore not very useful for early detection. Early detection of pests is of prime importance for stored products in order to minimize spread of an infestation to pest-free areas and to reduce the extent, risks, and cost of control measures (Burkholder, 1976).

Sticky traps have similar advantages and disadvantages as mechanical traps, but their simplicity and low cost has led to their increased use by pest control operators to monitor highly mobile pests such as cockroaches. For a nocturnal insect such as German cockroaches, sticky traps can reduce the need for night inspections, since traps placed in the daytime will catch cockroaches emerging at night. However, night inspections are useful in determining the overall distribution of active insects and in evaluating sanitation levels at a time when the premises are supposed to be clean. For best results sticky traps are placed in dark areas close to suspected cockroach feeding and harborage areas and where they will be least disturbed by cleaning measures. In practice, the use of sticky traps to indicate threshold levels for action has resulted in improved control of cockroaches with less use of insecticide and less labor (Kardatzke, Rhoderick, and Nelson, 1981).

Light traps are more attractive to certain insects than sticky or mechanical traps, but they have the disadvantage of high installation and maintenance costs. Another disadvantage is that light traps are only useful against flying insects. However, not all flying insects are equally attracted. For instance, even in a closely related group such as mosquitoes, light traps might be excellent for monitoring *Aedes taeniorhynchus* and *Psorophora columbiae* but useless for *Culex quinquefasciatus.*

Food attractants can be useful in luring insects to traps for monitoring purposes. The insects thus attracted are either killed by contact with an insecticide-treated surface or held by a sticky surface or mechanically. Considerable progress has been made in identifying and isolating components of food that are attractive to particular stored product insects (Burkholder, 1984; Freedman et al., 1982; Mikolajczak et al., 1984). For instance, researchers at the USDA Stored Product and Household Insects Laboratory in Madison, Wisconsin, have identified several components of wheat germ oil that are useful in larval monitoring of *Trogoderma* species. However, food attractants have several disadvantages as aids in pest monitoring. These disadvantages include an inability to attract insects over long distances, competition from other food, and the fact that only the feeding stages of a pest are attracted. In addition, the food attractant itself may sustain an infestation (Burkholder, 1981).

The greatest advances in insect monitoring relate to the use of insect pheromones (Burkholder, 1976). Pheromones are chemicals secreted by insects in order to modify the behavior of other insects belonging to the pheromones, and pheromones that stimulate mass attack or feeding or that mark territory boundaries (Barak and Burkholder, 1976; Cogburn, Burkholder, and Williams, 1984; Stockel and Sureau, 1981).

Research on pheromones has been most intense in the United States and Japan, and this research has already resulted in the isolation, identification, synthesis, and testing of pheromones for the monitoring of many stored product insects (Cogburn, Burkholder, and Williams, 1984; Vick et al., 1981; Williams et al., 1981). One of the key pioneers in this work has been Wendell Burkholder at the USDA Stored Product and Household Insects Laboratory in Madison, and many of the following comments about the use of pheromones are drawn from his work.

Sex pheromones and aggregation pheromones have been the prime objects of study, and their role in insect communication relates to the longevity and feeding requirements of adult insects. Adult insects that are short-lived (less than one month) and require no feeding for reproduction rely on sex pheromones for communication. These insects include moths and dermestid, bruchid, and anobiid beetles. These sex pheromones are usually produced by the female and attract males.

Long-lived adults that need to feed for reproduction rely on male-produced aggregation pheromones for long-distance communication. These insects include grain weevils, grain borers, flour and grain beetles, and both males and females respond to the aggregation pheromone. The effect of such aggregation pheromones is to attract adults to food sources where both feeding and breeding can then take place.

Alongside pheromone research has been the development of traps in which the pheromones or food attractants can be incorporated. A corrugated paper trap has been developed that simulates the cracks and crevices in walls sought as hiding places by many beetles (Barak and Burkholder, 1976; 1984; Burkholder, 1976; Williams et al., 1981). A commercially available *Trogoderma* trap is based on this design and it contains both a sex pheromone to lure adult males of the warehouse and khapra beetle and a food attractant to lure larvae. A similar *Tribolium* trap contains an aggregation pheromone and food attractant that lure both adult males and females of red and confused flour beetles. In both these traps the food attractant is oil based, and it serves the additional function of trapping and suffocating insects that fall into it.

A plastic grain probe trap is now commercially available that can be used alone or in which pheromones and oil lures can be inserted for extra attractiveness. Aggregation pheromones of the lesser grain borer and flour beetles have been successfully used with this trap, and oil lures have also been effective in attracting a wide variety of adult and larval beetles. These grain probe traps can be left in place for several days, and this provides much greater sensitivity in detecting light infestations.

A variety of funnel traps and adhesive traps baited with pheromones are available for trapping moths and beetles (Cogburn, Burkholder, and Williams, 1984; Williams et al., 1981). One popular commercial sticky trap incorporating sex pheromones attracts adult males of the Indian meal moth, almond moth, raisin moth, tobacco moth, and Mediterranean flour moth.

As with other traps, pheromone traps must be placed correctly. Moth traps are usually more effective if placed near ceilings, whereas beetles are more easily trapped on or near the floor (Smith, Burkholder, and Phillips, 1983). They can be placed where there is a history of infestation or in incoming goods areas, trucks, and rail cars to monitor for new infestations. In large areas large numbers of traps can be placed in a grid pattern at intervals of 25-50 ft. Then, as catches occur, the grid can be tightened to pinpoint problem areas, and the particular source of infestation can be confirmed by visual inspection. Individual pheromone traps should be placed in locations likely to be sought by the target insects and sheltered from damage by fork trucks or other machinery. In general, pheromone traps should be placed away from open doors and windows to avoid luring insects into a building from outside. In addition, it may be prudent to place traps where they will not be readily seen by government inspectors, who might use trap catches as the basis for enforcement action.

Trap locations should be identified on site maps to facilitate inspection and servicing. The numbers and types of pest caught in each location should be recorded to help plan and evaluate control measures. Traps based on pheromones or food attractants need routine servicing to replace exhausted attractants. Most pheromones need to be replaced every month, while food attractants are usually effective for one to two months.

Pheromone traps are especially useful for early detection of an attack. This is especially important with pests such as the lesser grain borer, where the adults are strong fliers and an infestation may become widespread before it is detected by conventional methods. Early detection of pests using pheromone traps is already widely established in agriculture and horticulture for such pests as codling moth, Japanese beetle, and boll weevils. Chapter 10 of this book discusses use of pheromone traps in pest management efforts against certain borers attacking woody ornamental plants. In structural pest control the increased use of such traps is similarly expected to result in improved pest management by enabling better targeting of pesticide and nonpesticide treatments (Barak, Shinkle, and Burkholder, 1977).

In the field of rodent management there have also been advances in pest monitoring (Davis et al., 1974). Rodent monitoring methods include trapping, dropping counts, and electronic counting. Increasingly, professionals are measuring tracking and feeding activity (Scalingi, 1981). Typically, measurements are taken immediately before implementing a rodent management program and then about two weeks later to determine the initial success of the program. Thereafter feeding and tracking activity might be measured at three-month intervals to help guide future control efforts.

Rodent-tracking activity is measured by using standard-sized patches (e.g., 6 × 12 in) of nontoxic tracking powder. These tracking patches are placed in areas of suspected rodent activity, and the numbers of rodent tracks are recorded. After each recording the tracking patch is smoothed out for the next assessment. Feeding activity is determined by measuring the consumption from nontoxic bait packages at bait points located in areas of suspected rodent activity. The amount of bait consumed can be accurately measured, but in most cases it will suffice to use a simple indexing system. In one commercially available rodent evaluation kit, developed with William Jackson of Bowling Green State University, Ohio, the following indices are suggested:

Tracking index	Feeding index
Index 1 = 0 tracks	Index 1 = bait package unopened
Index 2 = 1-4 tracks	Index 2 = bait package opened
Index 3 = 5-9 tracks	Index 3 = one quarter consumed
Index 4 = 10-19 tracks	Index 4 = one half consumed
Index 5 = 20-29 tracks	Index 5 = three quarters consumed
Index 6 = 30 or more tracks	Index 6 = wholly consumed

At each assessment such an index can be assigned for each tracking or baiting point and the result recorded for future comparisons and for estimating the success of control measures. With the increasing incidence of rodent resistance, particularly to warfarin, greater emphasis must now be placed on monitoring rodent response to control measures.

While census baiting primarily relates to rodent management there is one insect pest where pre- and post-treatment baiting is widely practiced. This is the Pharaoh ant, where baiting with small pieces of liver or drops of corn syrup and water solution provides an excellent means of monitoring infestations.

The above examples of monitoring might all be termed quantitative approaches, but there are other systems that are more subjective. The keeping of a log book on pest sightings is one such system. Log books are frequently maintained by clients of pest control operators, including food processors, hospitals, and apartment houses. These logs, while often incomplete and inaccurate, nonetheless provide useful clues as to pest activity in the premises, and in particular they can reveal trends in pest populations. Some pest control operators find the feedback from their clients so valuable that they do not rely on the client independently making entries in a log book. Instead they phone the client every month and ask for information on pest sightings. In this way they can keep their own log of customer comments.

CONSULTATION

The 1980s have been called the age of consulting in the urban pest management field (Moreland, 1982). The origins of this consulting age extend back to the 1972 amendments to the Federal Insecticide, Fungicide and Rodenticide Act, which laid the groundwork for more regulation of the pest control industry. With new regulations, increasingly irresponsible media coverage of pest control issues, and a more concerned and demanding general public, many pest control companies as well as institutions and industries with pest problems found they needed inputs from independent professionals. These consultants have various areas of specialization, including business and finance, food plant inspection, expert testimony, pest control in high technology environments, personnel training, new product evaluation, complaint investigation, and loss prevention.

Like consultants in other fields, consultants in the pest control field are valued for the extent to which they can save their clients time, money, and reputation. Increasingly, consultants are involved in reducing risks. Earlier in this chapter it was pointed out that the risks from pests were often not known to those who lived or worked in infested premises and that greater priority was given to pest management once these risks were understood. Consultants frequently help communicate these risks from pests, but consultation also involves weighing the risks of the various pest management programs. Such risks might include pesticide risks to people, pets, and property. Some of these risks may be obvious but others are not, and in a high-value situation it is best to seek advice before taking a chance.

The following examples of pesticide properties that are not widely known to users but that might cause problems in certain circumstances illustrates the kind of information that can be obtained from consultation with an expert:

Problem	Substance
Causes irritance, allergies	Pyrethrins and some pyrethroids
Corrodes copper and brass	Chlorpyrifos, phosphine
Corrodes iron and mild steel	Dichlorvos
Reacts with some carpet dyes	Acephate, dichlorvos, malathion
Highly toxic to fish	Pyrethroids
Highly toxic to dogs	Brodifacoum
No specific antidote	Pyrethrins, amidinohydrazone
Damaging to plastics	Aerosols with aggressive solvents such as methylene chloride.
Can taint certain foods	Lindane, pesticides with high content of xylene solvent.

In some cases the chemical risks are most unexpected. For instance, ethylene oxide, which is a standard fumigant for books, can react with water to form initially ethylene glycol and then diethylene glycol, which is a solvent for printers ink. This reaction would not normally be a problem, unless the books were wet at the time of fumigation.

In other cases a pest control operator may simply never have encountered a particular pest problem and not know how to deal with it. An infestation of German cockroaches inside a kidney dialysis machine would be one example. In this case formaldehyde might be the biocide of choice because it is used routinely for disinfecting the machine, and there are protocols for purging it from the machine for re-use. Another example might be cockroach control on aircraft, and an expert could give advice on what pesticides have passed the necessary aircraft corrosion tests. Of course, many pest control operators may not even be aware that there are specific approval procedures for aircraft pesticides but, as already mentioned, in high-value environments it pays to seek advice.

Some pest control operators routinely call in an outside expert whenever they tackle a new type of pest problem. In this way they extend the scope of their work with minimal risk. When using a consultant in this way it is best to first draft a pest management strategy and then ask the consultant to check it, make changes if necessary, and then endorse the final strategy. In effect, the pest control consultant acts as a sounding board, and his job is not to simply suggest novel answers to pest problems but appropriate answers, bearing in mind any special circumstances.

The answers may involve habitat modification, trapping, or physical exclusion of pests rather than a reliance on pesticides. In the integration of these various methods the urban pest management consultant is on the leading edge of technology. In the structural pest field it is now possible to think beyond pest management in or around individual buildings. As with large-scale mosquito management programs it is now possible to contemplate management of cockroaches, rodents, and other pests on a city scale (Anonymous, 1982). This approach, embodied in the concept of creating "urban pest-free zones" is already practiced in Europe (Story, 1977).

Whether working for city managers, pest control operators, food processors, institutional clients, or others, the independent consultant's role is primarily that of providing information. However, the decision to act on that information rests with the client.

REFERENCES

Anonymous. 1982. Budapest experiment controls cockroaches. *Pest Control.* **50**(1):76.

Apple, J. L., and R. F. Smith. 1976. *Integrated Pest Management.* Plenum, New York. 208p.

Barak, A. V., and W. E. Burkholder. 1976. Trapping studies with dermestid sex pheromones. *Environ. Entomol.* **5**(1):111-114.

Barak, A.V., and W. E. Burkholder. 1985. A versatile and effective trap for detecting and monitoring stored product Coleoptera. *Agric., Ecosyst., Environ.* **12:**207-218.

Barak, A.V., and P. K. Harein. 1982. Trap detection of stored-grain insects in farm-stored, shelled corn. *J. Econ. Entomol.* **75:**108-111.

Barak, A. V., M. Shinkle, and W. E. Burkholder. 1977. Using attractant traps to help detect and control cockroaches. *Pest Control* **45**(10):14-16, 18-20.

Beatson, S. H. 1972. Pharaoh's ants as pathogen vectors in hospitals. *Lancet,* February 19, pp. 425-427.

Breeland, S. G., and J. A. Mulrennan, Jr. 1983. Florida mosquito control: State's program is not a matter of chance. *Pest Control* **51**(2):16-24.

Burgess, N. R. H., and K. N. Chetwyn. 1979. Cockroaches and the hospital environment. *Nursing Times,* February 15, pp. 5-7.

Burkholder, W. E. 1976. Application of pheromones for manipulating insect pests of stored products. In *Insect Pheromones and Their Applications,* T. Kono and S. Ishii (eds.). Japan Protection Association, Tokyo, pp. 111-122.

Burkholder, W. E. 1981. Biomonitoring for stored-product insects. In *Management of Insect Pests with Semiochemicals,* E. R. Mitchell (ed.). Plenum, New York, pp. 29-40.

Burkholder, W. E. 1982. Reproductive biology and communication among grain storage and warehouse beetles. *J. Georgia Entomol. Soc.* **17**(2nd suppl., Oct.):1-10.

Burkholder, W. E. 1984. The use of pheromones and food attractants for monitoring and trapping stored-product insects. In *Insect Management for Food Storage and Processing,* F. Baur (ed.). American Association of Cereal Chemistry, St. Paul, Minn., pp. 69-86.

Cogburn, R. R., W. E. Burkholder, and H. J. Williams. 1984. Field tests with the aggregation pheromone of the lesser grain borer (Coleoptera: Bostrichidae). *Environ. Entomol.* **13:**138-142.

Davidson, J., and S. Gill. 1980. Urban integrated pest management. *Chemical Times and Trends.* July, pp. 29-31.

Davis, H., et al. 1974. *Urban Rat Surveys.* U.S. Department of Health, Education, and Welfare, Center for Disease Control, Atlanta, Georgia, 22p.

Donohoes, H. C., and D. F. Barnes. 1934. Notes on field trapping of Lepidoptera attacking dried fruits. *J. Econ. Entomol.* **27**(5):1067-1072.

Flint, M. L., and R. van den Bosch. 1981. *Introduction to Integrated Pest Management.* Plenum, New York, 240p.

Frankenfeld, J. G. 1948. *Staining Methods for Detecting Weevil Infestation in Grain,* ET 256. U.S. Bureau of Entomology and Plant Quarantine, July, 4p.

Freedman, B., K. L. Mikolajczak, D. R. Smith, Jr., W. F. Kwolek, and W. E. Burkholder. 1982. Olfactory and aggregation responses of *Oryzaephilus surinamensis* (L.) to extracts from oats. *J. Stored Prod. Res.* **18:**75-82.

Harris, K. L. 1950. Identification of insect contaminants of food by the micromorphology of the insect fragments. *J. Assoc. Off. Agric. Chem.* **33:**898-933.

Harris, K. L. 1960. *Microscopic-Analytical Methods in Food and Drug Control,* Technical Bulletin No. 1. U.S. Department of Health, Education, and Welfare, Washington, D.C., 256p.

Heuerman, R. F., and O. L. Kurtz. 1955. Identification of stored-product insects by the micromorphology of the exoskeleton, 1. Elytral patterns. *J. Assoc. Off. Agric. Chem.,* August.

Kardatzke, J. T., I. E. Rhoderick, and J. H. Nelson. 1981. How roach surveillance saves time, material and labor. *Pest Control* **49**(6):46, 47.

Kurtz, O. L., and K. L. Harris. 1955. Identification of insect fragments: Relationship to the etiology of the contamination. *J. Assoc. Off. Agric. Chem.* **38**(4):1010-1015.

Lanier, G. N. 1981. Pheromone-baited traps and trap trees in the integrated management of bark beetles in urban areas. In *Management of Insect Pests with Semiochemicals,* E. R. Mitchell (ed.). Plenum, New York, pp. 115-131.

Loschiavo, S. R. 1974. Laboratory studies of a device to detect insects in grain, and of the distribution of adults of the rusty grain beetle. *Cryptolestes ferrugineus* (Coleoptera: Cucujidae), in wheat filled containers. *Can. Entomol.* **106:**1309-1318.

Loschiavo, S. R. 1975. Field tests of devices to detect insects in different kinds of grain storages. *Can. Entomol.* **107:**385-389.

Loschiavo, S. R., and J. M. Atkinson. 1967. A trap for the detection and recovery of insects in stored grain. *Can. Entomol.* **99:**1160.

Loschiavo, S.R., and J. M. Atkinson. 1973. An improved trap to detect beetles (Coleoptera) in stored grain. *Can. Entomol.* **105:**437-440.

Mallis, A., ed. 1982. *Handbook of Pest Control,* 6th ed. Franzak and Foster, Cleveland, Ohio, 1,104p.

Metcalf, R. L., and W. H. Luckmann, eds. 1975. *Introduction to Insect Pest Management.* Wiley, New York. 592p.

Mikolajczak, K. L., B. W. Zilkowski, C. R. Smith, Jr., and W. E. Burkholder. 1984. Volatile food attractants for *Oryzaephilus surinamensis* (L.) from oats. *J. Chem. Ecol.* **10**(2):301-309.

Moreland, D. 1982. The expanding role of the consultant in the pest control industry. *Pest Control Technol.* **10**(1):57-60, 84.

Pratt, H. D., and R. C. Barnes. 1959. *Identification Keys for Common Mosquitoes of the United States.* U.S. Department of Health, Education, and Welfare, Communicable Disease Center, Atlanta, Georgia, 40p.

Scalingi, A. V. 1981. How to measure rodent infestation quotient. *Pest Control* **49**(8):13-16.

Smith, L. W., Jr., W. E. Burkholder, and J. R. Phillips. 1983. *Detection and Control of Stored Food Insects with Traps and Attractants: The Effect of Pheromone-Baited Traps and Their Placement on the Number of* Trogoderma *Species Captured,* Natick technical report 83/008. Natick, Mass., 13p.

Southwood, T. R. E. 1978. *Ecological Methods: With Special Reference to the Study of Insect Populations,* 2nd ed. Halsted Press, London, 524p.

Stockel, J., and F. Sureau, 1981. Monitoring for the Angoumois grain moth in corn. In *Management of Insect Pests with Semiochemicals,* E. R. Mitchell (ed.). Plenum, New York, pp. 63-73.

Story, K. O. 1977. *The Pest Free Zone: A New Pest Control Concept.* Proc. 41st annual Purdue University Pest Control Conference. 14p.

Truman, L. C., G. W. Bennett, and W. L. Butts. 1976. *Scientific Guide to Pest Control Operations,* 3rd ed. Purdue University/Harvest, Cleveland, Ohio, 276p.

Vick, K. W., J. A. Coffelt, R. W. Mankin, and E. L. Soderstrom. 1981. Recent developments in the use of pheromones to monitor *Plodia interpunctella* and *Ephestia cautella.* In *Management of Insect Pests with Semiochemicals,* E. R. Mitchell (ed.). Plenum, New York, pp. 19-28.

Weber, W. J. 1979. *Health Hazards from Pigeons, Starlings and English Sparrows.* Thomson, Fresno, Calif., 138p.

Weber, W. J. 1982. *Diseases Transmitted by Rats and Mice.* Thomson, Fresno, Calif., 182p.

Williams, H. J., R. M. Silverstein, W. E. Burkholder, and A. Khorramshahi. 1981. Dominicalure 1 and 2: Components of aggregation pheromone from male lesser grain borer *Rhyzopertha dominica* (F.) (Coleoptera: Bostrichidae). *J. Chem. Ecol.* **7**:759-780.

6

Potential for Biological Control in Urban Environments

Robert V. Flanders

Purdue University
West Lafayette, Indiana

Biological control deals with actual or potential relationships between natural enemies (parasites, predators, or pathogens) and their hosts or prey. In such relationships, natural enemies numerically affect populations of their hosts, so that lower densities occur than if the natural enemies were absent (DeBach, 1964*b;* Huffaker and Messenger, 1964*b;* Huffaker, Simmonds, and Laing, 1976). Naturally occurring biological control is a fundamental part of nature whose subtle effects protect man from an enormous number of potential pest species. Applied biological control is based on our recognition of the importance of these naturally occurring relationships and our understanding of them. Although other definitions exist (Wilson and Huffaker, 1976), applied biological control usually is defined as the study, importation, augmentation, and conservation of natural enemies for the regulation of densities of other organisms (DeBach, 1964*b*).

Nearly all biological control attempts and successes have occurred in agricultural (including silvicultural) situations. Consequently, most examples of biological control successes against urban pests have been by fortuitous expansion of successes against the same (e.g., cottony-cushion scale, black scale, woolly apple aphid, Comstock mealybug), or related (e.g., nigra scale, longtailed mealybug, dictyospermum scale) agricultural pests (Clausen, 1978). Despite their accidental occurrences, these successes indicate the possibly high potential for directed biological control efforts against other urban pests. However, the few attempts that have focused on urban pests (e.g., American and Australian cockroaches, European earwig, elm leaf beetle, European elm scale) have presumably relied on principles and practices that were primarily developed in and for agriculture. Since urban and agricultural environments possess some distinctly different characteristics (Frankie and Ehler, 1978; National Academy of Sciences, 1980; Sawyer and Casagrande, 1983), many existing biological control principles and practices may require modification in urban environments. Despite increasing interest in and attempts at biological control of urban pests, the principles and practices of biological control have not been adequately re-evaluated.

The purpose of this chapter is to evaluate the potential for and problems of biological control in urban environments. This evaluation will primarily be by comparison with biological control principles and practices developed for agriculture. Initially, some characteristics of urban environments that may affect natural and applied biological control will be identified and discussed. Biological control techniques will then be examined, emphasizing the applicability, potential, and research needs of each in urban areas. Finally, the current status of biological control will be discussed relative to the utilization of natural enemies by urban residents and organizations. Only arthropod pests (insects and mites) and their parasites (parasitoids), predators, or pathogens will be considered. Biological control of weeds, plant pathogens, and vertebrate pests will not be discussed nor will natural enemies that act as competitors. (For discussion of these see Huffaker and Messenger, 1976; Papavizas, 1981.) Several general references (DeBach, 1964*a,* 1974; Frankie and Koehler, 1983; Huffaker, 1971*a;* Huffaker and Messenger, 1976; Nay, 1976*a;* National Research Council, 1980) were used in writing this chapter and should be consulted by those readers desiring more background information.

ECOLOGICAL CONSIDERATIONS

General Concepts

Pests are species whose population densities are sufficiently high relative to the sensitivities of man to cause economic, aesthetic, social, or medical losses (Clark, et al., 1967; Rabb, 1970). Thus, in biological control, as in other pest control strategies, we are concerned with how densities are determined and how they can be modified. The density of a population (the number of individuals per unit area) is determined by interactions between the innate qualities of the species (e.g., fecundity, generation time, sex ratio, longevity) and those of its environment. These interactions modify innate qualities of individuals composing the population and affect extents to which immigration adds to and mortality and emigration subtract from density. The environment includes all biotic (e.g., populations of other species) and abiotic (e.g., heat, light, wind, moisture) factors that directly or indirectly affect the rates of addition to and subtraction from the density of a population (Clark et al., 1967; Huffaker and Messenger, 1964*a*, 1964*b*; Huffaker, Messenger, and DeBach, 1971; Huffaker, Simmonds, and Laing, 1976; May, 1976*a*; Pianka, 1974).

Natural enemies are biotic components of their host or prey populations' environments and usually affect their densities by influencing death rates. However, natural enemies also are affected by various environmental factors, which may differ from those affecting their hosts. Even when the same factors affect both natural enemy and host populations, the populations may respond differently. Consequently, mortalities in host populations caused by natural enemies may vary over time and space depending on the differential effects of various environmental factors (Clark et al., 1967; Huffaker and Messenger, 1964*a*, 1964*b*).

The extent to which a parasite or predator population affects the density of its host population depends primarily on its ability to locate susceptible host individuals (Huffaker and Messenger, 1964*b*; Huffaker, Messenger, and DeBach, 1971; Huffaker, Simmonds, and Laing, 1976; Nicholson, 1933). This ability or searching capacity depends on a multitude of environmental factors, including the population density of the host (Doutt, 1964*a*; Vinson, 1981). It is generally believed that the longer a predator or parasite and its host have coevolved, the more similar are their responses to abiotic environmental factors, the higher is the searching capacity of the natural enemy, and the lower the density at which the host's population will consequently be regulated by the natural enemy. Predators generally require higher host densities than parasites to maintain themselves in an area, since they must consume several prey individuals to complete development, while parasites only require one (Hagen, Bombosh, and McMurtry, 1976).

Virulent pathogens that cause high mortalities in their host populations usually are of greater interest in arthropod pest control than those that cause chronic diseases, (i.e., high host infection rates but low mortality rates (Jaques, 1983; Weiser, Bucher, and Poinar, 1976). However, naturally occurring virulent pathogens may only infrequently cause epizootics to occur in their host populations, since epizootic initiation and maintenance depend on dynamic relationships among densities of passively dispersing infective units (i.e., spores, conidia), susceptible hosts, and diseased hosts in an area (Anderson, 1982; Tanada, 1964). Spatial separations of diseased and susceptible hosts and weather conditions may also affect the ability of some pathogens to incite epizootics. Because of these relationships, naturally occurring virulent pathogens may allow their hosts to exhibit greater density fluctuations than do similarly effective parasites and predators (Hagen, Bombosh, and McMurtry, 1976). However, reductions of host densities may be dramatic when epizootics do occur.

Relative to these considerations, the presence, activity, and manipulation of natural enemies in urban areas depend on numerous past and present ecological and evolutionary factors. Not all these factors can be identified here for all urban situations, and only some of the more obvious ones will be discussed.

Community Composition and Stability

Ecological communities are assemblages of interdependent populations that coexist within an area (Brown, 1984; Pianka, 1974; Southwood and Way, 1970). Populations in a community are partitioned into trophic levels according to their position in the flow of energy through the community. Populations at one trophic level (e.g., herbivores) extract energy from populations at the next lower trophic level (e.g., green plants) and pass energy to populations at the next higher trophic level (e.g., carnivores: parasites, predators, and pathogens). Because of their energy dependencies, all populations in a community directly or indirectly interact so that each is a component of the environments of the others. Consequently, density means and fluctuations of populations in a community are dynamically interconnected, and amplitudes of density fluctuations are a measure of community stability (the lower the density fluctuations, the greater the stability). Current theory suggests that community stability depends on the complexity of food chains (species diversity) in the community and the degree of coevolution of the composing species (May, 1976*b;* Southwood and Way, 1970).

Ecological succession refers to directional changes in types and diversities of species in communities over time (Brown, 1984; Pianka, 1974). Succession begins with pioneering species invading new environments created by geologic or man-made disturbances, proceeds with continuous replacement and adaptation of species, and usually culminates in a relatively diverse and stable community. Annual plants rapidly colonize newly disturbed environments and are sequentially replaced by biennial plants and perennial shrubs and trees. Pioneer species (r-strategists: annual plants) typically possess life history strategies (e.g., high reproductive rates, high dispersal abilities, low competitive abilities) that enable them to exploit recent environmental disruptions, while their replacements are better suited to more stable situations (K-strategists: low reproductive rates, low dispersal abilities, and high competitive abilities). Thus successional replacement of plants occurs along an r-K continuum (Brown, 1984; Pianka, 1970, 1974). Life history strategies of herbivores and their natural enemies also can be categorized according to the r-K continuum, and successional replacement of these species in a community apparently coincides with equivalent plant replacements (Brown, 1984; Ehler, 1982; Force, 1974; Southwood, 1976). Therefore as communities naturally mature, all trophic levels may be sequentially filled by coadapted species exhibiting more or less equivalent life cycle strategies, eventually culminating in a stable community dominated by K-strategists.

Successional replacement of species in communities dominated by man (agricultural and urban) may be altered or stopped by environmental modifications enforced by man. These communities are either maintained at early successional stages by preserving the presence of annual plants that normally would be replaced through successional processes or maintained in mixed successional stages (annual plants mixed with biennials and perennials), that usually are temporally separated in natural communities (Price et al., 1980). Consequently, human-dominated communities may be composed primarily of r-strategists or of unnatural mixtures of r and K-strategists. In either case, the environmental modifications enforced by man may result in unstable communities.

Human-dominated (secondary) communities probably are best characterized as being composed of species that have been consciously or unconsciously selected for inclusion by man rather than by normal successional replacement (Polvolny, 1971). Many of these species originally were components of local natural (primary) communities but possess innate qualities that allowed then to immediately colonize (r-strategists) or eventually adapt (primarily K-strategists) to environments modified by man (Polvolny, 1971; Southwood and Way, 1970). Some species (primarily r-strategists) probably continue to rely on invasions from surrounding primary communities to maintain their presence in secondary communities. Thus secondary communities are composed of desirable plants and animals (domesticated populations) as well as species of variably adapted pests and natural enemies.

The ability of a species to invade and maintain itself in modified environments may depend on its trophic level. Those species that rely on the same resources consumed, cultivated, and accumulated by man (herbivores, scavengers) probably readily become components of secondary communities (Polvolny, 1971; Southwood and Way, 1970). However, natural enemies of these species, even if they are innately capable of surviving in these modified environments, must wait until their hosts (resources) become sufficiently adapted and abundant to maintain their presence. Thus, natural enemies may temporally lag behind their hosts in invading and adapting to new environments. Such delays may cause short-term instabilities during the formation of secondary communities as well as long-term instabilities if adaptive advantages gained by hosts continue to affect the density regulation abilities of natural enemies. Temporary resurgences of some pests to higher than previous densities following insecticide applications (temporary environmental modifications imposed by man) in agricultural fields are examples of the short-term consequences of such temporal lags (Ripper, 1956). Natural enemies must not only wait for dissipation of insecticide residues to recolonize fields but also for sufficient recolonizations by their hosts, which consequently allows host population densities to resurge. Similarly, the more common development of pesticide resistant pests than natural enemies also has been attributed to trophic dependencies (Croft and Strickler, 1983; Georghiou, 1972). The development of resistant natural enemy populations must be preceded by host resistance, a form of adaptation, even though innate abilities may be equal.

The ability of a species to adapt to modified environments also may partially depend on the degree of genetic variability that exists within its source population (Remington, 1968; Scarza, 1983; Whitten, 1970). If natural enemies are assumed to possess narrower gene pools than their hosts because of trophic restrictions, then adaptabilities of natural enemies to modified environments may be lower than those of their hosts. Apparently no major comparative studies on genetic variabilities between host and natural enemy populations have been conducted, but several authors have implied that differences may occur (Georghiou, 1972; Huffaker, 1971*b;* Huffaker and Messenger, 1964*b;* Messenger, Wilson, and Whitten, 1976; Remington, 1968; Whitten, 1970). Consequently, natural enemies may not only be less diverse than their hosts in secondary communities but when present also may be less completely adapted. However, the ability of a natural enemy to rapidly invade and adapt to environments may also partially depend on whether it is more of an r- or K-strategist, r-strategists being more rapid invaders and having broader gene pools than K-strategists. In addition, natural enemies that attack only a few species of hosts (K-strategy) may possess narrower gene pools than those that attack many (r-strategy; Messenger, Wilson, and Whitten, 1976). Consequently, broadly host-specific natural enemies, or r-strategists, may be more diverse and better adapted than narrowly host-specific species, or K-strategists, in secondary communities (Conway, 1976). Since narrow host-specificity and the ability of a natural enemy to regulate its hosts at low densities are believed to be correlated attributes (Doutt, 1964*a;* Vinson, 1981), some hosts may exhibit higher densities in secondary than in primary communities. Furthermore, since predators and pathogens generally are less specific than parasites (Hagen, Bombosh, and McMurtry, 1976), they consequently may have relatively greater impacts on the densities of certain hosts in secondary communities than in primary.

Polvolny (1971) divided secondary communities into those of agriculture and of human residence (urban). He based this division on the relative intensities of selection pressures (environmental modifications) imposed by man that influenced the composition of these communities; moderate selection pressures in agricultural environments and high selection pressures (severe environmental modifications) in urban environments. Intermediate environments are recognizable, depending on types and severities of environmental modifications. With the assumed gradation from low to high selection pressures, the compositions of secondary communities may be expected to vary from relatively complex in agriculture to simple in urban areas. This variation suggests that natural enemies may be more diverse and effective in agricultural than in urban environments. However, studies by Ehler (1982) and Ehler and Frankie (1979) that compared natural enemy complexes between urban and natural environments indicated that they were qualitatively similar but that proportions of species present differed. Proportional differences may have resulted from differing abilities among the species to adapt to environmental differences between the sites. In the study by Ehler (1982) there also was evidence that host density suppression was greater in natural than urban environments, which could have resulted from functional or adaptational differences among parasites (r- vs. K-strategists) between the study sites.

Polvolny's (1971) discussion on the evolution of secondary communities also suggests that r-strategist pests may be relatively more numerous than K-strategists in urban rather than agricultural environments because of differences in their dispersal capabilities and preadaptations to disturbed environments. Conway (1976) has discussed the potential for biological control of pests relative to their position on an r-K continuum and has suggested that biological control probably has low potential against r-strategists but high potential against pests that exhibit intermediate life history strategies. However, Ehler and Miller (1978) have conducted studies in temporary agricultural environments that indicated r-strategist pests were likely to possess equivalent r-strategist natural enemies that were capable of regulating their densities. Furthermore, Ehler (1982) has suggested that the r versus K classification may be more appropriately applied to natural enemy complexes than to individual species and that the potential for biological control of r-strategists consequently may be equal to that of K-strategists. However, it does appear that pest species exhibiting extreme r-strategist characteristics, such as aphids, may be more difficult to control by biological control methods (Hagen, Bombosh, and McMurtry, 1976) and that the apparently high probability of their occurrence in urban environments may affect the potential for urban biological control.

Many other compounding factors may influence the composition of urban communities, the impact of naturally occurring biological control agents, and the potential for biological control methods. For example, the composition of some urban communities may also depend on their nearness to agricultural or primary communities (Southwood and Way, 1970). When an urban area is situated adjacent to a natural area, the composition of the urban community and the degree of naturally occurring biological control may largely mimic that of the primary community because of the ease with which the composing species can disperse between them. However, as the distance between areas increases, species compositions and diversities may diverge because of different natural enemy and host dispersal abilities. Other confounding factors may include age of urban area and types or degrees of environmental modifications.

Environmental Heterogeneity

Modern agricultural environments are characterized as large areas planted as monocultures. Desirable plants are maintained at abnormally high densities and with unnatural uniform spatial separations. Monocultures also exist in urban environments (lawns, street trees), but these are usually smaller than those in agriculture (Frankie and Ehler, 1978; Kielbaso and Kennedy, 1983; Owen, 1983). Urban environments are more typically characterized as heterogeneous polycultures, composed of both annual and perennial plants whose intra- and interspecific spatial separations usually are quite different from those occurring in natural communities. This unnatural heterogeneity may influence the abilities of and manners by which those natural enemies that are present in urban environments regulate their hosts' densities (Stanton, 1983).

For example, pest outbreaks on isolated residential plantings are often used as collection sites in natural enemy introduction programs (Bartlett and van den Bosch, 1964). This exploration strategy is based on observations that natural enemies that regulate their hosts at low densities frequently cause host populations to fragment and exhibit local outbreaks. Partially because of mortalities caused by natural enemies, natural enemy and host populations become spatially and temporally separated on patches of the hosts' resources. This separation may allow hosts to temporarily reach high densities on local resource patches until their parasites, predators, or pathogens locate them (Anderson, 1982; Andrewartha and Birch, 1954; Huffaker and Messenger, 1964*b;* Nicholson, 1933; Price et al., 1980; Tanada, 1964). Thus, although a natural enemy may effectively keep the average density of its host at a low level over a large area, localized outbreaks may occur.

Spatially isolated pest outbreaks have not been of major concern in the biological control of agricultural pests, since large monocultures tend to decrease the probability and severity of their occurrence. However, they may be of concern in the heterogeneous urban environments. For example, the natural enemies of cottony-cushion scale, *Icerya purchasi*, maintain this pest at low densities in commercial citrus groves in southern California, but local outbreaks of the scale commonly occur on isolated citrus trees and certain ornamental plants in residential areas (Quezada and DeBach, 1973). A similar situation was reported by Frankie et al. (1977) for certain gall-forming insects and their parasites on isolated ornamental oak trees in urban areas. This phenomenon may have a greater impact on young perennial shrubs and trees that are spatially isolated from older plants of the same species. In such cases the temporal lag between infestation by the pest and location of the patch by natural enemies may be too long relative to abilities of young plants to recover.

Since parasites and predators respond to various biotic and abiotic environmental factors to locate their hosts (Doutt, 1964*a;* Vinson, 1981), problems associated with spatial and temporal separations of hosts from natural enemies may be aggravated by modified and variable physical factors in urban environments (Frankie and Ehler, 1978; Sawyer and Casagrande, 1983). Such factors include physical structures and lights as well as microclimatic factors, such as temperature, moisture, and wind. Unfortunately, few studies have been conducted to determine the effects of such factors on natural enemies in urban environments, but some effects may be anticipated. For example, physical structures and barriers are dominant features of urban environments (Frankie and Ehler, 1978) and may affect the searching behaviors and abilities of certain parasites and predators. Such effects may be why some ornamental plant pests (spider mites, whiteflies, mealybugs) frequently achieve higher densities indoors than outdoors on the same plants. Inabilities of natural enemies to invade human dwellings and structures may frequently be the cause of such indoor pest problems. Even when natural enemies invade buildings with their hosts, indoor structural features, climate controls (heating, cooling, humidification, dehumidification), and lighting (including windows) still may differentially affect behaviors and biologies of pests and natural enemies so that pests attain higher densities than outdoors.

Exterior artificial lights in urban areas also may affect host-natural enemy relationships. Many nocturnally active insects are attracted to porch and street lights. If the natural enemies of these species are not similarly attracted, then pest immigrations may numerically overwhelm local natural enemy populations so that high pest densities develop (Cantelo et al., 1974). For example, higher densities of hornworm larvae, *Manduca* spp., frequently are observed on tomato plants in urban gardens than in nearby commercial fields, which may be due to residential lights attracting gravid females that then overwhelm local natural enemy populations with eggs and larvae. Such pest outbreaks may be aggravated in residential areas by the use of insect traps and killing devices that rely on strongly attractive ultraviolet lights. If these devices only trap or kill a portion of the attracted individuals then those that escape may generate local pest outbreaks. For example, Jamornmarn (1979) reported that blacklight traps were equally attractive to both males and females of the corn earworm, *Heliothis zea*, in the laboratory but that females were less likely than males to enter traps. Consequently, attracted females that do not enter traps outdoors may cause local pest outbreaks if insufficient natural enemies are present in the area. Apparently no studies have been conducted in urban areas to confirm such effects.

Human Factors

The most overwhelming feature of urban environments and the feature that most distinctly separates urban environments from agricultural is the presence of dense human populations. This feature and its impacts on urban pest management are discussed in other chapters and elsewhere (Frankie and Koehler, 1983; National Research Council, 1980; Sawyer and Casagrande, 1983). Obviously, the political, social, psychological, economic, and ecological factors associated with the presence of dense human populations in urban areas will affect biological control principles and practices in numerous ways, and not all effects can be anticipated for all situations (see Merritt, Kennedy, and Gersabeck, 1983). Only a few obvious human factors can be discussed here.

Attempts to reduce pest populations to minimal densities are perhaps more intense in urban than agricultural environments, since human contact usually is more frequent and human tolerances consequently may be lower (National Academy of Sciences, 1980; Sawyer and Casagrande, 1983). Such intolerances may reduce the potential for biological control of certain urban pests, since even the most effective natural enemies of the same pests in agricultural or undisturbed environments may be incapable of reducing densities to levels low enough to satisfy coexisting humans. In addition, responses, attitudes, and knowledge relative to the presence, density, and type of pests and their natural enemies may vary between individuals and situations (parks, backyards, homes, shopping malls) and must be considered in proposing, developing, and evaluating urban biological control projects (Olkowski and Olkowski, 1976; Olkowski et al., 1976, 1978; Sawyer and Casagrande, 1983).

Since biological control practices deal with pest and natural enemy populations, areas within which manipulations occur must be large enough to accommodate critical population interactions. However, urban political or legal boundaries (property lines, city blocks, park boundaries) may not be relevant population boundaries. Attempts to manipulate or conserve natural enemies in one area (garden, yard, park) may be affected by pest control attempts and other environmental manipulations in surrounding areas. In general, areas within which biological control practices are attempted must be sufficiently large to allow the populations being manipulated to exhibit some degree of spatial and temporal continuity. This continuity may be more difficult to define and maintain in urban environments than has been experienced by biological control workers in agricultural environments.

Because of the various psychological, sociological, political, and legal factors associated with dense coexisting human populations, the potential for successfully controlling individual pest species by biological control strategies usually will be more difficult to assess in urban than agricultural environments. In addition, biological control successes may be more difficult to define and attain in urban situations as compared to agricultural. However, definable successes against some urban pests have occurred, and the potential for future successes is promising.

BIOLOGICAL CONTROL METHODS

Importations

In their native regions, organisms are geographically restricted by various natural barriers (Bartlett and van den Bosch, 1964; Clausen, 1978; Wilson and Graham, 1983). However, the migrational and commercial movements of man and his necessities have allowed many organisms to breach their barriers and to expand their distributions. The more closely an organism is associated with man and his necessities, the more likely its distribution has been expanded (Polvolny, 1971). Consequently most human ectoparasites and coinhabitants of human residences, such as the German cockroach, the house fly, the head louse, the body louse, and the bed bug, are now nearly cosmopolitan in distribution. Movements of agriculturally important or aesthetically pleasing plants and animals also have resulted in widespread introductions of their pests into new geographic areas (DeBach, 1964*b;* Wilson and Graham, 1983). Recently discovered incipient infestations of the exotic Mediterranean fruit fly, *Ceratitis capitata*, in California and Florida illustrate the continuing high frequency of such introductions into North America.

Introduced species commonly achieve higher densities in invaded areas than in their native regions, and many exotic arthropod species in North America consequently are pernicious agricultural and urban pests (Bartlett and van den Bosch, 1964; Clausen, 1978; Wilson and Graham, 1983). In the regions where a species evolved, it usually possesses various coevolved natural enemies that regulate it at low densities. However, natural enemies and hosts frequently become disassociated during transport to and establishment in new regions (Wilson and Graham, 1983). Consequently, introduced species frequently attain high densities and subsequently become pests in invaded areas (Bartlett and van den Bosch, 1964; Ehler and Andres, 1983). When coevolved natural enemies are successfully re-associated with their hosts in invaded regions, exotic pest densities may be permanently reduced to innocuous levels (DeBach, 1964*b*, 1974; Wilson and Huffaker, 1976; Huffaker, Simmonds, and Laing, 1976). The re-association of natural enemies with their hosts has been the primary goal of biological control since first successfully demonstrated against cottony-cushion scale in the late 1890s (DeBach, 1974; Doutt, 1964*b*).

The importation of natural enemies also may be effective against certain native pests (Pimentel, 1963). Indigenous natural enemies of these pests may be incapable of effectively regulating their densities in disturbed or modified environments or are insufficiently effective relative to human standards. If the same or related species of pest possesses effective natural enemies in other geographic regions, then these natural enemies also may effectively suppress densities of the native pest when introduced. However, relatively few natural enemy importations have occurred against native agricultural or urban pests as compared to exotics (Clausen, 1978).

Techniques and considerations for natural enemy introductions against agricultural pests have been discussed extensively elsewhere (Bartlett and van den Bosch, 1964; DeBach and Bartlett, 1964; Fisher, 1964; Zwolfer, Ghani, and Rao, 1976), but modifications may be required against urban pests because of unique urban environmental characteristics and problems. For example, releases to colonize imported natural enemies of agricultural pests usually are made in fields where host plants are uniformly spaced and pest densities are uniformly high. Released natural enemies may reproduce rapidly with minimum reliance on their dispersal capabilities. Consequently, colonizing natural enemies usually do not become so diluted at release sites that critical behavioral and biological phenomena, such as mating, are sufficiently disrupted to hinder establishment. However, release strategies may have to be modified in urban environments to alleviate dispersal and dilution problems that may develop because of more widely spaced and unnaturally dispersed host habitats and populations. Because of these urban environmental characteristics, longer periods of time may also be required to confirm natural enemy establishments and to evaluate their effectiveness as compared to releases in agricultural environments (DeBach and Bartlett, 1964; DeBach and Huffaker, 1971; DeBach, Huffaker, and MacPhee, 1976).

Importations of natural enemies may have greater potentials in urban than in agricultural environments because of the greater likelihood of exotic pests being introduced into them, the greater diversity of exotic plant and animal hosts that may occur in them to support exotic pests, and the more severe and diverse environmental modifications that occur in them to disrupt the density suppression capabilities of indigenous natural enemies. However, this potential varies with the association the pest has with man, the sensitivity of man to the pest, and the location of the pest problem (home, yard, park, roadside, shopping mall). The importation of natural enemies may have little potential against obligate human ectoparasites or obligate co-inhabitants of human dwellings. Even if highly specific coevolved natural enemies of such pests were located, it may be that man's sensitivities to the pests may be so high that density reductions would still be insufficient. The only acceptable solution for many of these pest problems may be eradication, but natural enemies only reduce average densities and usually are incapable of eradicating their hosts (DeBach, 1964*b*, 1974). It appears that the potential for effectively suppressing pest densities by the importation of natural enemies increases as the frequency of contact between humans and the pest decreases. Thus, pests of plants are more amenable to biological control methods than are sanitary pests (cockroaches) and ectoparasites; exterior pests are more amenable than interior pests; and exterior pests of plants in areas such as parks and roadsides are more amenable than those in residential gardens.

As in agriculture, most natural enemy importation attempts and successes in urban environments have been against scale insects on perennial plants (European elm scale, Bermuda cedar scale, araucaria scale, barnacle scale, lecanium scale, golden oak scale; Clausen, 1978). A recent example is the apparently successful establishment of five natural enemies in California to suppress densities of two exotic scales *Pulvinariella mesembryanthium* and *P. delottoi* on the ice plant, a perennial groundcover commonly planted along urban roads (Tassan, Hagen, and Cassidy, 1983). DeBach, Rosen, and Kennett (1971) have discussed the attributes of scale insects that have led to the great number of importation successes against them. The life cycle strategies of these pests appear to be intermediate on the r-K continuum, which may indicate a high potential for biological control (Conway, 1976). However, pests of perennial plants other than scales also have apparently been successfully controlled by importing their natural enemies (European earwig, Cuban laurel thrips, pine aphid, hibiscus mealybug, comstock mealybug, oak leaf miner, European pine shoot moth; Clausen, 1978). A recent example is the apparently successful establishment in southern California of a natural enemy against the Nantucket pine tip moth, *Rhyacionia frustrana*, an introduced pest of ornamental pine trees (Scriven and Luck, 1978). However, when the high diversities and abundances of exotic perennial groundcovers, shrubs, and trees that are planted in urban areas (Owen, 1983; Kielbaso and Kennedy, 1983) and the consequent great number of exotic pests that probably attack them are considered, it is apparent that natural enemy importation efforts against pests of perennial urban plants have been minimal. For example, euonymus scale, *Unaspis euonymi*, has been a serious pest of ornamental euonymous in North American urban areas for more than a century. The scale is of Asian origin, and no effective natural enemies occur in North America (Gill, Miller, and Davidson, 1982). As with other ornamental plant pests (tea scale, rose chafer, imported willow leaf beetle), the importation of natural enemies to effectively suppress densities of euonymus scale appears promising, but no efforts have been made.

Importations of natural enemies against pests of annual ornamental plants in urban areas probably has less potential than against pests of perennial plants. Numerous spatial, cultural, and temporal disruptions occur in plantings of annual ornamentals (Owen, 1983), and pests of these plants, such as aphids, usually exhibit extreme r-strategist characteristics (Hagen, Bombosh, and McMurtry, 1976; Ehler and Miller, 1978; Southwood, 1976). These disruptions and pest characteristics may reduce the probabilities of locating equivalently adapted natural enemies (Conway, 1976). However, a recent study by Ehler and Miller (1978) in annual cotton fields indicated that certain natural enemies may be capable of suppressing such pests.

Some urban medical and veterinary pests have had exotic natural enemies established against them that apparently have resulted in at least partially effective control (black widow spider, American cockroach, American dog tick; Clausen, 1978). The prospects and problems involved in attempting to use natural enemies against such pests were reviewed by Bay et al. (1976) and Legner et al. (1974). The primary problem with such pests is the low densities to which their populations usually must be reduced relative to the sensitivities of man. This problem is critical when vectors of human diseases are considered (Bay et al., 1976).

As DeBach (1974) has stated, the successful importation and establishment of natural enemies usually depends on intense and tenacious efforts. It is apparent that such efforts in urban situations have been minimal. This minimal effort is in contrast to the environmental and economic benefits that may result from successful importation projects (DeBach, 1964*b*, 1974; Doutt and DeBach, 1964; Ervin, Moffitt, and Meyerdirk, 1983; Huffaker, Simmonds, and Laing, 1976; Simmonds, 1968; Wilson and Huffaker, 1976). The primary reason for this lack of effort appears to be insufficient funding. Since the importation of an effective natural enemy may permanently reduce a pest's densities to innocuous levels and thus eliminate the need for future interventions by man (DeBach, 1964*b*; Huffaker, Simmonds, and Laing, 1976), this method usually is not commercially exploitable. The funding for natural enemy introductions must consequently come from user groups or government organizations, and these apparently are difficult to define and organize for urban situations.

Periodic Releases

When parasites, predators, or pathogens are insufficiently numerous to naturally suppress densities of pests, periodic releases may be used to artificially increase their densities and subsequently reduce pest densities to innocuous levels. Periodic releases may be divided into those that are inoculative and those that are inundative, depending upon relative numbers of natural enemies released and how rapidly the target pests' densities are expected to be suppressed (DeBach and Hagen, 1964; Rabb, Stinner, and van den Bosch, 1976; Ridgway and Vinson, 1977). Inundative releases involve relatively high numbers of natural enemies that are expected to immediately suppress high pest densities. Inoculative releases are made with relatively small numbers of natural enemies against presently low but increasing pest densities. Unlike inundation, inoculative releases rely on the progeny of the released natural enemies to eventually retard or eliminate increases in pest densities. Releases for both immediate and long-term pest suppressions also are possible.

DeBach and Hagen (1964) have discussed various factors that should be considered when developing, implementing, and justifying periodic releases against agricultural pests, and most of their discussions also apply to urban pests. Although periodic releases against urban pests are environmentally and commercially attractive, their effectiveness depends on thoroughly understanding the specific pest-natural enemy relationship and relevant environmental influences. However, nearly all periodic releases that have been attempted against urban pests were originally developed for use in agriculture, and surprisingly few intervening studies to account for differences in environmental, economic, and social factors have been performed in their transfer to urban situations. In addition, some natural enemies that are currently being sold to urban residents for periodic releases are not yet or never will be implemented in agriculture because of their poor economics, unreliabilities, or ineffectiveness. For example, egg cases of predacious preying mantids commonly are sold to urban residents for release into their gardens, although apparently no equivalent programs have been developed for agriculture. Recent studies by Hurd and Eisenberg (1984) have indicated that such releases probably have little or no effect on densities of most urban pests. Perhaps the only positive effect of their release may be that some homeowners may sufficiently reduce pesticide applications to allow other natural enemies previously hindered by such applications to regain control of certain pest species, but this is insufficient justification for such releases.

Certain species of lady beetles, especially the convergent lady beetle, *Hippodamia convergens*, also are frequently collected as adults from overwintering aggregations and sold to consumers for release in urban gardens (DeBach, 1974; Hagen, 1962). Unlike preying mantids, lady beetles may effectively suppress high densities of certain pests, especially aphids and mites. However, the dispersal capabilities and behaviors of adults may diminish the effectiveness of periodic releases in certain urban situations.

Minimum host densities usually are required to establish and maintain these predators in an area (Hagen, Bombosh, and McMurtry, 1976; Hagen, 1962). Most releases by homeowners probably are against insufficient or nonexistent pest densities, and most of the released predators probably rapidly disperse from such release areas. Periodic releases of most lady beetles are difficult to economically or practically justify in urban gardens, and their beneficial activities usually can be more effectively utilized by methods that promote and conserve the presence of indigenous species.

Trichogramma spp. are parasites of the eggs of several species of Lepidoptera and are extensively reared and released to control several agricultural pests around the world (Ridgway et al., 1981; Veronin and Grinberg, 1981; Ridgway and Vinson, 1977). They are also frequently advertised in North America as biological control agents for release in residential gardens against pests like the cabbage looper, *Trichoplusia ni*, and the corn earworm (tomato fruitworm), *Heliothis zea*. When these parasites are inundatively released into urban gardens against such pests, they may kill sufficient numbers of eggs to effectively suppress subsequent damaging larval densities. However, the effective use of these parasites depends on several operational and biological parameters especially the proper timing of releases to coincide with the target pest's egg stage (Ridgway et al., 1981). If releases do not occur when a pest's population is primarily in the egg stage, parasites will appear ineffective despite their innate potentials. Since most homeowners respond only to pest damage, cannot predict damage before it occurs, or cannot recognize the eggs of pests, most *Trichogramma* releases into urban gardens probably are improperly timed and thus procedurally ineffective. This and other problems have hindered attempts to implement inundative releases of *Trichogramma* spp. in agriculture, and such problems are compounded in urban situations (Ridgway et al., 1981; Ridgway and Vinson, 1977).

Barrows and Hooker (1981) have recently investigated the potential for inoculatively releasing the parasite *Pediobius foveolatus* to suppress larval densities of the Mexican bean beetle, *Epilachna varivestis*, and the squash beetle, *E. borealis*, in urban vegetable gardens. The parasite was introduced into North America to suppress Mexican bean beetle densities in commercial soybean fields but was unable to overwinter. An inoculative release program was developed for soybeans using snap bean nurse-plots (Stevens, Steinhauer, and Coulson, 1975; Flanders, 1985). Attempts to use the parasite in urban gardens were an extrapolation, with little modification, of techniques primarily developed for agriculture, and as in agriculture, there are several questions remaining as to the effectiveness, economics, operation, and reliability of this release program (Flanders, 1983, 1985). More research is required to economically and practically justify *P. foveolatus* releases in urban gardens and to define the operational parameters of the release program if it is justified.

Many studies have been conducted on periodic releases of the parasite *Encarsia formosa* to suppress densities of the greenhouse whitefly, *Trialeurodes vaporariorum*, on ornamental plants in commercial greenhouses (Helgesen and Tauber, 1974, 1977; Tauber and Helgesen, 1981). Similar studies have been conducted on releases of various predators against such pests as spider mites (Burnett, 1979; Field and Hoy, 1984; Gould and Light, 1971; Hamlen and Lindquist, 1981; Hamlen and Poole, 1980; Huffaker, van de Vrie, and McMurtry, 1970; Simmonds, 1972; Wheeler, Stinner, and Henry, 1975) and aphids (Gurney and Hussey, 1970; Hamalainen, 1977; Markkula and Tiittanen, 1977; Scopes, 1969, 1970; Wyatt, 1970). It appears that these release programs could be readily modified for use by homeowners and managers of public interior plantscapes to control the same pests. For example, Olkowski, Daar, and Olkowski (1983) described the incorporation of periodic releases into a pest management program being developed for a conservatory. Steiner and Elliott (1983) listed several natural enemies available for release against several pests of interior plantscapes, but effectiveness of these releases into such environments were inadequately documented. In general, releases of natural enemies to control pests of interior plants appear to have potential, but more objective evaluation and implementation studies are required before they can be reliably recommended.

Slater (1984) and Slater, Hurlbert, and Lewis (1980) have recently reported on attempts to suppress densities of the brownbanded cockroach, *Supella longipalpa*, by releasing the oothecal parasite *Comperia merceti.* Although the authors claimed that releases effectively suppressed infestations in university offices and laboratories, more research is required on the effects of indoor climates and structures on this parasite and on the sociological and psychological effects of such releases in residences before their effectiveness and practicality can be completely evaluated.

Inundative releases of certain pathogens, especially viruses and bacteria, appear to have high potentials for the control of many urban pests (Jaques, 1983). In general, such releases artificially induce epizootics by increasing densities of a pathogen's infective units. Such releases circumvent problems encountered by naturally occurring pathogens in overcoming spatial and temporal heterogeneities of urban environments that may inhibit natural epizootics. Preparations of pathogens, their infective units, or their toxic by-products are used (Burges and Hussey, 1971; Jaques, 1983). The popularity of such releases in urban situations has been due to application strategies similar to those of synthetic insecticides, their environmental safety, and their high specificity.

The most widely used pathogen preparations are those composed of the toxic crystals produced by various strains of *Bacillus thuringiensis* (Jaques, 1983; Weiser, Bucher, and Poinar, 1976). The crystals paralyze intestinal tracts of certain species of Lepidoptera and Diptera when ingested by larvae.

The deterioration of the intestinal lining then allows various facultative pathogens, including *B. thuringiensis*, to invade the hemocoel, where they eventually cause the death of the host. Applications of these formulations are nearly identical to those of synthetic insecticides and are consequently easily substituted into existing urban pest control programs (Jaques, 1983; Olkowski et al., 1976, 1978; Pinnock and Milstead, 1971). However, unlike broad-spectrum synthetic insecticides, *B. thuringiensis* formulations are highly specific and possess short residual activities. Because of their specificities, these formulations conserve the presence of other natural enemies in the areas where they are applied and consequently have great potential for urban pest management programs.

Besides *B. thuringiensis*, the only other bacterial preparation currently available for use against an urban pest is that of *Bacillus popillae* for the control of Japanese beetle grubs, *Popillia japonica*, in lawns and turf (Burgess and Hussey, 1971; Jaques, 1983; Weiser, Bucher, and Poinar, 1976). The bacterium causes milky disease when spores are ingested by susceptible larvae. Unlike *B. thuringiensis, B. popillae* applications are inoculative, since spores usually are produced by the pathogen at application sites and continue to influence Japanese beetle infestations.

Various virus formulations also are available or being investigated for use against some urban pests and their future availability and use appears promising (Burgess and Hussey, 1971; Jaques, 1983; Weiser, Bucher, and Poinar, 1976). Several nematode parasites of arthropods (*Neoplectana carpocapsae, Reesimermis nielseni*) currently are being investigated for periodic releases against some agricultural pests, and future investigations may expand to certain urban pests (Jaques, 1983; Poinar, 1979).

Compared to other biological control agents and methods, periodic pathogen releases (including nematodes) may have greater potentials for the suppression of certain urban medical and veterinary pests, especially those that are aquatic in their immature stages (mosquitoes, blackflies). Similarly, the use of pathogens to control pests like termites (Hanel and Watson, 1983) and certain pestiferous social wasps appears promising, but considerable research is required.

Few augmentative release strategies, except those involving certain bacterial and viral pathogen formulations, have been developed to the extent where they can be reliably recommended to suppress urban pests. Even with those few programs that may be adequately developed, most homeowners and municipal employees probably are inadequately informed or trained to properly implement them. As with natural enemy importations, periodic releases appear promising for the ecologically and economically effective suppression of many urban pests, but research efforts and support have been inadequate.

Natural Enemy Conservation

Various ant species, especially the Argentine ant, *Iridomyrmex humilis*, may adversely affect otherwise effective natural enemies of various plant pests (Bartlett, 1961; Flanders, 1945, 1951; Olkowski et al., 1978; van den Bosch and Telford, 1964; Way, 1959). These ants seek populations of certain homopterous insects (aphids, soft scales, mealybugs) to feed on the honeydew they excrete. Ant-attended populations frequently increase to high densities because of interferences by ants with normally effective parasites and predators. The ants also may interfere with natural enemies of other pests that do not excrete honeydew but occur on the same plant (armored scales), thus also allowing their densities to increase. Copious amounts of honeydew produced by ant-attended populations and the consequent unsightly development of sooty-mold fungus frequently affect infested and surrounding trees and plants, as well as nearby objects, in urban situations (Olkowski and Olkowski, 1976; Olkowski et al., 1978).

Such ant-instigated-outbreaks frequently occur in urban situations, and the usual remedy is the application of a broad-spectrum insecticide, such as malathion, to the entire plant or tree (Olkowski and Olkowski, 1976; Olkowski et al.,1978). Such applications may only aggravate existing problems, since offending ant colonies usually are located away from the site and consequently are unaffected. Effective but ant-suppressed natural enemies also may be even more adversely affected by insecticide applications. In most cases, eliminating ant access to plants by employing mechanical or chemical barriers or spot treating offending ant colonies with insecticides are the most appropriate remedies. Such remedies will allow natural enemies to regain control of honeydew-producing pests, as well as other pests whose natural enemies also were previously affected by the ants (Flanders, 1945, 1951; Olkowski et al, 1978). However, in some situations the remedy may not be as simple. My own observations on weeping willow trees in Indiana have indicated that attendances of aphids by several ant species increase aphid densities, but that ants also prey on coexisting populations of imported willow leaf beetle, *Plagiodera versicolora*. Parasites of the willow leaf beetle are adversely affected by ants, but ant predation more effectively suppresses beetle densities than do the parasites. Elimination of ants decreased aphid densities, but tree defoliation increased because of higher imported willow leaf beetle densities (R. V. Flanders, unpublished data). Thus, in urban situations where ants are present, a complete appraisal of the situation is required.

Certain cultural practices (pruning, watering, fertilizing) may also affect relationships between natural enemies and their hosts and should be considered in the development of urban pest management programs (Rabb, Stinner, and van den Bosch, 1976; van den Bosch and Telford, 1964). The nutritional qualities of certain plants may be altered by the application of fertilizers and consequently affect herbivore-natural enemy relationships. The irrigation of plants also may differentially affect relationships between pests and their natural enemies through changes in microclimate. For example, spider mite population densities frequently increase to higher than normal densities on plants that are water-stressed (van de Vrie, McMurtry, and Huffaker, 1972). Apparently, higher survival or reproductive rates of mites on such plants allow them to increase more rapidly than their predators. The pruning of plants also may alter pest-natural enemy relationships by altering microclimates, plant growth characteristics, or plant nutritional characteristics. For example, Hall and Ehler (1980) have investigated populations of *Aphis nerii* on urban oleander bushes and have found that succulent regrowth resulting from frequent pruning and irrigation favored increases in aphid densities. In addition, the parasite *Lysiphlebus testaceipes* was more efficient at low aphid densities on less frequently irrigated and pruned plants.

Many studies have confirmed the importance of requisites such as nectar, alternate hosts, and refuges to maintain the presences or efficiencies of parasites and predators in agricultural fields (Rabb, Stinner, and van den Bosch, 1976; van den Bosch and Telford, 1964). Apparently no equivalent studies have been conducted in urban situations. The availability, types, and distributions of such requisites undoubtedly affect natural enemy efficiencies in urban environments and should be considered in the development of urban pest management programs.

Modifying Insecticide Usage

Applications of nonspecific persistent insecticides have repeatedly been shown to more adversely affect natural enemies than pests (Bartlett, 1964; Croft and Brown, 1975; DeBach, 1974; Doutt, 1964*c;* Doutt and Smith, 1971; Georghiou, 1972; Ripper, 1956). Not only may parasites and predators be more physiologically vulnerable to toxicants than their hosts, but their trophic dependencies also may make them more ecologically vulnerable. Thus insecticide applications may not only cause more severe mortalities in natural enemy populations but may also spatially and temporally disrupt natural enemy-host relationships. Consequently, target pests that possess only partially effective natural enemies may resurge to higher densities following insecticide applications than they exhibited before (pesticide-induced resurgences), or the densities of previously innocuous pests that possess highly effective natural enemies may increase to damaging levels following insecticide applications (pesticide-induced upsets).

Several examples of insecticide-induced upsets of arthropods in urban environments have been reported. Luck and Dahlsten (1974, 1975) have documented outbreaks of pine needle scale, *Chionaspis pinifoliae*, in South Lake Tahoe, California, that resulted from the adverse effects of mosquito-fogging operations on the scale's natural enemies. Modifications in the mosquito control program allowed the parasites to regain their influence on pest populations and the return of pest densities to acceptable levels (Roberts, Luck, and Dahlsten, 1973). Similarly, Merritt, Kennedy, and Gersabeck (1983) have found that insecticide applications to control pestiferous flies on Makinac Island, Michigan, disrupted parasites and predators of European fruit lecanium, *Parthenolecanium corni*, causing scale outbreaks to occur on residential shade and fruit trees. DeBach and Rose (1977) reported outbreaks of citrus red mite, *Panonychus citri*, purple scale, *Lepidosaphes beckii*, and woolly white fly, *Aleurothrixus floccosus*, in residential and commercial citrus trees that resulted from insecticidal eradication attempts against the Japanese beetle. Ehler and Endicott (1984) recently reported on outbreaks of walnut aphid, *Chromaphis juglandicola*, and black scale, *Saisettia oleae*, in urban areas that were attributed to destruction of their natural enemies by malathion-bait sprays against the Mediterranean fruit fly in northern California. In all these cases, pest upsets occurred as a result of pesticide applications over relatively large urban areas. In urban residential areas, resurgences and upsets may be less frequent or at least less severe, since insecticide applications usually are more spatially restricted because of polycultures and property boundaries. However, the growing popularity of commercial companies that contract with homeowners to prophylactically apply broad-spectrum insecticides on lawns and gardens may dramatically increase intensities and frequencies of urban pest upsets and resurgences. Such applications may adversely affect host-natural enemy relationships in treated as well as adjacent yards and should be discouraged.

Since most arthropods in urban environments are maintained at low densities by indigenous natural enemies (DeBach, 1974; Hagen, van den Bosch, and Dahlsten, 1971), this natural control must be preserved when considering insecticide applications against those pests that do not possess similarly effective natural enemies. Insecticides that are narrowly species specific within and between trophic levels and that are nonpersistent should be employed whenever possible (Messenger, Biliotti, and van den Bosch, 1976; Newsom, Smith, and Whitcomb, 1976; Stevenson and Walters, 1983). The benefits of switching from broadly to narrowly specific insecticides have been discussed by Olkowski et al. (1976) for urban shade tree pests. Highly specific and environmentally nondisruptive pathogen formulations, such as *B. thuringiensis*, are most desirable when available for urban pest control. However, broad-spectrum nonspecific insecticides are still most frequently used against urban pests because of current marketing strategies and consumer perspectives and knowledge (Barrows, DeFilippo, and Tovallali, 1983; Levenson and Frankie, 1983).

When only nonspecific insecticides are available for the effective control of an urban pest, selective application strategies that preserve natural enemy presences may still be possible. In some situations, insecticides can be applied when natural enemies are physically or behaviorally nonsusceptible to their toxic effects (Bartlett, 1964; Newsom, Smith, and Whitcomb, 1976). For example, Flanders, Bledsoe, and Edwards (1984) have found that if nonspecific insecticides were applied against the Mexican bean beetle when *P. foveolatus* populations were primarily in their immature stages and protected within the remains of their hosts, then low parasite mortalities occurred. However, if the toxic residues of an insecticide persisted for a long time in the area, then adult parasites were rapidly killed when they emerged from the protective remains of their hosts. Consequently, short residual activity, like high specificity, is a preferred characteristic of insecticides (Newsom, Smith, and Whitcomb, 1976). Unfortunately, long residual activity usually is considered a desirable characteristic for insecticides by most urban residents and pesticide applicators, since it implies long term control (high market appeal for manufacturers) and reduces the need to time applications to when susceptible pest stages are present (high probability of success for untrained applicators) (Barrows, DeFilippo, and Tovallali, 1983).

In general, synthetic insecticides presently are an undesirable but necessary component of modern urban environments. In many situations, synthetic insecticide applications and biological control could be compatible pest control strategies, but little effort has been directed at developing such approaches, especially in urban environments. Urban residents and pesticide applicators should be informed of the ecological problems associated with insecticides, and urban researchers should develop programs that minimize biological and ecological disruptions that may result from their applications.

BIOLOGICAL CONTROL AND URBAN PEST MANAGEMENT

Any observant urban gardener is well aware of the multitude of plant and animal species that society coexists with. Fortunately, only a few of these species actually attain densities sufficient to be troublesome. Thus, in attempting to ameliorate the problems associated with the few species that are pests, we must always be aware of the multitude of potential pests that are maintained at innocuous densities by effective natural controls. When any pest control operation is attempted in any situation, its effects must not be viewed solely with regard to the target pest population but within the context of the entire ecological community. Naturally occurring biological control should be considered the framework within which all pest control practices should be intertwined. This philosophy forms the basis of the integrated pest management (IPM) concept (Corbet and Smith, 1976; Sawyer and Casagrande, 1983; Huffaker, Simmonds, and Laing, 1976).

Urban IPM programs should be developed from base-line information on naturally occurring species and their interactions in areas where control operations (especially pesticide applications) have not been and are not being applied. Pertinent pest control strategies should then be attempted to determine how each affects the natural occurrences and interactions of species, especially natural enemies. The least disruptive but effective strategies should then be incorporated into the IPM program. Such ideal scenarios have rarely been used in the development of agricultural or urban pest control programs that are currently being labeled as integrated. In other words, the philosophy and reality of IPM have yet to be completely meshed. In most current IPM programs, reductions in pesticide usage through the development of treatment (economic) thresholds and substitutions of selective and nonpersistent pesticides for nonselective and persistent insecticides have been understandably emphasized. These efforts have helped reduce some adverse environmental impacts that have resulted from previous pesticide application programs but have been the result of short-term, crisis-oriented studies that have contributed little to our understanding of the ecological systems involved. Since biological control strategies require long-term research commitments, few have been incorporated into modern IPM programs, especially in urban environments.

Biological control strategies, when available or sufficiently developed, tend to be among the least environmentally disruptive pest control alternatives. Since importation, augmentation, and conservation of natural enemies are tactics based on existing or missing components of natural control, they tend to support or strengthen interactions in ecological communities. Thus, biological control strategies ameliorate human environmental disruptions and return ecological systems to more natural and stable states. However, biological control should not be expected to solve all pest problems. As previously discussed, biological control probably has low potentials against human ectoparasites, obligatory coinhabitants of human residences, and disease vectors. In general, the more remotely located a pest situation is from areas of human residence and frequent human contact, the higher the potential for biological control may be. Even in outdoor situations biological control may not be capable of solving all problems. For example, urban situations with multiple pests, as also occurs in agriculture, may not be immediately amenable to biological control approaches because of the high host-specificities of many natural enemies. Additionally, biological control programs usually require longer periods of time to develop than other pest control strategies and consequently cannot be immediately instituted to solve new problems. Even if a biological control strategy has potential against a newly erupted pest problem, it is likely that other intervening strategies must be adopted before a biological control solution can be instituted. Various other problems or disadvantages may occur when attempting to incorporate biological control strategies into urban pest situations (Merritt, Kennedy, and Gersabeck, 1983; Sawyer and Casagrande, 1983). These problems should not be construed as insurmountable to urban biological control efforts, since similar problems have been overcome in agriculture.

Other than previous natural enemy importations that have resulted in permanent establishments of exotic natural enemies and some bacterial and viral periodic strategies, there presently are very few reliable or objectively evaluated biological control strategies available for use by urban residents and organizations. Despite high potentials for biological control in numerous urban situations, research support and efforts have been insufficient. Perhaps with the increasing realization that urban situations possess distinct and serious pest problems, biological control efforts will be intensified. Until such intensifications occur, endorsements and exploitations of insufficiently developed biological control strategies must be discouraged. Failures of such programs may not only ingrain distrust in urban residents and organizations for the particular program and agent but also for the entire concept of biological control. Future biological control efforts could be seriously hindered by such developments.

As in agriculture, urban biological control programs must be developed by trained specialists in universities and government research organizations. Once these programs are developed, responsibility for delivery and maintenance will shift to other groups. This shift in responsibility usually does not occur in importation programs, since established natural enemies permanently perpetuate themselves in the environment, but does occur in periodic release and conservation programs. Extensive university, governmental, and commercial delivery and maintenance systems are well developed for agricultural pest control programs, but such systems generally are less extensive for urban situations. This state of affairs may pose problems for implementing urban periodic release and conservation programs. Conservation programs appear to be the most easily implemented, since existing urban delivery and maintenance systems, which primarily deal with pesticide recommendations, could easily incorporate recommendations on natural enemy conservation tactics. Periodic releases of natural enemies against urban pests appear to be the most difficult to implement and maintain. Tauber and Helgesen (1981) have discussed problems and considerations for implementing periodic releases of *Encarsia formosa* in commercial greenhouses, and many of their discussions also apply to urban situations. Since periodic releases involve a marketable product (the natural enemies), commercial involvements in the implementation and maintenance of urban programs are likely. For example, pathogen formulations have been readily incorporated into urban pest management programs primarily because of commercial interests. Similar involvements are likely to occur for parasite and predator release programs when sufficiently developed. However, the abilities and training of urban applicators may ultimately determine the success of periodic releases, which may dictate the need for extensive training and monitoring programs. Other potential problems with periodic releases may include product standardization, shipping and storage methods, and rearing programs.

In summary, the more frequent and diverse environmental modifications and disruptions that occur in urban areas may more severely affect naturally occurring biological control agents than in agriculture. However, such effects may indicate that urban biological control efforts have potentials equal to or greater than those experienced in agriculture, since the primary objective of applied biological control is to reduce or eliminate such adverse effects to benefit natural enemy occurrences and efficiencies. Biological control methods that have primarily been developed for use in agriculture situations may need modification to overcome difficulties associated with urban environments. With the high number of exotic pests that are likely to occur in urban environments, natural enemy importation appears to have the greatest potential, compared to the other methods, but research support is a major problem. Compared to importations, natural enemy conservation strategies, especially those related to insecticide use, may be more easily and quickly developed and implemented for many urban pest situations. Periodic releases, especially of parasites and predators, appear to be the most difficult to develop and implement, but commercial interests may aid and hasten their effective use against certain urban pests. In general, past urban biological control efforts and support have been minimal and must be increased to take maximum advantage of the apparently high potential of biological control techniques to suppress numerous urban pest problems.

ACKNOWLEDGMENTS

Critical comments and suggestions on the original manuscript by G. W. Frankie (University of California, Berkeley), L. E. Ehler (University of California, Davis), and J. W. Yonker (Purdue University, W. Lafayette, Ind.) have been appreciated.

REFERENCES

Anderson, R. M. 1982. Theoretical basis for the use of pathogens as biological control agents of pest species. *Parasitology* **84**:3-33.

Andrewartha, H. G., and L. C. Birch. 1954. *The Distribution and Abundance of Animals.* University of Chicago Press, Chicago.

Barrows, E. M., and M. E. Hooker. 1981. Parasitization of the Mexican bean beetle by *Pediobius foveolatus* in urban vegetable gardens. *Environ. Entomol.* **10:**782-786.

Barrows, E. M., J. S. DeFilippo, and M. Tovallali. 1983. Urban community gardener knowledge of arthropods in vegetable gardens in Washington, D.C. In *Urban Entomology: Interdisciplinary Perspectives,* G. W. Frankie and C. S. Koehler (eds.). Praeger, New York, pp. 107-126.

Bartlett, B. R. 1961. The influence of ants upon parasites, predators and scale insects. *Ann. Entomol. Soc. Am.* **54:**543-551.

Bartlett, B. R. 1964. Integration of chemical and biological control. In *Biological Control of Insect Pests and Weeds,* P. DeBach (ed.). Chapman and Hall, London, pp. 489-511.

Bartlett, B. R., and R. van den Bosch. 1964. Foreign exploration for beneficial organisms. In *Biological control of Insect Pests and Weeds,* P. DeBach (ed.). Chapman and Hall, London, pp. 283-304.

Bay, E. C., C. O. Berg, H. C. Chapman, and E. F. Legner. 1976. Biological control of medical and veterinary pests. In *Theory and Practice of Biological Control,* C. B. Huffaker and P. S. Messenger (eds.). Plenum, New York, pp. 457-479.

Brown, V. K. 1984. Secondary succession: Insect-plant relationships. *Bioscience* **34:**710-716.

Burges, H. D., and N. W. Hussey, eds. 1971. *Microbial Control of Insects and Mites.* Academic Press, New York.

Burnett, T. 1979. An acarine predator-prey population infesting roses. *Res. Popul. Ecol.* **20:**227-234.

Cantelo, W. W., J. L. Goodenough, A. H. Baumhover, J. S. Smith, Jr., J. M. Stanley, and T. J. Henneberry. 1974. Mass trapping with blacklight: Effects on isolated populations of insects. *Environ. Entomol.* **3:**389-395.

Clark, L. R., P. W. Geier, R. D. Hughes, and R. F. Morris. 1967. *The Ecology of Insect Populations in Theory and Practice.* Methuen, London, 232p.

Clausen, C. P., ed. 1978. *Introduced Parasites and Predators of Arthropod Pests and Weeds: A World Review,* ARS Agricultural Handbook 480. U.S. Department of Agriculture, Washington, D.C.

Conway, G. 1976. Man versus pests. In *Theoretical Ecology: Principles and Applications,* R. M. May (ed.). W. B. Saunders, Philadelphia, pp. 257-281.

Corbet, P. S., and R. F. Smith. 1976. Integrated control: A realistic alternative to misuse of pesticides? In *Theory and Practice of Biological Control,* C. B. Huffaker and P. S. Messenger (eds.). Academic Press, New York, pp. 661-682.

Croft, B. A., and A. W. A. Brown. 1975. Responses of arthropod natural enemies to insecticides. *Annu. Rev. Entomol.* **20:**285-335.

Croft, B. A., and K. Strickler, 1983. Natural enemy resistance to pesticides: Documentation, characterization, theory and application. In *Pest Resistance to Pesticides,* G. P. Georghiou and T. Saito (eds.). Plenum, New York, pp. 669-702.

DeBach, P., ed. 1964*a*. *Biological Control of Insect Pests and Weeds.* Chapman and Hall, London.

DeBach, P. 1964*b*. The scope of biological control. In *Biological Control of Insect Pests and Weeds,* P. DeBach (ed.). Chapman and Hall, London, pp. 3-20.

DeBach, P. 1974. *Biological Control by Natural Enemies.* Cambridge University Press, London.

DeBach, P., and B. R. Bartlett. 1964. Methods of colonization, recovery and evaluation. In *Biological Control of Insect Pests and Weeds,* P. DeBach (ed.). Chapman and Hall, London, pp. 407-428.

DeBach, P., and K. S. Hagen. 1964. Manipulation of entomophagous species. In *Biological Control of Insect Pests and Weeds,* P. DeBach (ed.). Chapman and Hall, London, pp. 429-458.

DeBach, P., and C. B. Huffaker. 1971. Experimental techniques for evaluation of the effectiveness of natural enemies. In *Biological Control,* C. B. Huffaker (ed.). Plenum, New York, pp. 125-148.

DeBach, P., and M. Rose. 1977. Environmental upsets caused by chemical eradication. *Calif. Agric.* **32:**8-10.

DeBach, P., C. B. Huffaker, and A. W. MacPhee. 1976. Evaluation of the impact of natural enemies. In *Theory and Practice of Biological Control,* C. B. Huffaker (ed.). Academic Press, New York, pp. 255-285.

DeBach, P., D. Rosen, and C. E. Kennett. 1971. Biological control of coccids by introduced natural enemies. In *Biological Control,* C. B. Huffaker (ed.). Academic Press, New York, pp. 165-194.

Doutt, R. L. 1964*a*. Biological characteristics of entomaphagous adults. In *Biological Control of Insect Pests and Weeds,* P. DeBach (ed.). Chapman and Hall, London, pp. 145-167.

Doutt, R. L. 1964*b*. The historical development of biological control. In *Biological Control of Insect Pests and Weeds,* P. DeBach (ed.). Chapman and Hall, London, pp. 21-42.

Doutt, R. L. 1964*c*. Ecological considerations in chemical control: Implications to nontarget invertebrates. *Bull. Entomol. Soc. Am.* **10:**83-88.

Doutt, R. L., and P. DeBach. 1964. Some biological control concepts and questions. In *Biological Control of Insect Pests and Weeds,* P. DeBach (ed.). Chapman and Hall, London, pp. 118-142.

Doutt, R. L., and R. F. Smith. 1971. The pesticide syndrome—Diagnoses and suggested prophylaxes. In *Biological Control,* C. B. Huffaker (ed.). Plenum, New York, pp. 3-15.

Ehler, L. E. 1982. Ecology of *Rhopalomyia californica* Felt (Diptera: Cecidomyiidae) and its parasites in an urban environment. *Hilgardia* **50:**1-32.

Ehler, L. E., and L. A. Andres. 1983. Biological control: Exotic natural enemies to control exotic pests. In *Exotic Plant Pests and North American Agriculture,* C. L. Wilson and C. L. Graham (eds.). Academic Press, New York, pp. 395-418.

Ehler, L. E., and P. C. Endicott. 1984. Effect of malathion-bait sprays on biological control of insect pests of olive, citrus, and walnut. *Hilgardia* **52:**1-47.

Ehler, L. E., and G. W. Frankie. 1979. Arthropod fauna of live oak in urban and natural stands in Texas, II. Characteristics of the mite fauna (Acari). *J. Kan. Entomol. Soc.* **52:**86-92.

Ehler, L. E., and J. C. Miller. 1978. Biological control in temporary agroecosystems. *Entomophaga* **23:**207-212.

Ervin, R. T., T. J. Moffitt, and D. E. Meyerdirk. 1983. Comstock mealybug (Homoptera: Pseudococcidae): Cost analysis of a biological control program in California. *J. Econ. Entomol.* **76:**605-609.

Field, R. P., and M. A. Hoy. 1984. Biological control of spider mites on greenhouse roses. *Calif. Agric.* **38**(3-4):29-32.

Fisher, T. W. 1964. Quarantine handling of entomophagous insects. In *Biological Control of Insect Pests and Weeds,* P. DeBach (ed.). Chapman and Hall, London, pp. 305-327.

Flanders, R. V. 1983. Evaluating Mexican bean beetle biocontrol by *Pediobius foveolatus. IPM Pract.* **5**(7):1, 7-8.

Flanders, R. V. 1985. Biological control of the Mexican bean beetle: Potentials for and problems of innoculative releases of *Pediobius foveolatus.* In *World Soybean Research Conference III: Proceedings,* R. Shibles (ed.). Westview Press, Boulder, Colo., pp. 685-694.

Flanders, R. V., L. W. Bledsoe, and C. R. Edwards. 1984. Effects of insecticides on *Pediobius foveolatus* (Hymenoptera: Eulophidae), a parasitoid of the Mexican bean beetle (Coleoptera: Coccinellidae). *Environ. Entomol.* **13:**902-906.

Flanders, S. E. 1945. Coincident infestation of *Aonidiella citrina* and *Coccus hesperidum,* a result of ant activity. *J. Econ. Entomol.* **38:**711-712.

Flanders, S. E. 1951. The role of the ant in the biological control of homopterous insects. *Can. Entomol.* **83:**93-98.

Force, D. C. 1974. Ecology of insect host-parasitoid communities. *Science* **184:**624-632.

Frankie, G. W., and L. E. Ehler. 1978. Ecology of insects in urban environments. *Annu. Rev. Entomol.* **23:**367-387.

Frankie, G. W., and C. S. Koehler, eds. 1983. *Urban Entomology: Interdisciplinary Perspectives.* Praeger, New York, 496p.

Frankie, G W., D. L. Morgan, M. J. Gaylor, M. J. Benskin, J. G. Clark, W. E. Reed, and H. C. Hamman. 1977. *The Mealy-Oak Gall on Ornamental Live Oak in Texas,* Research Bulletin. Texas Agricultural Experiment Station. College Station, Texas.

Georghiou, G. P. 1972. The evolution of resistance to pesticides. *Annu. Rev. Ecol. Syst.* **3:**33-168.

Gill, S. A., D. R. Miller, and J. A. Davidson. 1982. *Bionomics and Taxonomy of the Euonymous scale,* Unaspis euonymi *(Comstock), and Detailed Biological Information on the scale in Maryland (Homoptera: Diaspididae),* Miscellaneous Publication 969. Maryland Agricultural Experiment Station, College Park, Md.

Gould, H. J., and W. I. Light. 1971. Biological control of *Tetranychus urticae* on stock plants of ornamental ivy. *Plant Pathol.* **20:**18-20.

Gurney, B., and N. W. Hussey. 1970. Evaluation of some coccinellid species for the biological control of aphids in protected cropping. *Ann. Appl. Biol.* **65:**451-458.

Hagen, K. S. 1962. Biology and ecology of predaceous Coccinellidae. *Annu. Rev. Entomol.* **7:**289-326.

Hagen, K. S., S. Bombosh, and J. A. McMurtry. 1976. The biology and impact of predators. In *Theory and Practice of Biological Control,* C. B. Huffaker and P. S. Messenger (eds.). Academic Press, New York, pp. 93-142.

Hagen, K. S., R. van den Bosch, and D. L. Dahlsten. 1971. The importance of naturally-occurring biological control in the western United States. In *Biological Control,* C. B. Huffaker (ed.). Plenum, New York, pp. 253-293.

Hall, R. W., and L. E. Ehler. 1980. Population ecology of *Aphis nerii* on oleander. *Environ. Entomol.* **9:**338-344.

Hamalainen, M. 1977. Control of aphids on sweet peppers, chrysanthemums and roses in small greenhouses using the ladybeetles *Coccinella septempunctata* and *Adalia bipunctata* (Col., Coccinellidae). *Ann. Agric. Fenn.* **16:**117-131.

Hamlen, R. A., and R. K. Lindquist. 1981. Comparison to two *Phytoseiulus* species as predators of two spotted spider mites on greenhouse ornamentals. *Environ. Entomol.* **10**:524-527.

Hamlen, R. A., and R. T. Poole. 1980. Effects of a predacious mite on spider mite populations of *Dieffenbachia* under greenhouse and interior environments. *Hortic. Sci.* **15**:611-612.

Hanel, H., and J. A. L. Watson. 1983. Preliminary field tests on the use of *Metarhizium anisopliae* for the control of *Nasutitermes exitiosus* (Hill) (Isoptera: Termitidae). *Bull. Entomol. Res.* **73**:305-313.

Helgesen, R. G., and M. J. Tauber. 1974. Biological control of greenhouse whitefly, *Trialeurodes vaporariorum* (Aleyrodidae: Homoptera), on short-term crops by manipulating biotic and abiotic factors. *Can. Entomol.* **106**:1175-1188.

Helgesen, R. G., and M. J. Tauber. 1977. The whitefly-*Encarsia* system: A model for biological control in short-term greenhouse crops. In *Pest Management in Protected Culture Crops,* F. R. Smith and R. E. Webb (eds.). ARS-NE-85, U.S. Department of Agriculture, Washington, D.C., pp. 71-73.

Huffaker, C. B., ed. 1971*a*. *Biological Control.* Plenum, New York.

Huffaker, C. B. 1971*b*. The ecology of pesticide interference with insect populations. In *Agricultural Chemicals: Harmony or Discord for Food, People and The Environment,* J. E. Swift (ed.). University of California, Division of Agricultural Science, Berkeley, pp. 92-104.

Huffaker, C. B., and P. S. Messenger. 1964*a*. Population ecology—Historical development. In *Biological Control of Insect Pests and Weeds,* P. DeBach (ed). Chapman and Hall, London, pp. 45-73.

Huffaker, C. B., and P. S. Messenger. 1964*b*. The concept and significance of natural control. In *Biological Control of Insect Pests and Weeds,* P. DeBach (ed.). Chapman and Hall, London, pp. 74-117.

Huffaker, C. B., and P. S. Messenger, eds. 1976. *Theory and Practice of Biological Control.* Academic Press, New York.

Huffaker, C. B., P. S. Messenger, and P. DeBach. 1971. The natural enemy component in natural control and the theory of biological control. In *Biological Control,* C. B. Huffaker (ed.). Plenum, New York, pp. 16-67.

Huffaker, C. B., F. J. Simmonds, and J. E. Laing. 1976. The theoretical and empirical basis of biological control. In *Theory and Practice of Biological Control,* C. B. Huffaker and P. S. Messenger (eds.). Academic Press, New York, pp. 41-78.

Huffaker, C. B., M. van de Vrie, and J. A. McMurtry. 1970. Ecology of tetranychid mites and their natural enemies: A review, 2, Tetranychid populations and their possible control by predators: An evaluation. *Hilgardia* **40**:391-458.

Hurd, L. E., and R. M. Eisenberg. 1984. Experimental density manipulations of the predator *Tenodera sinensis* (Orthoptera: Mantidae) in an old-field community, 2, the influence of mantids on arthropod community structure. *J. Anim. Ecol.* **53**:955-967.

Jamornmarn, S. 1979. Potential for using host plant selection by corn earworm adults as a management tool for control of larval damage. Ph.D. diss., Purdue University, West Lafayette, Ind.

Jaques, R. P. 1983. The potential of pathogens for pest control. *Agric. Ecosys. Environ.* **10**:101-126.

Kielbaso, J. J., and M. K. Kennedy. 1983. Urban forestry and entomology: A current appraisal. In *Urban Entomology: Interdisciplinary Perspectives,* G. W. Frankie and C. S. Koehler (eds.). Praeger, New York, pp. 423-440.
Legner, E. F., R. D. Sjogren, and I. M. Hall. 1974. The biological control of medically important arthropods. *Critical Reviews in Environ. Control* **4:**85-113.
Levenson, H., and G. W. Frankie. 1983. A study of homeowner attitudes and practices toward arthropod pests and pesticides in three U.S. metropolitan areas. In *Urban Entomology: Interdisciplinary Perspectives,* G. W. Frankie and C. S. Koehler (eds.). Praeger, New York, pp. 67-106.
Luck, R. F., and D. L. Dahlsten. 1974. Bionomics of the pine needle scale *Chionaspis pinifoliae* (Fitch) and its natural enemies at South Lake Tahoe, California. *Ann. Entomol. Soc. Am.* **67:**309-316.
Luck, R. F., and D. L. Dahlsten. 1975. Natural decline of a pine needle scale (*Chionaspis pinifoliae* Fitch) outbreak at South Lake Tahoe, California, following cessation of adult mosquito control with malathion. *Ecology* **56:**893-904.
Markkula, M., and K. Tiittanen. 1977. *Use of the Predatory Midge* Aphidoletes aphidimyza *(Rond.) (Diptera, Cicidomyiidae) against Aphids in Glasshouse Cultures,* ARS-NE-85. U.S. Department of Agriculture, Washington, D.C., pp. 41-44.
May, R. M., ed. 1976*a*. *Theoretical Ecology.* W. B. Saunders, Philadelphia.
May, R. M. 1976*b*. Patterns in multi-species communities. In *Theoretical Ecology: Principles and Applications,* R. M. May (ed.). W. B. Saunders, Philadelphia, pp. 34-76.
Merritt, R. W., M. K. Kennedy, and E. F. Gersabeck. 1983. Integrated pest management of nuisance and biting flies in a Michigan resort: Dealing with secondary pest outbreaks. In *Urban Entomology: Interdisciplinary Perspectives,* G. W. Frankie and C. S. Koehler (eds.). Praeger, New York, pp. 277-299.
Messenger, P. S., E. Biliotti, and R. van den Bosch. 1976. The importance of natural enemies in integrated control. In *Theory and Practice of Biological Control,* C. B. Huffaker and P. S. Messenger (eds.). Academic Press, New York, pp. 543-563.
Messenger, P. S., F. Wilson, and M. J. Whitten. 1976. Variation, fitness, and adaptability of natural enemies. In *Theory and Practice of Biological Control,* C. B. Huffaker and P. S. Messenger (eds.). Academic Press, New York, pp. 209-231.
National Research Council. 1980. *Urban Pest Management.* Report prepared by the Committee on Urban Pest Management, Environmental Studies Board, Commission on Natural Resources. National Academy Press, Washington, D.C., 272p.
Newsom, L. D., R. F. Smith, and W. H. Whitcomb. 1976. Selective pesticides and selective use of pesticides. In *Theory and Practice of Biological Control,* C. B. Huffaker and P. S. Messenger (eds.). Academic Press, New York, pp. 565-591.
Nicholson, A. J. 1933. The balance of animal populations. *J. Anim. Ecol.* **2:**132-178.
Olkowski, H., and W. Olkowski. 1976. Entomophobia in the urban ecosystem: Some observations and suggestions. *Bull. Entomol. Soc. Am.* **22:**313-317.
Olkowski, W., S. Daar, and H. Olkowski. 1983. IPM for a conservatory and greenhouse: A case history. *IPM Pract.* **5**(8):4-6,9.
Olkowski, W., H. Olkowski, R. van den Bosch, and R. Hom. 1976. Ecosystem management: A framework for urban pest control. *Bioscience* **26:**384-389.
Olkowski, W., H. Olkowski, A. I. Kaplan, and R. van den Bosch. 1978. The potential for biological control in urban areas: Shade tree insect pests. In *Perspectives in*

Urban Entomology, G. W. Frankie and C. S. Koehler (eds.). Academic Press, New York, pp. 311-347.

Owen, J. 1983. Effects of contrived plant diversity and permanent succession on insects in English suburban gardens. In *Urban Entomology: Interdisciplinary Perspectives*, G. W. Frankie and C. S. Koehler (eds.). Praeger, New York, pp. 395-422.

Papavizas, G. C., ed. 1981. *Biological Control in Crop Production*. Allanheld, Osmun and Co., Totowa, N.J.

Pianka, E. R. 1970. On r-, and K-selection. *Am. Nat.* **104:**592-597.

Pianka, E. R. 1974. *Evolutionary Ecology*. Harper and Row, New York.

Pimentel, D. 1963. Introducing parasites and predators to control native pests. *Can. Entomol.* **95:**785-792.

Pinnock, D. E., and J. E. Milstead. 1971. Biological control of California oakmoth with *Bacillus thuringiensis*. *Calif. Agric.* **25**(10):3-5.

Poinar, G. O., Jr. 1979. *Nematodes for Biological Control of Insects*. CRC Press, Boca Raton, Florida.

Polvolny, D. 1971. Synanthropy. In *Flies and Disease: Ecology, Classification and Biotic Associations*, B. Greenberg (ed.). Princeton University Press, Princeton, N. J., pp. 16-54.

Price, P. W., C. E. Bouton, P. Gross, B. A. McPheron, J. N. Thompson, and A. E. Weiss. 1980. Interactions among three trophic levels: Influence of plants on interactions between insect herbivores and natural enemies. *Annu. Rev. Ecol. Syst.* **11:**41-65.

Quezada, J. R., and P. DeBach. 1973. Bioecological and population studies of the cottony-cushion scale, *Icerya purchasi* Mask., and its natural enemies, *Rodolia cardinalis* Muls. and *Cryptochaetum iceryae* Will., in southern California. *Hilgardia* **41:**631-688.

Rabb, R. L. 1970. Introduction to the conference. In *Concepts of Pest Management* R. L. Rabb and F. E. Guthrie (eds.). North Carolina State University, Raleigh, pp. 1-5

Rabb, R. L., R. E. Stinner, and R. van den Bosch. 1976. Conservation and augmentation of natural enemies. In *Theory and Practice of Biological Control*, C. B. Huffaker and P. S. Messenger (eds.). Academic Press, New York, pp. 233-254.

Remington, C. L. 1968. The population genetics of insect introduction. *Annu. Rev. Entomol.* **13:**415-426.

Ridgway, R. L., and S. B. Vinson, eds. 1977. *Biological Control by Augmentation of Natural Enemies*. Plenum, New York.

Ridgway, R. L., J. R. Ables, C. Goodpasture, and A. W. Hartsack. 1981. *Trichogramma* and its utilization for crop protection in the USA. In *Use of Beneficial Organisms in the Control of Crop Pests*, J. R. Coulson (ed.). Entomological Society of America, College Park, Md., pp. 41-48.

Ripper, W. E. 1956. Effect of pesticides on balance of arthropod populations. *Annu. Rev. Entomol.* **1:**403-438.

Roberts, F. C., R. F. Luck, and D. L. Dahlsten. 1973. Natural decline of a pine needle scale population at South Lake Tahoe. *Calif. Agric.* **27:**10-12.

Sawyer, A. J., and R. A. Casagrande. 1983. Urban pest management: A conceptual framework. *Urban Ecol.* **7:**145-157.

Scarza, R. 1983. Ecology and genetics of exotics. In *Exotic Plant Pests and North*

American Agriculture, C. L. Wilson and C. L. Graham (eds.). Academic Press, New York, pp. 219-238.

Scopes, N. E. A. 1969. The potential of *Chrysopa carnea* as a biological control agent of *Myzus persicae* on glasshouse chyrsanthemums. *Ann. Appl. Biol.* **64:**433-439.

Scopes, N. E. A. 1970. Control of *Myzus persicae* on year-round chrysanthemums by introducing aphids parasitized by *Aphidius matricariae* into boxes of rooted cuttings. *Ann. Appl. Biol.* **66:**323-327.

Scriven, G. T., and R. F. Luck. 1978. Natural enemy promises control of Nantucket pine tip moth. *Calif. Agric.* **32:**19-20.

Simmonds, F. J. 1968. The economics of biological control. *PANS* **14:**207-215.

Simmonds, S. P. 1972. Observations on the control of *Tetranychus urticae* on roses by *Phytoseiulus persimilis. Plant Pathol.* **21:**163-165.

Slater, A. J. 1984. Biological control of the brownbanded cockroach, *Supella longipalpa* (Serville) with an encrytid wasp, *Comperia merceti* (Compere). *Pest Mange.* **3**(4):14-17.

Slater, A. J., M. J. Hurlbert, and V. R. Lewis. 1980. Biological control of brownbanded cockroaches. *Calif. Agric.* **34**(8-9):16-18.

Southwood, T. R. E. 1976. Bionomic strategies and population parameters. In *Theoretical Ecology: Principles and Applications,* R. M. May (ed.). W. B. Saunders, Philadelphia, pp. 26-48.

Southwood, T. R. E., and M. J. Way. 1970. Ecological background to pest management. In *Concepts of Pest Management,* R. L. Rabb and F. E. Guthrie (eds.). North Carolina State University, Raleigh, pp. 6-29.

Stanton, M. L. 1983. Spatial patterns in the plant community and their effects upon insect search. In *Herbivorous Insects,* S. Ahmad (ed.). Academic Press, New York, pp. 125-157.

Steiner, M. Y., and D. P. Elliott. 1983. *Biological Pest Management for Interior Plantscapes.* Alberta Environmental Centre, Vegreville, Alberta.

Stevens, L. M., A. L. Steinhauer, and J. R. Coulson. 1975. Suppression of Mexican bean beetle on soybeans with annual inoculative releases of *Pediobius foveolatus. Environ. Entomol.* **4:**947-952.

Stevenson, J. H., and J. H. H. Walters. 1983. Evaluation of pesticides for use with biological control. *Agric. Ecosys. Environ.* **10:**201-215.

Tanada, Y. 1964. Epizootiology of insect diseases. In *Biological Control of Insect Pests and Weeds,* P. DeBach (ed.). Chapman and Hall, London, pp. 548-578.

Tassan, R. L., K. S. Hagen, and D. V. Cassidy. 1983. Imported natural enemies established against ice plant scales in California. *Calif. Agric.* **36**(9-10):16-17.

Tauber, M. J., and R. G. Helgesen. 1981. Development of biological control systems for greenhouse crop production in the U.S.A. In *Use of Beneficial Organisms in the Control of Crop Pests,* J. R. Coulson (ed.). Entomological Society of America, College Park, Md., pp. 37-40.

Van den Bosch, R., and A. D. Telford. 1964. Environmental modification and biological control. In *Biological Control of Insect Pests and Weeds,* P. DeBach (ed.). Chapman and Hall, London, pp. 459-488.

Van de Vrie, M., J. A. McMurtry, and C. B. Huffaker. 1972. Ecology of tetranychid mites and their natural enemies: A review, 3, Biology, ecology, and pest status, and host-plant relations of tetranychids. *Hilgardia* **41:**343-432.

Veronin, K. E., and A. M. Grinberg. 1981. The current status and prospects of *Trichogramma* utilization in the U.S.S.R. In *Use of Beneficial-Organisms in the Control of Crop Pests,* J. R. Coulson (ed.). Entomological Society of America, College Park, Md., pp. 49-51.

Vinson, S. B. 1981. Habitat location. In *Semiochemicals,* D. A. Nordlund, R. L. Jones, and W. J. Lewis (eds.). Wiley, New York, pp. 51-77.

Way, M. J. 1959. Mutualism between ants and honeydew-producing Homoptera. *Annu. Rev. Entomol.* **8:**307-344.

Weiser, J., G. E. Bucher, and G. O. Poinar. 1976. Host relationships and utility of pathogens. In *Theory and Practice of Biological Control,* C. B. Huffaker and P. S. Messenger (eds.). Academic Press, New York, pp. 169-185.

Wheeler, A. G., Jr., B. R. Stinner, and T. J. Henry. 1975. Biology and nymphal stages of *Deraeocoris nebulosus* (Hemiptera: Miridae), a predator of arthropod pests on ornamentals. *Ann. Entomol. Soc. Am.* **68:**1063-1068.

Whitten, M. J. 1970. Genetics of pests in their management. In *Concepts of Pest Management,* R. L. Rabb and F. E. Guthrie (eds.). North Carolina State University, Raleigh, pp. 119-137.

Wilson, C. L., and C. L. Graham, eds. 1983. Exotic Plant Pests and North American Agriculture. Academic Press, New York.

Wilson, F. and C. B. Huffaker. 1976. The philosophy, scope, and importance of biological control. In *Theory and Practice of Biological Control,* C. B. Huffaker and P. S. Messenger (eds.). Academic Press, New York, pp. 3-15.

Wyatt, I. J. 1970. The distribution of *Myzus persicae* (Sulz.) on year-round chrysanthemum, 2, Winter season: The effect of parasitism by *Aphidius matricariae* Hal. *Ann. Appl. Biol.* **65:**31-41.

Zwolfer, H., M. A. Ghani, and V. P. Rao. 1976. Foreign exploration and importation of natural enemies. In *Theory and Practice of Biological Control,* C. B. Huffaker and P. S. Messenger (eds.). Academic Press, New York, pp. 189-207.

7

Nonpesticidal Components Essential to Pest Management

F. Eugene Wood

University of Maryland
College Park, Maryland

In an urban environment, the development of natural resources often implies the conversion of natural habitat to other uses (Mueller-Dombois, Kartawinata, and Handley, 1983). For example, development may take the form of using a woodlot for an office building, a marsh for an industrial site, or forestland for a housing development. In any of these events, the stable mix of species in communities that existed before development are replaced by a more homogeneous group of species and their environment. In other words, when we change or develop a natural area we substitute a single less variable crop or structure and eliminate many of the natural species that formerly coexisted within that environment. We see this clearly when a monoculture or single crop replaces the many plant and animal species of a natural community. Those species displaced include parasites, predators, pollinators, varied plant communities, fungal and bacterial associations, and numerous plant feeders. A few of the species whose environmental requirements remain satisfied in the developed area may still persist there and perhaps even become pests. More likely however, nonnative, or introduced species whose environmental regimens are met by the modified situation will likely invade and become pests (Frankie and Ehler, 1978).

All crop species were once wild species. By selection and breeding they have been modified for man's use. The variation that allowed at least some individuals of the wild species to survive under adverse conditions is eliminated in favor of the uniformity of a genetic strain. This produces a monoculture synchronized for planting and harvesting. Selection for high productivity, size, and conformity are other positive traits of a developed crop. To maximize these selected traits, pesticides, fertilizers, special feeds, irrigation, and modified cropping practices must be added (Bottrell, 1979).

A major shortcoming of this design of crop utilization is frequently the failure to adequately address subsequent pest control needs. Pest species find nearly unlimited, nutritious food as well as favored habitats in our agricultural monocultures. Urbanization mirrors the attempts of agriculture in simplifying a formerly variable habitat along with its checks and balances and substituting uniformity augmented by the constant infusion of outside resources. The negative aspects of urban development intensify when the pace of urbanization is rapid. Settlement (or resettlement) projects are becoming more common around the world as populations increase (Weatherly, 1983). Without proper planning, these instant cities accumulate all of the conditions for pest problems by default. It will tax all of the resources of urban planning, including incorporation of urban pest management considerations, to assure that a large increase of immigrants will not overburden the available facilities and lead to worse conditions than those left behind. It will be immensely important to see that one community planning or design shortcoming will not be the exclusion of pest control.

While settlement projects in underdeveloped nations do not parallel urbanization in developed nations, comparisons exist. Building design, turf maintenance, urban gardens, ornamental plantings, indoor landscaping, refuse disposal, warehousing, food and product distribution and supply, sanitation and health facilities, and recreation facilities are elements in most population centers. Development of new population centers within or adjacent to urbanized areas is common in the developed world. In the United States during World War II, housing and support facilities for workers in war plants were settlement projects. Many are now public housing projects. Low-cost housing, industrial plants and high tech centers, low income housing, "new towns," urban renewal, and shopping complexes are a few other examples of development with urgent need for pest control in design and maintenance with pest management in mind.

Maintenance of pest control programs in the recent past implies the use of pesticides and often only the use of pesticides. While pesticides are credited with vastly effective and efficient control of pests, when we consider the enormous populations of most pests, we note that relatively few of the total ultimately succumb to the use of synthetic organic pesticides. With the enormous fecundity and short generations of algae, German cockroaches, house flies, weeds, domestic rodents, and countless other species of agricultural and public health pests, humans would be helpless depending on any man-made defense as their only protection. Biological, genetic, physical, and other natural environmental factors keep plant and animal populations below the numbers that would otherwise inundate people and their crops. In fact, humans usually supply the factors that allow pests to develop to threatening proportions. While pesticides are acknowledged as the most powerful tool widely available for use in pest management (Metcalf, 1982), sole reliance on them elicits an even more powerful reaction from the pests, that is pesticide resistance.

RESISTANCE

Resistance of pests to pesticides is ultimately more influential in changing control programs than even the fear of environmental contamination. This influence is due to the short-term economic pressures on the users or producers. The first instance of pest resistance to pesticides, that of the San Jose scale's resistance to lime sulfur in 1914, was followed by the resistance of ten other pest species to hydrogen cyanide, lead arsenate, sodium arsenite, and other inorganic pesticides over the next 12 years. This relatively slow development of resistance accelerated after World War II when modern synthetic organic pesticides became widely used. The Colorado potato beetle, for instance, was controlled by Paris Green in 1865, later by lead arsenate, and by DDT in 1946. This important crop pest became resistant to DDT in 1949 and then became resistant to all of the 17 or more pesticides employed in its control in the subsequent 33 years (Metcalf, 1982). This same pattern of sequential pesticide resistance has been duplicated in our efforts to control the house fly, German cockroach, and other pest populations. Patterns of multiple resistance have even compounded to the point where the selection of resistance to certain pesticides has conferred a cross-resistance or the ability to detoxify compounds never previously used in control programs for those species (Collins, 1973, 1975; Nelson and Wood, 1982; Rust and Rierson, 1978).

While universally used pesticides have delivered enormous benefits (chlordane use for subterranean termite control is a notable example), they have frequently been overused, created environmental problems, succumbed to resistance, or have even precipitated the outbreak of secondary pests ("new" pests that were not problems until their parasites and predators were eliminated). Pesticides are not inherently bad in this regard. These same shortcomings can be elicited through the intense use of any single component control program.

The shortcomings of pesticide use vividly point up the dynamism of insects and their relatives. It is not surprising that this dominant form of life on earth would allow only a fleeting moment of success to any single threat to its existence. We can surely expect this pattern to continue. While weed resistance to herbicides has been slow in coming and its advent underrated by many pest control practitioners, it is inevitable and may be devastating to some cropping practices. Perhaps the surest way to identify a candidate species for the development of pesticide resistance would be to list a pest problem where a single component control agent appears to be completely successful now and for the foreseeable future.

The most obvious and appropriate response to the failure of control programs based on the use of a single control component is a multicomponent approach. This means seeking control alternatives and companions to pesticides. The key to successful alternatives and companions will be found in the factors that keep pest population from increasing to high levels when they are in their natural environment (Labeyrie, 1978). Finding these factors entails a great deal of basic research and time—two reasons frequently in short supply in the area of urban development. Nevertheless, multicomponent controls are becoming increasingly important, and we must seek economical ways of using them routinely as part of the pest control process. The pest control units using them will be the preferred contractors where pest management methods are looked on as positive or progressive pest control.

NONPESTICIDAL COMPONENTS OF URBAN INTEGRATED PEST MANAGEMENT (IPM)

Insect Growth Regulators

Since the outer covering or external skeleton of insects and their relatives does not grow, it must be periodically molted for the individual to increase in size and advance in maturity. Insects continue to go through immature molts because juvenile hormones suppress the characters found in the adult, the stage at which growth stops. If the hormone that maintains the insect in a pre-adult or juvenile state can be manipulated, it could cause the insect to mature differentially or die because it cannot successfully emerge from an immature stage. Various juvenile hormone "mimics" or synthetic insect growth regulators (IGRs) are being investigated. Two of these IGRs are currently in use. One, methoprene, is used to stop the development of fleas, flies, Pharoah ants, mosquitoes, whiteflies, mealybugs, and soft scales. Another, hydroprene, is used for cockroach control. Both have been tested on other insects, including termites (Haverty and Howard, 1979; see discussion in chapter 13 of this book). These and many other IGRs are being investigated for the control of cockroaches, ants, stored product pests, and yellowjackets. Growth regulators do not affect humans and other vertebrates in the same way they do insects and other arthropods, and thus a great margin of safety is attained.

Growth regulators act slowly, since they generally do not kill the pest until it molts into the adult stage. This is a drawback where immediate reduction of the population is needed or where the damaging stage is the immature or larval stage. This shortcoming can be compounded when an IGR treatment might even extend that stage. IGRs can, however, be used in conjunction with contact pesticides, and they can also be used on pest populations by factoring them into the economic (or aesthetic) threshold of a pest management scheme. For urban pests that build up in one season and are naturally suppressed in other seasons, the application of an IGR might retard the population increase until the unfavorable season occurs, thus maintaining the population at an acceptable level. As an example, after treatment, a cat flea population would be held below the human biting level (the adults are prevented from emerging from the pupal stage) until the lower humidity and temperatures of fall and winter naturally suppress population growth.

A great advantage of using hormones (e.g., the juvenile hormone) to suppress the development of an insect pest, is that there would be very little impetus for selection against a biological product that is part of the normal life cycle. This theoretical foundation is circumvented by the fact that physiological control products are mimics that elicit an action like the natural product. They are synthetically compounded in laboratories, and until every component of the natural compound is exactly reproduced, the potential for resistance will be present—and naturally expressed in due time. Laboratory experiments using a multiresistant strain of the house fly have already illustrated this process with methoprene (Georghiou, Lee, and DeVries, 1978).

Chitin Inhibitors

Chitin is a large and important part of the exoskeleton of arthropods. It is also a component of the insect egg and the inner anatomy of insects. Chitin is insoluble in water, alcohol, dilute acids and bases, and mammalian digestive enzymes. In nature it is broken down by some bacteria and fungi, some insects, and a few other animals. With this in mind, it is easy to see that inhibiting the development of this important substance will also affect insect egg development, molting, and pupation. The chitin inhibitor diflubenzuron has been used to control mosquitoes, defoliating caterpillars, and beetles (Sacher, 1971). Since chitin is not a selective substance in nature, widespread inhibition of its formation would indiscriminately affect thousands of insect species; therefore extensive application of chitin inhibitors must be well planned. Application in selected areas holds much promise for pest management.

Pheromones

Hormones, substances that are generally more familiar than pheromomes, are chemical compounds secreted inside organisms that stimulate the physiological reactions of various organs. Juvenile hormones, for instance, maintain insects in the pre-adult state, and certain quantities will not allow them to become mature. Pheromones have been defined by some as hormones produced for external use. In the broad sense, they are chemical signals emitted by animals that elicit a reaction when they are perceived by another. In the strict sense the definition is narrowed to odor communication within a species. However defined, chemical communication appears to be the primary method of communication among insects (Pasteels, 1977; Shorey, 1977).

With pheromones, we see the more discrete or restrictive end of a continuum of pest management options. On the broad, widely applicable end we have pesticides, many of which poison an extensive number of species. The classical synthetic organic pesticides attack a system that many species have in common—the animal nervous system. Next, chitin inhibitors are more restrictive in that they directly affect only the animals that incorporate a particular chemical (chitin) in their exoskeletons. Even though over a million species of arthropods are affected, many other large groups are not. The control scope of growth regulators is narrower still, since they affect certain groups of species that have the vulnerability of particular hormonal systems. Finally, pheromones can be so restrictive that they can elicit a reaction from only a single species or even one sex of a single species. The logical extension of this is to affect one stage of one sex of a species. This narrowing continuum of pest management tools will enable us to excise a discrete portion of a pest population, leaving the rest of the ecosystem intact. The research and use of odor communication for control of agricultural, forestry, stored product, and urban plant pests has resulted in significant interest and advancement in pest management methods (Roelofs, 1979; Trammel, Roelofs, and Glass, 1974).

Pheromones of the important stored product pests have been thoroughly researched and identified (Burkholder, 1982, 1984). The pheromones identified for stored product pests will frequently attract a complex of species. Seven pheromone groups are available as attractants to baits or traps and for use as monitoring devices or even controls (Shapas, Burkholder, and Boush, 1977). The communication and reproductive strategies of these pests have been identified as falling within two types (Burkholder and Ma, 1985). The first type includes species where the adults are short-lived, less than one month. These species frequently do not feed as adults and their principal communication is devoted to reproduction; thus, the development of sex pheromones is essential. The second type includes species with long-lived adults—those living more than one month. The adults of this latter type must feed, and the pheromones they produce are likely to be aggregation pheromones that bring both sexes together. Understanding these two types of pheromones might have usefulness in pest management programs involving pests with complete metamorphosis. A pest with a short adult lifespan—the exposed, flightless female gypsy moth, for instance—needs to mate quickly, and pheromones selected for calling members of the opposite sex are advantageous. Species with long-lived adults cannot live only on stored fat so they must forage and feed. The adults of insects with complete metamorphosis and often the mature larvae generally disperse from the area of their larval development. An aggregation pheromone will direct or assemble the adults to areas where they can feed and engage in behavior preparatory to mating.

Other pheromone-mediated behavior points toward the importance of these chemicals and the potential IPM benefits in their manipulation. The sex pheromone of the American cockroach is very active (Bell et al., 1977) and appears to be attractive to several cockroach species. Although the ecological requirements of those species differ, and their ranges seldom overlap, the pheromone of one species can be used to attract the males of all. The complexity of insect pheromone communication can perhaps be exemplified by the collected studies of mass attacks of susceptible trees by bark beetles (Bordon, 1974). A few female pioneer beetles find a host tree, attracted by identifying odors. Pines attractive to *Dendroctonus* tend to be older and often under stress from dry weather. The females begin feeding, releasing the aggregation pheromone frontalin. Frontalin and a terpene from the tree combine to make a very strong attractant. The mass attack begins, but male bark beetles are attracted in greater numbers than females. As the males feed, they produce verbenone, which inhibits the reception of the aggregation combination. As the verbenone value rises and the sap flow from the tree slows, producing less terpene, additional beetles are not recruited, the sex ratio becomes even, and the mass attack on the host tree is terminated, but pioneer beetles continue to be attracted to the general area. The short-range communication of sex pheromones and sound initiate mating and egg laying. The mass bark beetle attack may also attract associated species such as wood borers, which also take advantage of the weakened tree.

The workers of some species of *Reticulitermes*, subterranean termites, follow the trails of each other in their tunnels in soil and wood. The termite trail-following pheromone was found to be identical to an extract of wood decayed by a brown-rot fungus (Matsumura, Tai, and Coppel, 1969). These pheromones are renewed by each worker as long as they are positively reinforced by other workers. When the reinforcement is interrupted, the trailing breaks down, and others are not led into a changed or dangerous area. While the correct amount of pheromone sends clear commands, releases above the attractant threshold levels can be disruptive. This use of pheromones confuses the recipient and discourages mating and communication (Roelofs, 1979).

Sex pheromone trapping of clearwinged moths, the larvae of which bore into many shrubs and trees, exemplify how pheromones can be used to coordinate pest management strategies. The isolation of pheromone system components of the lesser peach tree borer and the peach tree borer allowed for the manipulation of the chemical lure so that it could be used with a sticky trap. The pheromone was contained in a series of hollow microfibers attached inside the trap and was dispersed over a seven-month period. Weekly trap analysis of the adult male moths outlined peak flight periods (Neal and Eichlin, 1983). From pest flight times population levels can be estimated and egg-laying periods can be calculated. Pest management decisions can be based on this information.

The efficacy of pheromone trapping of clearwing moths has cleared up problems in insect biology and taxonomy. Where a presumed single species had been collected sporadically through the summer using classical insect collection methods, the collection data from pheromone traps showed that, in fact, two species whose flight periods did not overlap actually occupied the same range. A clear understanding of the biology of pest species allows the recommendation of discrete strategies, rather than the maintenance of a pesticidal covering of plants with the hope that the pests will arrive when the pesticide is still toxic.

Biological Control

The life of every species at some time runs contrary to that of another. The variation in the communities of a natural ecosystem is the result of selection for the competitiveness of each species and against the total advantage of one alone. Inherent in this are abiotic limits and innumerable factors involving species competition and defense. Plant species compete for favorable space and light or adapt to the reality that allows them less. Animals adjust to harborage and food limitations but continue to compete. The imperative of reproduction then becomes a sink for the energy each species produces, but it is also the ultimate driving force for continuous competition and adaptation. The success of a species is resisted by adverse natural factors or opposing species. The result of species being contrary to one another is biological control. (See chapter 6 for more extensive discussion of this topic.)

Manipulating antagonistic insect species began in a formal way in 1888 when the vedalia lady beetle was introduced to control the cottony-cushion scale (Stehr, 1982). This successful endeavor is still a classical study in biological control. Less obvious but just as important, the vedalia beetle program established a scientific breadth to the control of insect pests. Until then, insect control was accomplished by rather primitive mechanical controls, marginally effective pesticidal baits and dusts, and entomological commissions. This success came at a propitious time when the fledgling U.S. Department of Agriculture was first being funded by Congress and when U.S. experiment stations were in their infancy.

Predators and Parasites. A general understanding of the term predator is that it is a free living organism that kills and eats its prey. Tigers, foxes, ladybeetles, and praying mantids act in this way. Predators in buildings and other modern human habitation consist of spiders, centipedes, and a few less visible and less well-known mites and ants. Some vertebrates have been suggested as indoor predators, such as geckos and even snakes. Recommendations of these species as predators of household pests are seldom met with enthusiasm; yet many predators are kept inside as pets, even spiders and snakes. There seems to be no place for free-living undomesticated animals inside human structures.

Predators outside urban dwellings, however, are tolerated for the perceived good they do, that is, attack and kill pests of yards and gardens. For predators to be helpful in pest management, they need harborage and prey for food. The less variable the plant community is spatially and in numbers of species, the more restricted the predators will be.

Parasites are generally beneficial as are predators except that they tend to be smaller and less noticeable. Some parasites cause diseases in humans, so the term needs to be clear in the content of its use. This is not always easy when dealing with the general public. It is not even easy as it is applied in the context of pest management, since, technically speaking, parasites do not directly kill their hosts. In this strict use, "parasites" that do kill their hosts are called parasitoids. For our general use we will assume that the parasites involved in urban pest management programs either directly or indirectly contribute to the death of their hosts.

Parasites of pests in homes or buildings are often assumed to be pests, and indeed when the term pest is judged in the aesthetic sense, they are. The most common parasites in dwellings are small wasps. Wasp is another term that is easily misunderstood. Several families of parasitic Hymenoptera parasitize different species of cockroaches. One encyrtid wasp, *Comperia merceti*, is well known as an egg parasite of the brownbanded cockroach (Roth and Willis, 1960) and is even used for control in certain situations where it is acceptable (Slater, Hurlbert, and Lewis, 1980). Trials have been made using parasitic wasps as part of an urban pest management program where American and smokey brown cockroaches were problems (Piper and

Frankie, 1978). The pest management strategy was to reduce the cockroach density outside houses, precluding their entry. Parasites have even been used in short-term pest control. One parasite of the Mexican bean beetle cannot overwinter in the United States and is kept in the laboratory throughout the winter and released each spring in soybean fields (Stevens, Steinhauer, and Coulson, 1975). Like predators, parasites are much more appreciated outside but are never kept inside—even as pets.

For pest management programs involving natural enemies in urban yards and gardens, certain alternatives should be considered. Plants should be selected that will not be under stress at their planting site. Pests build up faster, and stressed plants succumb quickly. A diversity of plants should be maintained, and some tolerance to injury should be allowed. A majority of perennial plants is preferable to large numbers of annuals. Parasite and predator populations can increase over the years in perennials, and they allow fewer weeds. To maintain a stable environment that allows for predator and parasite establishment, avoid disruptive cultural practices and preserve areas where parasites, predators, and inoffensive prey can find refuge. Avoid pesticide use (Croft, 1977). Predators with little dispersal ability are easily killed, and prey reduction stresses parasites and predators and eliminates a generation of parasitized prey.

Plant Resistance. Plants appear to be at the mercy of insects, but if that were true, few would persist long enough to set seed, and disasters caused by defoliating insect epidemics would be much less spectacular and traumatic than they are. Perhaps the most obvious accommodation plants have made to insects is the evolution of flowers. Their intricate adaptations in color and form compete to assure insect pollination. Flowering plants are responsible for the increase of insects dependent on pollen and nectar. At the same time many of these insects will also be plant feeders in some stage of their life cycle and will consume leaves, roots, stems, and seeds. Regardless of the many possible evolutionary pathways available, once a plant feeder has adapted to a host, changes in the host plant elicit adaptation of the pest. This co-evolutionary relationship is expressed by the preponderance of some pest groups, including whiteflies, thrips, jumping plant lice, and aphids, each of which is specific to single genera of plants (Kogan, 1977).

The vulnerability of crops to new pests is exemplified by the introduction of the grape phylloxera from North America to France in the early 1860s precipitating a calamity that affected all of Europe. The larvae of that small beetle feeding on the roots of European grapevines brought the continent's wine industry to the point of disaster by 1880. The recommendation of USDA entomologist, C. V. Riley, to use resistant North American rootstocks on which were grafted European grapevine scions was the reprieve that saved the wine industry (Mallis, 1971).

Plant resistance to insects is expressed in many ways, partially because certain plants that are less damaged or less infested than others under the same circumstances are judged resistant. It is not easy to demonstrate that the absence of feeding is due to a kind of plant defense or resistance by the plant. The "action" of nonfeeding is inherent in both the pest and the host.

Some factors that affect plant feeding are ecological. For instance, weather might delay or might shorten plant development so that a vulnerable or preferred stage of growth is delayed or passes, and the pest cannot take advantage of it. Certain cultural practices such as the use of planting dates, fertilizers, and irrigation can mitigate these factors. Other factors that affect plant feeding are genetic (Pal and Whitten, 1974). Morphological characteristics such as hairy leaves, thorns, shape, or even color, such as the yellow hue which attracts aphids, are important in host finding, host recognition, and host suitability for oviposition. An area of major importance in plant insect interactions is that of genetically controlled chemical compounds or metabolites that influence pest and sometimes plant behavior. These compounds stimulate the pests' sense of smell or taste. They are important in finding or recognizing hosts or in repelling pests from plants. Some pests are arrested or their movement stopped until further stimuli are received. Certain compounds initiate feeding or stimulate the continuation of feeding or oviposition. Others deter feeding or inhibit it or inhibit oviposition. Finally, many toxic metabolites are being discovered that can debilitate an insect or even kill it (Kogan, 1982).

Microbial Control. Microorganisms including viruses, fungi, bacteria, rickettsiae, and protozoa are among the primitive forms of life established on earth long before the advent of insects and their arthropod relatives. Modern species of all these life forms, including arthropods, live together, and while some are dependent on each other, others are pathogenic in their associations. Most insects and most other animals have intestinal microorganisms, and while specific advantages cannot always be attributed to them, their elimination often presages problems. Insects and mites abound in environments that are also friendly to microorganisms, such as the upper soil levels, the humus and thatch layers, made of decomposing plant parts, and in and around all decaying organic matter.

Termites are well known for their need of certain protozoa and bacteria to break down the cellulose in the wood they consume. Subterranean termites also appear to receive cues from some wood-infesting fungi and indeed more readily consume wood infested with fungi. Clothes moths, which evolved feeding on the fur of animal carcasses long before humans began processing it for clothing, still need vitamins produced only by microorganisms and cannot live on cleaned wool or feathers. Stored product pest populations accelerate where fungal and bacterial infestations are expanded by increased moisture, and fruit flies, *Drosophila*, are found on fermenting fruit because their larval food is the yeast organisms that cause the fermentation.

While some microorganisms are beneficial or even necessary to some insects, many are debilitating or pathogenic. Disease-causing microorganisms usually persist in numbers sufficient to keep up a low level of reproduction either outside or inside a host, and many forms even have resistant stages; spores, for example, can persist for certain periods as inactive organisms. Insect pathogens are most noticeable when they are infecting their hosts at a high level, producing an epizootic that can reduce an insect population to such low numbers that years might be needed for recovery. Manipulating insect pathogens or their by-products so that they can be used to predictably limit insect populations is the goal of research in insect pathology.

In pest management, insect pathogens can be used in several ways (Maddox, 1982). One way is by the maximum utilization of naturally occurring diseases. For example, several viruses fatal to armyworms and other caterpillars occur annually in some areas. By close monitoring, taking into consideration the total caterpillar population and that part of the population already infected, decisions can be made on whether the population will be controlled by the pathogen or whether a pesticide application may be needed. Another way of using insect pathogens is by the introduction of pathogens into insect populations as permanent mortality factors. This use is exemplified by milky disease of the Japanese beetle. *Bacillus popilliae* was discovered in New Jersey in 1933 in dying beetle grubs that had a white appearance due to the spindle-shaped spores of the bacterium. Tests showed that when the bacteria-infected Japanese beetle grubs died, the bacterial spores were taken in by other foraging beetle grubs. These grubs also became spore producers and in this way spores were generally distributed in the uppermost soil levels. The *Bacillus* spores, which resist excessive heat, cold, desiccation, and moisture, persist in the soil for many years and are a permanent control for the Japanese beetle and several other species of white grub.

The final pathogen use as a pest management strategy is the application of insect pathogens as microbial insecticides for the temporary control of an insect pest. Processes necessary for the development of antibiotics for medicine have substantially contributed to the improvement of the technology of insect pathogen research. Mass fermentation methods, tissue culture techniques, and synthetic diet research have aided in the isolation and production of pathogens that can be applied as insecticidal sprays for immediate pest control for a limited period of time (Angus, 1977). Several illustrations immediately come to mind. *Bacillus thuringiensis (B.t.)* and the Nuclear Polyhedrosis Viruses have been used successfully to control many lepidopterous larvae. At least one other strain of *B.t.* has been found to control mosquitoes and blackflies, and many other strains are known to occur. With progress in biotechnology and genetic engineering, the development of insect pathogens may be entering a new era.

The use of insect pathogens in urban pest management has a number of advantages. The transfer from agriculture to urban flower and vegetable gardening is relatively direct. *B.t.* for urban gypsy moth control is often preferred over other pesticides because of the perceived hazards of synthetic organic pesticides to people and wildlife. The use of insect pathogens for pest control in urban structures is advancing less rapidly. Agricultural application is less direct than household application, and registration requirements are more rigorous because of the use of viruses and other pathogens for use inside.

Attractants

Attracting or luring animals so as to catch them or kill them or use them in some other way appears to be a phenomenon discovered or selected for by almost all animal groups. Many insects, for instance, display all or parts of themselves to attract mates and even predators. Some fish lure their prey with a special luminous structure, bird displays are legendary, and mammals attract others for sex, food, and even play. Attractants can consist of pheromones advertising sex, the trail home, olfactory signals that identify nutrients, small presents proffered by spiders and flies begging for attention, cricket songs, and firefly glimmerings, or they can even be bits of oddly arranged feathers or plastic worms dangled by fishermen in vain hope. With such potential it is surprising that attractants are not available to manipulate every economic species, pestiferous or beneficial. In reality many lures and baits are available for limited use in local areas. Attracting ants on a kitchen counter can sometimes be done with the foodstuffs they are infesting—sugars or fats. For instance, Pharaoh ants can be controlled with a mixture of boric acid and mint apple jelly. Boards painted schoolbus yellow and covered with oil will attract and stick whiteflies and winged aphids in the garden or greenhouse.

Attractants in general are used for detection or monitoring, for removal or control trapping, and for mating confusants (Metcalf and Metcalf, 1982).

When attractants are used to monitor populations, they often eliminate "cover sprays" of pesticides. Where effective attractants and accurate trapping evaluations are available, pest population levels can be measured at proper time intervals and pest management decisions based on those data. For instance, monitoring data can be used to time pesticidal applications for controlling codling moth larvae, using the adult males' appearance in pheromone baited traps. Quarantined pests such as the Mediterranean fruit fly can be detected and controlled at an early stage (Beroza et al., 1961), and gypsy moth advance can be tracked and evaluated for local pesticidal control or parasite releases. Even infested warehoused commodities can be detected (Burkholder and Ma, 1985) and removed or traced back to the origin of infestation. Survey traps using attractants such as those employed for the Japanese beetle were so impressive that they have been placed on the consumer market. Although the beetle trap catches are ineffective in controlling reinfestation or adult suppression, the attraction is so spectacular that many believe it is helping.

Using attractants in conjunction with removal trapping is the principal motive for investigating most attractants. The words "attractant," "bait," and "lure" imply that those materials will induce or entice a pest from some distance and, by controlling its actions, lead it to its capture or death.

In urban settings and especially in a home or commercial structure, economic (or aesthetic) thresholds are often very low. Hospital management or sanitarians or food-processing plant managers are reluctant to acknowledge any pest level. Therefore the capture of a pest in a monitoring trap is viewed as evidence of a pest infestation (Burkholder and Ma, 1985). At the same time cover sprays to control pests inside structures are difficult to maintain effectively and are thought of as unhealthy. Nontoxic attractants and discrete or orderly placed traps that will intercept every pest are the imaginary ideal of most urban pest management clients. Unfortunately, that degree of control can seldom be obtained by current attractants. Even so, where attractants have been well worked out, the future looks bright. Mass trapping when pest populations are low appears to be the key to control using attractants (Burkholder and Ma, 1985). Where good sanitation and maintenance are emphasized, low populations may be suppressed or eliminated in this way (Burkholder, 1981).

Another pest management strategy, combining the use of attractants and pathogens, has been found useful. Protozoan pathogen spores were placed in conjunction with a pheromone. Male dermestid beetles attracted to the pheromone were contaminated with the gregarine and later transferred the spores to females (Shapas, Burkholder, and Boush, 1977).

Even in apple orchards where little fruit damage is tolerated, red-banded leaf roller populations were managed using pheromone traps, to the point that fruit injury was maintained at an acceptable level (Trammel, Roelofs, and Glass, 1974).

The last type of attractant use, as mating confusants, has been tested in extensive experiments but has not shown much promise to date. Environmental saturation with sex pheromones reduced mating success of a gypsy moth but not enough to measurably affect the next generation. Likewise the use of sex pheromones to disrupt mating of the pink bollworm did not compete economically with pesticide applications (Metcalf and Metcalf, 1982).

Repellents

Repellents are substances that are not acceptable to pests and disturb their sensory receptors or are sensed as alarm signals. The repellency of pesticides (Ebeling, Reierson, and Wagner, 1967, 1968*a*, 1968*b*; Ebeling, Wagner, and Reierson, 1966) is one example, and it becomes more important as a control limitation the longer pests have to stand on a treated surface to become poisoned.

Some repellents have been used since ancient times, the use of smoke and cedar oils for instance. Bordeaux mixture, used as a fungicide on plants, also repels many insects. Creosote was poured in ditches as a barrier to chinch bugs migrating from maturing fields of wheat to corn. Creosote on wood is a fungicide and also a termite repellent. Naphthalene repels moths from woolens, carpet beetles from insect collections, and raccoons from chimneys. There are new foliage repellents to deter deer feeding on forest and orchard seedlings (Henderson, 1983) as well as oily or sticky tactile repellents for birds. Since repellents need not kill pests, they can have a low toxicity and usually can be used safely around man and animals (Committee on Plant and Animal Pests, 1969). The only modern mosquito repellent effective for application to humans is diethyl toluamide (Deet). These chemicals are not likely to be toxic to parasites and predators but will likely be repellent. The major drawback to repellents tends to be their application. Repellent glues are sticky and run when they get hot, crystals must be put in place, and liquids must cover an entire area to be protective. Furthermore, once the repellents are disrupted the pests quickly return.

Colorless dyes applied to wool during the dyeing process give a permanent protection from clothes moths and carpet beetles (Pence, 1959, 1966). Many synthetic cockroach repellents have been tested (Burden, 1975; Burden and Eastin, 1960; Mallis, Easterlin, and Miller, 1961; McGovern, Burden, and Berzoa, 1975); some are used on cases in bottling plants, but little other utilization has been made. Repellents that can safely be used on or incorporated in the cases of thermostats, smoke detectors, and computers are needed, since cockroaches often live in those protected spaces.

Recently, a natural "dispersion-inducing substance" has been reported for the German cockroach (Suto and Kumada, 1981). Natural deterrents and repellents may provide the basis for a new generation of synthetic repellents. Long residual repellent substances that can safely be injected into soil to protect structures or that can be applied to surfaces such as wood or concrete to exclude rodents, wood-destroying pests, crickets, millipedes, and other structure-invading pests would be welcome for use in urban pest management programs.

Traps and Trapping

It is hard to come to terms with the idea of using traps in a time where physical and mechanical controls represent the old and seemingly primitive ways of pest control, before the advent of modern pesticides. This reluctance is further complicated by the prevalent idea that we in the developed world are too technologically advanced to return to methods that do not include the guaranteed success of modern chemistry. In the United States and western Europe, an entire generation of farmers, pest control professionals, and consumers have based their concepts of pest control on the application of synthetic organic pesticides. Rodent resistance to repeat feeding of anticoagulent baits (Jackson, Spear, and Wright, 1971), and insect resistance to many residual pesticides left urban pest control to fall back on fewer chemicals. This brought about a renaissance in various kinds of traps.

Vertebrate Trapping. Traps were a very early tool of humans. Snares, pitfalls, and deadfall traps were developed by aboriginal hunters.

The first steel trap was designed in 1540 (Snetsinger, 1983), and modern improvements range from multiple catch traps for rodents to light traps with attached electrocuters for flying insects. Traps were first designed for catching vertebrates, and they still have the most direct and efficient usefulness in that field. Perhaps since humans share their senses more directly with vertebrates, vertebrates are easier to understand than other pests such as arthropods or fungi. Vertebrate infestations are also generally less intense and pervasive than insect pests. While they may occur in colonies of hundreds or flocks of thousands, their body size alone imposes limits that invertebrates easily exceed. Still, size is nearly always intimidating, and visible rodents are more alarming than visible cockroaches (Wood et al., 1981). The importance of vertebrates as pests influences the importance of traps in vertebrate pest management.

Even though trap use requires more maintenance than baits or sprays, they have many advantages. Some examples are quick reduction of high rodent populations, especially mice; certain disposal—once mice or squirrels or birds are trapped they can clearly be discarded, eliminating potential odor problems and secondary pest infestations; traps can be retrieved and reused after the trapping period—poisonous baits are used up or more often left in place, frequently attracting stored product pest infestations or becoming food for cockroaches; traps are less hazardous to the environment than chemicals; effective trapping requires understanding pest biology and behavior; and traps do not contribute to pesticide resistance (Frishman, 1982*b*).

Traps and trapping are well documented (Baumgartner, 1982; Bjornson, Pratt, and Littig, 1968; Brown, 1960; Marsh and Howard, 1982; National Pest Control Association, 1982; Timm, 1983). The types or choice of traps, how they are used, where they are placed (Knote, 1984), and how they are baited is, of course, critical and to a great extent dependent on experience as well as training. Other important elements that support the trapping activity are sanitation, reduction of harborage, client education, evaluation, and follow-up, such as monitoring and program maintenance.

Glueboards. Another type of vertebrate trap is the glueboard. Glueboards have become much more effective with recent formulations of sticky or tacky surface materials. Modern glues can be more easily handled and will not liquefy in the heat. These nonrepellent surfaces catch mice and other small vertebrates but are less effective for rat control.

Cockroach Trapping. The impetus to trap cockroaches is usually for the evaluation of a pest management procedure, to identify an infested dwelling unit, or perhaps to pinpoint a high population area within a unit, but it is seldom to control a cockroach infestation. It stands to reason that pests that are reared by the garbage-can-full (as cockroaches are in research laboratories) will also be able to overrun any sort of trap.

Commercial cockroach traps are sold to consumers and to pest control operators. They are essentially disposable cardboard covers of small gummy surfaces that stick cockroaches fast when they enter them. These traps are more appropriate for smaller cockroaches. Large species, especially the American cockroach, can pull themselves off the sticky surface and escape. Sticky traps are also frequently used to monitor population reduction in the evaluation of new pesticide applications (Moore and Granovsky, 1984). Traps used for most research purposes are usually made of various-sized jars with different barriers and attractants that are themselves differentially efficacious (Owens and Bennett, 1983; Reierson and Rust, 1977). There are even some traps consisting of containers with electrically energized barriers that are used to collect large numbers of live cockroaches (Ballard and Gold, 1983; Wagner, Ebeling, and Clark, 1964).

Some of the same limitations are true of cockroach trapping as are true of rodent trapping, or for that matter, trapping any pest population with adequate food, water, harborage, and a high reproductive potential. Where large pest populations are infesting a favorable environment, removal trapping for control is out of the question. Now the question should be: can mass trapping or removal trapping control cockroaches in low populations? Trapping can eliminate low populations of rodents, and there are indications the same can be done with some stored product pests (Burkholder, 1981). While the sex attractant of the American cockroach is available for attracting males of that species, there are no baits or lures that will predictably attract other cockroach species, particularly the German cockroach. Many tests have been made to find acceptable baits to use in cockroach traps (Frishman, 1982*a;* Wileyto and Boush, 1983); while each test designates a superior bait, there is no agreement between tests. Given the general feeding habits of cockroaches, it is not surprising that no one food odor predominates. The lack of a food bait indicates that cockroaches enter traps only as they forage. Traps can, however, be located in places more likely to be stuck on than they would be in others. German cockroaches favor traveling along the intersection of two plane surfaces, i.e., the floor and the wall (Ebeling, 1975). They are also likely to inspect an incline (R. V. Carr, personal communication); they prefer particular sized spaces (3/16 inch), and they tend to hesitate and bunch up in corners. These behavioral characteristics can be used as guides for trap placement, and the addition of moist or yeasty baits seems to make traps more attractive. However, there is no indication that populations are reduced to acceptable levels using traps, and there even seems to be a population threshold below which trap catches are rare. Even so, cockroach trapping is an important part of a pest management program. Traps can be used to identify dwelling units with large populations, or focus units (Akers and Robinson, 1985). These populations migrate directly into other apartments when construction elements, i.e., plumbing, are favorable (Owens and Bennett, 1982).

Traps are also useful in ongoing monitoring. However, translating cockroach trap counts into a pest management action level is difficult. The U.S. Navy uses an average of 2, and the National Park Service uses an average of 2.5 cockroaches per trap night as action levels (Ruggiero, personal communication). Until a reliable method of estimating the entire cockroach population in a structure is available, action levels will have to be set by experience and aesthetic needs in individual situations. A beginning has been made in modeling German cockroach populations (Grothaus et al., 1981), but until an accurate measurement of the hidden population can be made, routine treatments and "complaint spraying" will continue to be the predominant practices in cockroach control.

Fly Papers and Strips. A few flies in summer are easily excused, but the sudden appearance of one fly buzzing in a room on a warm winter day can cast grave doubt on an otherwise efficient pest management program. Attic flies are those flies that wind up in attics, ceiling voids, cupolas, and roof top equipment rooms in every type of building from homes to hospitals. Houseflies, face flies, cluster flies, blow flies, and others hide in crevices for protection during the winter. Their quest can begin as early as August, and once settled they remain quiet until springtime when they leave hiding to resume doing things flies do.

Urban pest management programs assume the management of all pests in an area of operation, and this definitely includes attic flies in winter as well as filth flies in summer. Hanging sticky fly strips are effective for intercepting the flies before they have time to enter lower floors. Fly strips should be routinely placed in buildings where attic fly entry cannot be prevented. The strips operate best near windows because the flies, when active, move to the morning light and try to escape. Flies are also attracted to verticals such as light cords and ceiling supports, and this tendency makes hanging strips attractive to land on. Some fly strips take advantage of this and use printed vertical lines under the tacky surface. Others are manufactured as sticky orange tubes with printed fly silhouettes in behaviorally attractive aggregations, intended to draw flies to the resting surface. The commonest fly strip is the classical rolled strip that is pulled spiraling from a short cylinder. Putting the strips in places where flies will be active is most important. Placement should be made after attic temperatures cool down because the tacky surface runs and drips in extreme heat.

Light Traps. Blacklight traps are commonly used to monitor populations of moths as they build up on agricultural crops. These traps are very useful for assessing peak flight periods or population levels that help make pest control decisions. Many insects are able to detect images that reflect in the ultraviolet or "blue" portion of the light spectrum. For instance, certain flowers reflect ultraviolet light, which insects see but humans do not (Jones and Buchmann, 1974). Conversely, many insects do not detect images that reflect light at the red end of the spectrum or that end opposite the blue. For these reasons, blue light and visible light toward the blue end of the spectrum attract many insects. Many flies, such as houseflies and cluster flies, and some beetles, such as cigarette beetles and carpet beetles, can be found at eastern windows, where they are attracted to the distant early morning light. Later in the day, with bright light coming from all angles, these insects are no longer directionally oriented. Thus, different species of insects are attracted by different light wave-lengths and intensities. Mosquito species are attracted to a weak source of visible light. That is why light traps used by the Public Health Service and mosquito control agencies use very weak flashlight-size bulbs, combined with small fans to suck mosquitoes into the capture bag of the trap. In urban neighborhoods electrocution grids are often attached to blacklights. Midges or chironomids, which look very much like mosquitoes, are attracted to a blacklight source. So are some beetles and night flying moths, and it is usually these larger insects that make the prolonged sizzle that satisfies insect electrocuter owners.

Blacklight fly traps for indoor use may have fans that blow attracted insects into holding containers or onto sticky surfaces. Still others have attached electrocuters. Indoor blacklight fly traps are for use at night when inside building lights are off. They are intended to attract and kill flies that come in during the day. These traps should be evaluated by inspecting the trap catch. If the target pests are not being captured, the traps are useless or are in the wrong place. Blacklight traps cannot compete with other light sources, such as vending machines and neon signs. They should be placed where they will not attract outside insects to doors or windows. Light traps should never be placed outside entrances where attracted insects can crawl inside in the morning when the lights are off or ineffective.

Mercury vapor lights also attract flying insects because of the ultraviolet waves they emit. These lights should not be placed where they will draw insects toward buildings. Rather they should be placed away from buildings, where flying insects will not cause problems, and the less attracting sodium vapor lamps should be used near structures (Gilbert, 1984).

Screening

Window screens are as common on houses in the developed world as the windows they cover. This has not always been true, however, even in the United States. In the early part of the twentieth century, state university extension services promoted screening as a major program in rural areas for the exclusion of flies and mosquitoes. The resistance to screening, because of expense and the imagined obstruction of airflow, was difficult to overcome even in a vulnerable national period of typhoid and yellow fever transmission. Screens were finally accepted, and wire screening was not only fitted to windows but was placed on doors of pantries and kitchen cabinets and even made into small screened cages in which to place babies for naps in the shade.

For several decades after the Second World War, the efficiency of building cooling and heating was greatly improved. In this same period, window screens fell into disuse, principally because the windows of air-conditioned buildings were seldom opened. Homes built in the decades of the 1950-1970s utilized window screens made for ease of attachment and removal, and half inch gaps at the bottoms were accepted as normal. Window screens became tokens to building codes rather than pest excluders. Commercial buildings were centrally heated and cooled, and the positive pressure was vented through gaps specifically placed between easy access doors. Pests such as cockroaches, crickets, millipedes, mice, and any swarm of flying insects attracted to shopping mall lights had immediate access to stores. New emphasis is now being placed on screened openings in homes since energy has become costly and since nonmetal screening, which is durable and easy to work with, has become available.

Routine screening of attic and crawl space ventilators to exclude insects and vertebrates allows for moisture reduction without infestation of the structure. Weatherproofing homes and commercial buildings has narrowed gaps around doors, and exclusion of pests rather than residual spraying is seen as a positive effort.

A recent commercial innovation is the enclosing of entire shopping malls. This process reduces the number of openings to the outside but creates a new space for pest buildup—the common areas of malls. Invasion of businesses from this area will bring different pests in contact with each store, but exclusion still remains the first line of defense.

CO_2 Fumigation

The use of CO_2 for fumigation is certainly environmentally sound, and several tests have shown it to be an efficient method of pest control (Barnhart, 1963; Cantwell and Neidhardt, 1978). Several variables such as temperature and gas circulation must be controlled, but for the difficult elimination of cockroaches in hospital food carts, CO_2 in the form of dry ice has proven effective. Other pests are also susceptible to small space CO_2 fumigation (Cantwell, Tompkins, and Watson, 1973). Perhaps this simple procedure will find a place in integrated pest management, where the time and effort spent controlling small pest populations in portable artifacts and packages is often out of proportion to the overall program.

Ultrasonic Devices

Devices that emit ultrasound designed to affect rodents are either purchased to fit in a pest management scheme or in many cases unwisely bought as a single control method. Theoretically, the devices should produce ultrasound that repels rodents from areas at which the sound waves are directed. This causes the rodents to retreat to areas shielded from the sound waves. Toxic baits, traps, and other control methods are placed in these sound shadows or shielded refuges, thereby effecting control.

One test conducted in an egg-laying establishment found that, while the ultrasound repellent effect was not absolute in all areas, where good sanitation and construction occurred, rodent activity decreased (Jackson, 1980). Another test using Philippine rats indicated that well-fed, sound-naive rats were repelled from food for ten days because of an ultrasonic device 30 cm (about 1 ft) from the food. The device was operating at 20 kHz and 118 dB. When the rats' food was restricted the device was ineffective. All tested devices were ineffective when the dB level was lowered from 118 dB to 88 dB, simulating a 10.7 m increase in distance from the target area. Intensities under 90 dB are considered ineffective, but it should also be noted that high intensities of 120 dB will produce sound-induced seizures and death in Norway rats (Shumake et al., 1982).

Several problems are indicated when ultrasonics are considered for rodent control. For instance, sound shadows are established by any obstruction, and the intensity of the ultrasound decreases very rapidly with distance. In large warehouses or storage rooms, sound shadows could be nearly continuous. Furthermore, rats normally avoid clear, unobstructed areas and seek protected places, somewhat negating the need for ultrasound interdiction. Rats will also habituate to many situations (including ultrasound) and become very bold when they are hungry. Birds appear to be unaffected in ultrasonic device tests (Martin and Martin, 1984).

There are no indications that ultrasonic devices, as a single or lone method, will control rodents, and regardless of the desire for a "magic bullet," ultrasonic devices are useless for the control of insect pests (D. A. Reierson and Federal Trade Commission, personal communications).

Sterility and Genetic Control

The use of sterile male releases to suppress the screwworm fly opened a new window on pest control and even pest eradication. The screwworm program involved the massive release of sterile male flies, which competed with native fertile males, to inseminate females, which only mated once. The program was successfully tested on an island population of screwworms. The test was followed by the release and ultimate eradication of the fly in the southwestern United States. The screwworm eradication program is a milestone in pest control, comparable to the vedelia beetle releases and the discovery of synthetic organic pesticides. It was successful at a time when the limitations of pesticides had become obvious, and it demonstrated the need and potential of biological and technological research. The championing and acceptance of integrated pest management was supported in part by the confidence inspired by the popular screwworm eradication program.

Principles of pest population management have been broadened a great deal in recent years (Knipling, 1979), and the application of genetic manipulation has provided several examples of pest management besides the screwworm. These include programs to control the Mediterranean fruit fly, the olive fly, and mosquitoes (National Research Council, 1980).

Genetic manipulation programs usually involve many years of research

(Pal and Whitten, 1974). Before cytogenetic research can begin, the organisms involved must be studied in nature and their biology understood in order to establish laboratory colonies. The organism will need to be mass reared, which involves artificial diets or available hosts, methods of handling large numbers of organisms, emphasizing a particular stage or sex or clone, protection from diseases, guarding against the selection of limiting laboratory attributes, and many other parameters (Metcalf and Metcalf, 1982).

While the release of genetically altered pests in agricultural or forest areas may be applauded, releasing pests in any condition in urban areas must be accompanied by well-prepared educational or mass communication programs. Population density alone magnifies communication problems. A release program involving entire agricultural regions may not affect as many people as one involving an urban apartment housing complex. Misunderstood alarm about "irradiated" insects can be initiated by a minority, but with the communications media available and the close proximity of people in urban neighborhoods, individual feelings can be translated into group action in a matter of hours.

Genetic engineering will facilitate genetic pest control. The replacement of a normal allele with one that is lethal when certain environmental conditions occur, such as low temperatures or reduced daylight, may be possible in the future. Chromosomal translocations have produced German cockroach males that are genetically sterile but equally competitive with normal males (Ross and Cochran, 1973, 1976). While people are not likely to bear the further burden of additional cockroaches released in their apartments, they would not likely notice sterile male releases in an apartment building garbage room as part of a total IPM program.

Finally, a condition known as *daughterless* has been discovered in a species of parasitic wasp. The male-transmitted condition is expressed by the production of male offspring instead of female offspring (Werren, Skinner, and Charnou, 1981). The same species also expresses a *son-killer* trait, which causes the production of female-biased sex ratios through the mortality of males. The *son-killer* trait is associated with bacterial infections in the adult wasps. It is suspected that the *daughterless* trait also is caused by an infectious agent, possibly transferred to the female reproductive tract during copulation (J. H. Werren, personal communication). Pathogen-mediated sterility may be of value in the management of susceptible pests.

Organic Pest Management

A major polarization exists among many who practice pest control, either as lay persons or professionals. The argument is whether to use organic pest control materials and methods or synthetic pesticides and fertilizers. At times individuals of each group assumes or accepts a position normally thought to be that of the other, but generally such ideas do not develop into balanced philosophies. Claims of economics, expediency, human health and diet, even biblical authority are invoked by each group to refute the other.

The important lesson to be learned from the "organic versus synthetic" argument is to not exclude useful and responsible pest control methods from pest management programs because of pressure from one pole or the other. Certain procedures are without prejudice. A list would include understanding as much about the pests as possible; obtaining an acceptable level of sanitation; promotion of plant vigor; the safe use of pesticides; regard for environment and stewardship of the land; and basic ethics.

Urban and agricultural pest management share the same principles, but there are areas where emphasis differs. Plant pest management in urban areas involves many different plants on a small acreage. The plants may be vegetables, shrubs, trees, and turf, each type represented by many species. Organic procedures are more acceptable in this kind of situation. Agricultural pest management on the other hand, involves large acreages of a monoculture, or a single crop of plants—or animals—with efficient production as the criterion for success. Accepting that agricultural practices should not operate at the expense of harming people or the environment, pest management practices often must involve the use of pesticides as well as limiting their unnecessary use. Synthetic organic pesticides were developed because there was an important need, and this need remains true today.

There are many organic pest management practices that should be used in the urban landscape, for example, mulch around tree and shrub bases to keep mowers away from them; prune cleanly to keep from tearing bark and giving borers entry places; place plants in areas that suit their biological needs—put shade loving plants in the shade and maintain plant vigor by using the correct fertilizer, mulching, and watering when needed—plants that are under stress are susceptible to pest attack. There are many other examples in all gardening books—those that support organic methods and those that support synthetic pesticides and fertilizers.

There are recommendations for vegetable gardening using organic methods for nearly every garden plant and nearly every pest insect (Yepsen, 1976). Most universities and governmental agencies have at least some organic recommendations, and all land grant universities have advice on good horticultural practices. There is little university or governmental research on organic pest control because agricultural production has a necessary priority over urban gardening and urban pest management. This is not to say that the research is all on pesticides. In fact most is not, but pesticide testing and recommendations are an important part of agricultural pest management.

Of the organic pest management recommendations, most fall into cultural control, physical or mechanical control, biological control, and nontoxic pesticides. Some examples of each are as follows.

Cultural Controls. Plant mixed crops. The more varied the plants, the less chance there is of major pest infestations on one plant species (Cantelo and Webb, 1983). Use resistant plants. They may sometimes have a lower potential yield, but diseases, for instance, will not affect them. Plant perennials when possible; parasites and predators are less disrupted at harvest and weeds are discouraged. Time planting dates to miss problem pests when plants are most vulnerable, and use repellent plants if they are known.

Physical or Mechanical Controls. Use roof shingles or foil on soil surface around cole crop seedings to keep cabbage maggots from laying eggs. Use collars such as paper cups around plant stems for cutworm barriers. Cover wire wickets over spinach and chard rows with plastic screening for spinach leafminers. Air out, decrease moisture for slugs. Pick off hornworms, bagworms, and imported cabbage worm eggs under the outer leaves of cabbage. Prune off fall webworm webs at the tips of branches, tear out tent caterpillar webs, and pick off caterpillars. Use burlap barriers on trees for gypsy moth larvae. Mulch with black plastic or newspapers for weeds.

Biological Controls. Spray with *Bacillus thuringiensis* for caterpillars. Plant trap crops away from gardens and transfer predators and parasitized pests. Glue ladybeetle and other predator wings together with a drop of clear fingernail polish to keep them from flying away.

Nontoxic Pesticides. Apply dormant oils to smother scale and aphid eggs. Apply during dormant season, and parasites will not be affected. Apply insecticidal soaps or detergents for soft-bodied insects. Beer is used to attract and drown slugs.

Efficacious organic plant protection methods are safe to use, economical, and very satisfying for attentive plant growers.

Folk Nostrums and Panaceas

There are innumerable cures, preventives, and remedies for pests that have been handed down, generally through folklore. When these pallatives sound reasonable, they are frequently accepted without question. While some of the old tales are absolutely correct, most cannot be confirmed without well-controlled experiments.

A sampling of these nonchemical nostrums followed with possible explanations are as follows. Cucumber peels repel or kill cockroaches. When tested, the peels dried out and the German cockroaches thrived. Baking soda in rat food will cause gas to build up, presumably on the rodent's acid stomach, and since rodents cannot belch or "burp", they will explode! Actually rats cannot regurgitate; gas is another matter. Plaster of Paris fed to rats will harden in their intestines and kill them. Norway rats fed large amounts of plaster of Paris with their diet only produced chalky white fecal pellets (W. B. Jackson, personal communication). Plaster of Paris with sugar or flour eaten by cockroaches will harden in their guts and kill them. Tiny plaster of Paris casts of cockroach guts have never been observed where this was tried. Osage orange fruit will kill or repel cockroaches. Cockroaches reared in dishes with Osage orange fruit were not affected at all. Those hit with Osage oranges died. Small cotton balls attached to screen doors will keep flies away. This has been tested and is apparently not true. However, when stuffed in a hole in the screen the cotton will keep the flies out.

SUMMARY

Where possible, all of the nonpesticide management practices should be put in place on the outside landscape and inside the structures first, then pesticide application, where needed and when needed, will be much more effective. Nonpesticide practices are usually of lasting value and upkeep is usually minimal.

Outside

An ornamental pest management program (Holmes and Davidson, 1984) is a systematic procedure involving landscape map construction, identification of the plants, proposals for plant maintenance, and periodic monitoring of the premises, which includes pheromone trapping, spot treating when necessary, and reporting. A pest management program of this type can be performed on an urban lot or over an entire community (Raupp and Noland, 1984).

Some landscape/pest management alterations might be to change part of the grade to create different drainage, mulch trees and shrubs to conserve moisture and protect trunks, replace stressed plants with better suited or resistant plants, and properly prune to shape plants and make them stronger and less susceptible to wind and ice damage. Some trees might be removed to arrest pest problems, like female boxelder trees, which produce an annual late summer horde of boxelder bugs. Or, tree removal will take out dead trees or trees with holes, which are habitat for biting midges and mosquitoes and for wood rot which weakens the trunk or limbs allowing them to snap. It will also open up areas to light to increase the vigor of surrounding plants.

Incorrect grading and plantings around buildings can directly contribute to pest problems in and of the structure. Most moisture problems of buildings are caused by improper grading or settling. When the grade drains moisture toward the building (the predominant problem with historical structures), fungi or wood rots become established, damaging the structures themselves and predisposing wooden members to termites, carpenter ants, old house borers, and anobiid powderpost beetles. The grade, directed toward the building, also washes soil against the foundation or even the sill, exacerbating the moisture problem and bringing termites, ants, and other invading pests closer to wood and the interior. Leaking gutters and improperly directed downspouts complete the scenario that turns crawl spaces into murky, miasmic swamps.

Improper landscaping around buildings can also cause pest infestation problems. Thick overgrown foliage next to the foundation obstructs ventilation and encourages moisture retention below a structure. Proper ventilation and a plastic soil moisture barrier will often correct humidity problems under a building. Foliage next to the foundation affords food, in the form of honeydew from scales and aphids, and access to mold-infested wood for carpenter ants. Vines can perform the same functions. Organic mulch of plants next to a building fosters millipedes, black vine weevils, crickets, and flies and provides cockroach harborage in the warmer climates. Mulch, vines, and thick shrubs give cover to rodents and, most important, deny access for inspection and maintenance. No plants should be closer than 3 ft from a foundation for the previously described reasons and because the roof overhang will dry them out. Concrete foundations can be responsible for increasing basic soil conditions and daytime warmth, which in the winter encourages growth and nighttime freezing. Breaking all wood-soil contact, installing barrier strips of gravel covered plastic, as well as caulking doors and window frames and draining window wells will provide an early defense line against pest infiltration. Insulated pipes from air conditioner compressors can breech the house foundation and bring subterranean termites into wooden floor joists. Utility lines and overhanging tree limbs at the roof area are also conduits, bringing carpenter ants and squirrels. Caulking around service lines will keep out those acrobatic pests, as well as flying pests such as bats, honeybees, yellowjackets, flies, and elm leaf beetles.

A final consideration outside buildings is garbage disposal and the placement of trash containers. Available garbage left overnight encourages rat problems, and cat, raccoon, and skunk scavenging. Providing for trash protection and timely removal will obviate much rodent bait use as well as unnecessary pesticide application for flies and yellowjackets. The placement of garbage cans and dumpsters is critical around homes as well as around hospitals and restaurants. The natural tendency is to aim for easy access from the building. Placement of dumpsters near doors allows for quick disposal, but it also invites birds, flies, yellowjackets, honeybees, and rodents to entryways, and from there they have access to all parts of the building. Replacement of dumpsters and cans must be made, keeping in mind that building personnel will need reasonable access or they will outmaneuver the changes or create other problems. Consider dumpster and garbage can size and numbers. Oversized dumpsters and many cans indicate that garbage pick-up doesn't happen as often as it should. Undersized facilities will result in container overflow. Entryways are also commonly propped open for deliveries. Air curtains, screened doors, and hanging plastic strips will help if they are not circumvented by building and delivery personnel. Inspection and observation of personnel practices are essential to evaluating these "cultural" problems and take time to be properly understood.

Inside

Sanitation, control of moisture, and reduction of harborage are always necessary considerations in urban pest management programs.

Sanitation. Inspections should reveal the sanitation level of human food preparation areas, which sustain cockroaches and rodents. These areas include kitchens, coffee areas, dining areas, and places where food is taken that might not be obvious at first. Food is taken to patient rooms in hospitals and sanitariums, but it is also taken to nursing stations and offices. It is brought in by visitors and horded by patients. In homes food is taken to dens, playrooms, bedrooms and fed to pets almost anywhere.

Other food that pests consume includes old rodent bait and other stored products such as milled cereals, spices, dried fruits, and whole grains. Silverfish, grain mites, and psocids feed on starches, usually infested by molds in places of high humidity. Carpet beetles feed on high protein grain, and wool, feathers, and fur in the form of clothing, tapestries, carpets, taxidermy mounts, dead animals, insect specimens, and bat guano. Clothes moths also feed on many of the same materials. Subterranean termites and wood borers feed on trash wood beneath buildings and paper products inside buildings. Drywood termites, powderpost beetles, wood borers, and carpenter ants infest wooden artifacts and firewood. Flea larvae feed on dried blood from adult fleas. Flies infest garbage, animal filth, overripe fruit, sink traps, and floor drains. Spiders feed on all of the indoor pests that they trap in their webs or pounce on at windows. All these sanitation problems must be eliminated, or pesticidal treatment will fail.

Moisture Control. Moisture inside may consist of free water from dripping faucets, dishes in the sink, commodes, condensation, pet dishes, leaky roofs, and overflowing gutters. Adequate moisture for pests is also provided by humidity in the air. Building maintenance, better housekeeping, ventilation, and dehumidification are necessary pest management options, and in many cases moisture control will stop the pest infestation without the use of pesticides.

Reduction of Harborage. Quite often reduction of harborage inside means the elimination of clutter. Sometimes harborage is unsanitary accumulation, or possibly it is created by openings in building walls, but quite often it is simply surface area and voids created by an over accumulation of unused household furniture, stuffed kitchen cupboards, littered office rooms, and neglected storerooms. Reduction of harborage is more difficult to explain to lay public than sanitation and moisture problems.

CONCLUSION

Pest management is the management of all pests in an infested area (Luckmann and Metcalf, 1982). It seldom implies the eradication of a pest, except perhaps, for a limited time or in a limited space. More often than not, urban pest management research has not established a survey method capable of telling whether low populations of pests are even present. A commodity that cannot be overused in pest management is education. Education of pest management practitioners and their subsequent education of program recipients is essential. If pest management programs are to remain flexible as dynamic target pests adjust to them and as research provides new answers, education will be the vehicle for it all.

REFERENCES

Akers, R. C., and W. H. Robinson. 1981. Spatial patterns and movements of German cockroaches in urban, low-income apartments. *Proc. Entomol. Soc. Wash.* **83:**168-172.

Angus, T. A. 1977. Microbial control of arthropod pests. In *Proceedings of XV International Congress of Entomology.* Entomological Society of America, College Park, Md., pp. 473-477.

Ballard, J. B., and R. E. Gold. 1983. Field evaluation of two traps used for control of German cockroach population. *J. Kans. Entomol. Soc.* **56:**506-510.

Barnhart, C. S. 1963. Dry ice fumigation. *Pest Control* **31:**30.

Baumgartner, D. M. 1982. *Animal Damage Control in Washington,* Extension Bulletin 1147. Cooperative Extension Service, Washington State University, Pullman.

Bell, W. J., S. B. Vuturo, S. Robinson, and W. A. Hawkins. 1977. Attractancy of the American cockroach sex pheromone (Orthoptera: Blattidae). *J. Kans. Entomol. Soc.* **50:**503-507.

Beroza, M., N. Green, S. I. Gertler, L. F. Steiner, and D. H. Miyashita. 1961. New attractants for the Mediterranean fruit fly. *J. Agric. Chem.* **9:**361-365.

Bjornson, B. F., H. D. Pratt, and K. S. Littig. 1968. *Control of Domestic Rats and Mice.* Public Health Service, Communicable Disease Center, Atlanta, Georgia.

Bordon, J. H. 1974. Aggregation pheromones in the Scolytidae. In *Pheromones,* M. C. Birch (ed.). American Elsevier, New York, pp. 135-160.

Bottrell, D. G. 1979. *Integrated Pest Management.* Council on Environmental Quality, Washington, D.C., 120p.

Brown, R. Z. 1960. *Biological Factors in Domestic Rodent Control,* Public Health Service Publication 773.

Burden, G. S. 1975. Repellency of selected insecticides. *Pest Control* **43:**16,18.

Burden, G. S., and J. L. Eastin. 1960. Laboratory evaluation of cockroach repellents. *Pest Control* **28:**14.

Burkholder, W. E. 1981. Biomonitoring for stored-product insects. In *Management of Insect Pests with Semiochemicals: Concepts and Practice,* E. R. Mitchell (ed.). Plenum, New York, pp. 29-40.

Burkholder, W. E. 1982. Reproductive biology and communication among grain storage and warehouse beetles. *J. Georgia Entomol. Soc.* **17**(2nd suppl., Oct.):1-10.

Burkholder, W. E. 1984. The use of pheromones and food attractants for monitoring and trapping stored-product insects. In *Insect Management for Food Storage and Processing,* F. Baur (ed.). American Association of Cereal Chemistry, St. Paul, Minn., pp. 69-82.

Burkholder, W. E., and M. Ma. 1985. Pheromones for monitoring and control of stored-product insects. *Ann. Rev. Entomol.* **30:**257-272.

Cantelo, W. W., and R. E. Webb. 1983. An examination of selected companion plant combinations and how such systems might operate. *J. Wash. Acad. Sci.* **73:**100-106.

Cantwell, G. E., and K. Neidhardt. 1978. Fumigation of hospital food carts to control German cockroaches. *Pest Control* **49:**14-16.

Cantwell, G. E., G. J. Tompkins, and P. N. Watson. 1973. Control of the German cockroach with carbon dioxide. *Pest Control* **41:**40, 42, 48.

Collins, W. J. 1973. German cockroach resistance, 1, Resistance to diazinon includes cross-resistance to DDT, pyrethrius, and propoxur in a laboratory colony. *J. Econ. Entomol.* **66:**44-47.

Collins, W. J. 1975. Resistance in *Blattella germanica* (L.) (Orthoptera, Blattidae): The effect of propoxur selection and nonselection on the resistance spectrum developed by diazinon selection. *Bull. Entomol. Res.* **65:**399-403.

Committee on Plant and Animal Pests. 1969. *Insect Pest Management and Control,* Publication 1695. National Academy of Sciences, Washington, D.C.

Croft, B. A. 1977. Resistance in arthropod predators and parasites. In *Pesticide Management and Insecticide Resistance,* D. L. Watson and A. W. Brown (eds.). Academic Press, New York, pp. 377-393.

Ebeling, W. 1975. *Urban Entomology.* Division of Agricultural Sciences, University of California.

Ebeling, W., D. A. Reierson, and D. E. Wagner. 1967. Influence of repellency on the efficacy of blatticides, 2, Laboratory experiments with German cockroaches. *J. Econ. Entomol.* **60:**1375-1390.

Ebeling, W., D. A. Reierson, and D. E. Wagner. 1968*a*. Influence of repellency on the efficacy of blatticides, 3, Field experiments with German cockroaches with notes on three other species. *J. Econ. Entomol.* **61:**751-761.

Ebeling, W., D. A. Reierson, and D. E. Wagner. 1968*b*. Influence of repellency on the efficacy of blatticides, 4, Comparison of four cockroach species. *J. Econ. Entomol.* **61:**1213-1219.

Ebeling, W., R. E. Wagner, and D. A. Reierson. 1966. Influence of repellency on the efficacy of blatticides, 1, Learned modification of the behavior of the German cockroach. *J. Econ. Entomol.* **59:**1374-1388.

Frankie, G. W., and L. E. Ehler. 1978. Ecology of insects in urban environments. *Annu. Rev. Entomol.* **23:**367-387.

Frishman, A. M. 1982*a*. Cockroaches. In *Handbook of Pest Control,* 6th ed., A. Mallis (ed.). Franzak and Foster, Cleveland, Ohio, pp. 101-153.

Frishman, A. M. 1982*b*. Rats and mice. In *Handbook of Pest Control,* 6th ed., A. Mallis (ed.). Franzak and Foster, Cleveland, Ohio, pp. 5-77.

Georghiou, G. P., S. Lee, and D. H. DeVries. 1978. Development of resistance to the juvenoid methoprene in the house fly. *J. Econ. Entomol.* **71:**544-547.

Gilbert, D. 1984. How light affects insects. *Pest Control Technol.* **12:**42.

Grothaus, R. H., D. E. Weidhass, D. G. Haile, and G. S. Burden. 1981. Superroach challenged by microcomputer: Think growth rate not percent kill in controlling the German cockroach. *Pest Control* **49**(1,2,3).

Haverty, M. I., and R. W. Howard. 1979. Effects of insect growth regulators on subterranean termites: Induction of differentiation, defaunation, and starvation. *Ann. Entomol. Soc. Am.* **72:**503-508.

Henderson, R. F. 1983. Moles. In *Prevention and Control of Wildlife Damage,* R. M. Timm (ed.). Cooperative Extension Service, University of Nebraska, Lincoln, pp. D53-D61.

Holmes, J. J., and J. A. Davidson. 1984. Integrated pest management for arborists: Implementation of a pilot program. *J. Arboric.* **10:**65-70.

Hrdy, I., and J. Krecek. 1972. Development of superfluous soldiers induced by juvenile hormone analogues in the termite, *Reticulitermes lucifugus santonensis. Insectes sociaux* **19:**105-109.

Jackson, W. B. 1980. Ultrasonics protect egg farm. *Pest Control* **48:**28, 38.

Jackson, W. B., P. J. Spear, and C. G. Wright. 1971. Resistance of Norway rats to anticoagulent rodenticide confirmed in the United States. *Pest Control* **39:**13-14.

Jones, C. E., and S. L. Buchmann. 1974. Ultraviolet floral patterns as functional orientation cues in hymenopterous pollination systems. *Anim. Behav.* **22:**481-485.

Knipling, E. F. 1979. *The Basic Principles of Insect Population Suppression and Management,* Handbook 512. U.S. Department of Agriculture, Washington, D.C.

Knote, C. E. 1984. Attack mice problems on building exteriors. *Pest Control* **52:**46-52.

Kogan, M. 1977. The role of chemical factors in insect/plant relationships. In *Proceedings of XV International Congress of Entomology.* Entomological Society of America, College Park, Md., pp. 211-227.

Kogan, M. 1982. Plant resistance in pest management. In *Introduction to Pest Management,* R. L. Metcalf and W. H. Luckmann (eds.). Wiley, New York, pp. 93-134.

Labeyrie, V. 1978. The significance of the environment in the control of insect fecundity. *Ann. Rev. Entomol.* **23:**69-90.

Luckmann, W. H., and R. L. Metcalf. 1982. The pest-management concept. In *Introduction to Pest Management,* 2nd ed., R. L. Metcalf and W. H. Luckmann (eds.). Wiley, New York, pp. 1-32.

McGovern, T. R., G. S. Burden, and M. Beroza. 1975. *n*-alkanesulfonamides as repellents for the German cockroach. *J. Med. Entomol.* **12:**387-389.

Maddox, J. V. 1982. Use of insect pathogens in pest management. In *Introduction to Pest Management,* 2nd ed., R. L. Metcalf and W. H. Luckmann (eds.). Wiley, New York, pp. 175-216.

Mallis, A. 1971. *American Entomologists.* Rutgers University, New Brunswick, N.J.

Mallis, A., W. C. Easterlin, and A. C. Miller. 1961. Keeping cockroaches out of beer cases. *Pest Control* **29**(6):32-35.

Marsh, R. E., and W. E. Howard. 1982. Vertebrate pests. In *Handbook of Pest Control,* A. Mallis (ed.), 6th ed., Franzak and Foster, Cleveland, Ohio, pp. 791-861.

Martin, L., and P. C. Martin. 1984. Research indicates propane cannons can move birds. *Pest Control* **52:**52.

Matsumura, F., A. Tai, and H. C. Coppel. 1969. Termite trail following substance, isolation and purification from *Reticulitermes virginicus* and fungus-infected wood. *J. Econ. Entomol.* **62:**599-603.

Metcalf, R. L. 1982. Insecticides in pest management. In *Introduction to Pest Management,* 2nd ed., R. L. Metcalf and W. H. Luckmann (eds.). Wiley, New York, pp. 217-278.

Metcalf, R. L., and R. A. Metcalf. 1982. Attractants, repellents, and genetic control in pest management. In *Introduction to Pest Management,* 2nd ed., R. L. Metcalf and W. H. Luckmann (eds.). Wiley, New York, pp. 279-314.

Moore, W., and T. Granovsky. 1984. Interpreting cockroach sticky trap catches. *Pest Control Technol.* **12:**64-72.

Mueller-Dombois, D., K. Kartawinata, and L. L. Handley. 1983. Conservation of species and habitats: A major responsibility in development planning. In *Natural Systems for Development,* R. A. Carpenter (ed.). Macmillan, New York, pp. 76-91.

National Pest Control Association. 1982. *Bird Management Manual.* NPCA, Dunn Loring, Va.

National Research Council. 1980. *Urban Pest Management.* Report prepared by the Committee on Urban Pest Management, Environmental Studies Board, Commission on Natural Resources. National Academy Press, Washington, D.C., 272p.

Neal, J., and T. D. Eichlin. 1983. Seasonal response of six male sesiidae of woody ornamentals to clearwing borer (Lepidoptera: Sesiidae) lure. *Environ. Entomol.* **12:**206-209.

Nelson, J., and F. E. Wood. 1982. Multiple and cross-resistance in a field-collected strain of the German cockroach (Orthoptera: Blattellidae). *J. Econ. Entomol.* **75:**1052-1054.

Owens, J. M., and G. W. Bennett. 1982. German cockroach movement within and between urban apartments. *J. Econ. Entomol.* **75:**570-573.

Owens, J. M., and G. W. Bennett. 1983. Comparative study of German cockroach

(Dictoptera: Blattellidae) population sampling techniques. *Environ. Entomol.* **12**:1040-1046.

Pal, R., and M. J. Whitten. 1974. *The Use of Genetics in Insect Control.* Elsevier, Amsterdam.

Pasteels, J. M. 1977. Evolutionary aspects in chemical ecology and chemical communication. In *Proceedings of XV International Congress of Entomology.* Entomological Society of America, College Park, Md., pp. 281-293.

Pence, R. J. 1959. A new approach to fabric pest control. *Soap Chem. Spec.* **35**:65-68, 105-106.

Pence, R. J. 1966. Use of antimetabolities for pest control. In *Encyclopedia of Chemistry,* 2nd ed., G. L. Clark (ed.). Reinhold, New York, pp. 139-152.

Piper, G. L., and G. W. Frankie. 1978. Integrated management of urban cockroach populations. In *Perspectives in Urban Entomology,* G. W. Frankie and C. S. Koehler (eds.). Academic Press, New York, pp. 249-266.

Raupp, M. J., and R. M. Noland. 1984. Implementing landscape plant management programs in institutional and residential settings. *J. Arboric.* **10**:161-169.

Reierson, D. A., and M. K. Rust. 1977. Trapping, flushing and counting German cockroaches. *Pest Control* **45**:40-44.

Robbins, W. E. 1972. *Hormonal Chemicals for Invertebrate Pest Control.* National Academy of Sciences, Washington, D.C.

Roelofs, W. L. 1979. *Establishing Efficacy of Sex Attractants and Disruptants for Insect Control.* Entomological Society of America, College Park, Md.

Ross, M. H., and D. G. Cochran. 1973. German cockroach genetics and its possible use in control measures. *Patna. J. Med.* **47**:325-337.

Ross, M. H., and D. G. Cochran. 1976. Sterility and lethality in crosses involving two translocation heterozygotes of the German cockroach. *Experientia* **32**:445-447.

Roth, L. M., and E. R. Willis. 1960. *The Biotic Associations of Cockroaches.* Smithsonian Miscellaneous Collection, Washington, D.C.

Rust, M. K., and D. A. Reierson. 1978. Comparison of the laboratory and field efficacy of insecticides used for German cockroach control. *J. Econ. Entomol.* **71**:704-708.

Shapas, T. J., W. E. Burkholder, and G. M. Boush. 1977. Population suppression of *Trogoderma glabrum* by using pheromone luring for protozoan pathogen dissemination. *J. Econ. Entomol.* **70**:469-474.

Shorey, H. H. 1977. The adaptiveness of pheromone communication. In *Proceedings of XV International Congress of Entomology.* Entomological Society of America, College Park, Md. pp. 294-307.

Shumake, S. A., A. L. Kolz, K. A. Crane, and R. E. Johnson. 1982. Variables affecting ultrasound repellency in Philippine rats. *J. Wildl. Manage.* **46**:148-153.

Slater, A. J., M. J. Hurlbert, and V. R. Lewis. 1980. Biological control of brownbanded cockroaches. *Calif. Agric.* **34**(8-9):16-18.

Snetsinger, R. 1983. *The Ratcatcher's Child: The History of the Pest Control Industry.* Franzak and Foster, Cleveland, Ohio.

Stehr, R. W. 1982. Parasitoids and predators in pest management. In *Introduction to Pest Mangement,* 2nd ed., R. L. Metcalf and W. H. Luckmann (eds.). Wiley, New York, pp. 135-174.

Stevens, L. M., A. L. Steinhauer, and J. R. Coulson. 1975. Suppression of Mexican

bean beetle on soybeans with annual inoculative releases of *Pediobius foveolatus*. *Environ. Entomol.* **4:**947-952.
Suto, C., and N. Kumada. 1981. Secretion of dispersion-inducing substance by the German cockroach. *Blattella germanica* (L). *Appl. Entomol. Zool.* **16:**113-120.
Timm, R. M., ed. 1983. *Prevention and Control of Wildlife Damage.* Cooperative Extension Service, University of Nebraska, Lincoln.
Trammel, K. W., W. L. Roelofs, and E. H. Glass. 1974. Sex pheromone trapping of males for control of red-banded leafroller in apple orchards. *J. Econ. Entomol.* **67:**159-170.
Wagner, R. E., W. Ebeling, and W. R. Clark. 1964. An electric barrier for confining cockroaches in a large rearing or field collecting can. *J. Econ. Entomol.* **57:**1007-1009.
Weatherly, W. P. 1983. Insect pest outbreaks. In *Systems for Development,* R. A. Carpenter (ed.). Macmillan, New York, pp. 211-238.
Werren, J. H., S. W. Skinner, and E. L. Charnou. 1981. Paternal inheritance of a daughterless sex ratio factor. *Nature* **293:**467-468.
Wileyto, E. P., and G. M. Boush. 1983. Attraction of the German cockroach, *Blattella germanica* (Orthoptera: Blattellidae), to some volatile food components. *J. Econ. Entomol.* **76:**752-756.
Wood, F. E., W. H. Robinson, S. K. Kraft, and P. A. Zungoli. 1981. Survey of attitudes and knowledge of public housing residents toward cockroaches. *Bull. Entomol. Soc. Am.* **27:**9-13.
Yepson, R. B., Jr., ed. 1976. *Organic Plant Protection.* Rodale, Emmaus, Pa.

8

Education, Information Transfer, and Information Exchange

Gordon W. Frankie
University of California, Berkeley

James I. Grieshop
University of California, Davis

J. Kenneth Grace
University of California, Berkeley

Jack B. Fraser
University of California, Berkeley

The field of urban pest management is as much people oriented as it is pest oriented (National Research Council, 1980; Extension Committee on Organization and Policy [ECOP], 1981; Frankie and Ehler, 1978; Frankie and Koehler, 1983; Frankie, Fraser, and Lewis, 1982; Frankie et al., 1981; Sawyer and Casagrande, 1983). Although most pest organisms are easily identified, there are often human constraints associated with pest infestations that are not easily recognized, much less dealt with in an effective manner. Most of these constraints can be traced to relevant human attitudes and actions, and to identify and address these constraints represents a major step toward dealing with the people problems of urban pest management (UPM).

Effective education and information transfer are logically the principal means for dealing with people problems, and both processes are often acknowledged to be important components of UPM (National Research Council, 1980; Extension Committee on Organization and Policy, 1981; Todaro, 1984). Yet the literature reveals that very little has been written about these components (Farace, 1980; National Research Council, 1980; Worf, 1981). Further, with few exceptions (Fear et al., 1983) little effort has been made to develop fundamental approaches and tools for dealing with education and technology transfer needs in UPM. One notable example of this kind of deficiency is found in a recent UPM report developed by a special committee of the National Research Council (1980). In this 273-page document, only 4 pages directly addressed the issues of education and technology transfer.

Selected processes and channels that provide for the practical and theoretical transfer of UPM technology from research and extension specialist to practitioner to urbanite are examined and evaluated in this chapter. Much of our discussion centers on the potential for using a marketing approach for new education and information transfer efforts in UPM. In adopting a marketing framework, we recognize that information transfer involves more than a simple unidirectional movement process. It also involves an exchange relationship among those who develop and distribute the information and those who ultimately use it.

The paper is divided into four sections: (1) descriptions of terms used in the paper, (2) application of marketing principles and practices for transferring/exchanging UPM information, with descriptions of relevant case histories, (3) traditional and possible future educational approaches in client and practitioner education, and (4) a set of specific recommendations.

DESCRIPTION OF TERMS

For our purposes here, we view UPM broadly as the management of pest populations at levels that are acceptable to particular urban groups. The methods employed for management may be simple and unilateral (such as the one-time application of a pesticide); they may involve a combination of chemical or nonchemical means or both; or they may be fashioned into an integrated program. Regardless of the direct methods employed, to be characterized as management we maintain that thought and planning must predate the decision to take action against a given pest. A "thoughtful" program includes understanding the biology, behavior, and ecology of the target species and the nature of its negative impact on the clientele.

Effective education and communication (here considered to be nearly synonymous) use organized, deliberate, and sustained processes to transmit information to bring about changes in attitudes and practices (Cremin, 1978; Schramm, 1973). At the core of this definition is the concept of transmission or transfer. A transfer approach to education emphasizes the unidirectional movement of a product (such as UPM information) from a source (the UPM practitioner) to a receiver (the audience or client). In the commercial field this approach corresponds to selling. In contrast, commercial marketing and marketers advocate a broader view: the marketer sells a product and analyzes what the marketplace is ready to absorb. In this view, emphasis shifts from production to marketing (Kotler, 1984).

Marketing is the exchange of goods and services for other goods, services, or money. Marketing management is the "analysis, planning, implementation and control of programs developed to bring about desired exchange with target audiences" (Kotler, 1984). Such management occurs when people become conscious of opportunities to gain from a more careful planning of exchange relationships. Using the above distinctions, education, including UPM education, can thus be specified as "education as selling" and "education as marketing." In the former, an educator tries to "peddle the output of the factory." Regarding "education as marketing," an educator tries to understand the needs and wants of the market (the users).

UPM EDUCATION AS MARKETING

Application of Marketing Theory to Education

Education (as an exchange process and not simply a transfer process) and educational management (as analysis, planning, implementation, and control of programs) are analogous to marketing and marketing management. The UPM educator/marketer, in our view, must actively engage in exchange relationships, with an eye to what the user/audience wants. Education becomes less a matter of transfer and more a matter of exchange of information and products: for UPM marketers and educators it is vital to understand the audience if they are to develop appropriate educational and technical information (Peters and Waterman, 1982).

There are two principal urban UPM audiences: the urban public in general; and the specific and private audience composed of owners of pest management firms, managers of restaurants and other businesses, as well as institutional agencies, such as hospitals, military bases, and schools. Within each of these audiences there is variation in the type of information needed (or sought) and the ways in which this information is used. Whereas some members of an audience want information to solve immediate problems, others accumulate information they think may later be useful. It is important to become sensitive to the user and the need for information exchange (Peters and Waterman, 1982). To know the audience is a basic marketing principle.

Marketers use the concept of "audience segmentation" as a means for understanding the audience. Audience segmentation is a recognition that a given audience is usually not monolithic but consists of a series of subaudiences that must be reached or educated with different approaches (Kotler and

Zaltman, 1971). Marketing audience segmentation traditionally focuses on demographic factors such as age, sex, and residence. However, it is now widely acknowledged that effective segmentation also requires input from the psychological and behavioral sciences. More specifically, markets are known to be influenced by differences in buyer attitudes, motivations, values, usage patterns, aesthetic preferences, and degree of susceptibility (Yankelovich, 1964).

Within UPM, audience segmentation can lead to improved understanding of the two major audience categories, their motives, how they operate, act and react, variations within audience type, and so on. In short, "know thy audience" is advice to be followed—by educators as well as by marketers.

Marketers view their problem as one of developing the right *product,* backed by the right *promotion,* put in the right *place,* and at the right *price* (Kotler, 1972). UPM promoters and practitioners must also be concerned with these marketing principles. What is the product in UPM? It is not always a management technique; it can also be knowledge. In some cases, the product might be an educational campaign providing advice on how to avoid pest problems in the home or hospital. UPM educators must try to create specific tangible products and services that can be marketed and purchased. UPM practitioners also must be concerned with promotion. In many cases, a communication/persuasion strategy, using a variety of promotional tools such as advertising, publicity, and personal selling, can be utilized. Place is the third element of the marketing approach. There must be adequate and compatible distribution and response channels so that those who want to purchase the educational materials or services have a way to obtain them. Outlets accessible to general and specific audiences must be arranged. Price represents the cost that the target consumer will accept to obtain the product. Price does not necessarily always refer to monetary cost; it can also include opportunity, energy, and psychological costs. Price in UPM education is clearly an issue for those responsible to the public, such as Cooperative Extension workers.

Marketing UPM

The marketing approach has been used extensively by the corporate UPM sector. UPM-related commercial products are routinely tested before they are released to consumers. Information associated with product development and testing usually bears a proprietary stamp and is therefore unavailable for use by other concerns, private or public. Although use of marketing techniques has been demonstrated in the corporate UPM arena, these techniques have received relatively little attention by workers in public institutions. Given the potential of this approach and the need to more effectively educate and transfer and exchange UPM technology to wider audiences, it seems only logical that public institutions, especially land grant universities, should explore the use and application of marketing techniques.

In the following sections we review two current marketing research studies in UPM that are being supported by public institutions. These case histories reveal the diverse techniques available in marketing, all of which are designed to improve the process of transferring and exchanging technology. Two extension studies, an extension/commercial endeavor and a public health project that partially reflect the marketing approach are examined in later sections.

California Consumers and Pesticide Information. Several U.S. national surveys of public attitudes and practices toward pests and pesticides (referenced in Frankie et al., 1981) indicate that most urbanites (47-55%) who buy pesticides purchase them in grocery stores. Using this information as a base, Frankie, Koehler, Grace, and Hesketh (in prep.) entered into a cooperative arrangement with the Safeway grocery chain to advertise a free University of California publication on pesticides in selected stores in northern California. The goals of this pilot project were twofold: to use an in-place corporate structure to advertise (promote) Cooperative Extension information (the product); and to survey, via a questionnaire (the place), the recipients of the free information (price) to assess their reactions to the publication.

The pilot study was conducted in 26 Safeway stores in the San Francisco Bay area during the months of July and August 1983. Postage-paid coupons were placed on pesticide shelves to advertise the free Extension publication. Almost 400 requests were received; 210 people returned the questionnaire that accompanied the request form and publication. Requests and returned questionnaires were viewed as a clear demonstration of the feasibility of using this outreach method.

Regarding the second goal, a great deal of information was generated by the survey: the ways people generally acquire information about pests and pesticides and ways they transfer it to others; why people send for the information; topics not adequately covered in the publication; and specific shortcomings. These data are currently being used to construct a more comprehensive questionnaire that will be used in a future and larger program to distribute and assess the same information, with the ultimate goal of producing consumer-tested information on pesticides. Both the completed and planned studies actively use marketing principles to exchange information concerning urban pesticide use.

California Freeways: Iceplant Scales and UPM. Another type of UPM marketing approach is currently being tested by Peter Wilton and his colleagues at the University of California, Berkeley. Using a large data base on the biology, ecology, and management of two scale species on iceplant along California freeways (Tassan, Hagen, and Cassidy, 1983; Washburn and Frankie, 1985), Wilton and colleagues have developed a marketing scheme to test and deliver a package of tailor-made UPM information on scale identification, monitoring, and management to freeway workers and their supervisors in the California Department of Transportation (Caltrans). The initial phase of the research consisted of a survey designed to assess needs, interests, knowledge, expertise, and competing priorities of selected decision makers in the Caltrans organization (a form of audience segmentation and analysis). The descriptive survey provided information on the kinds and amount of UPM information needed by the various employee levels. It also allowed for an assessment of the mode of packaging the UPM information. (Initially, top level administrators of Caltrans and the university had decided to package the UPM information in a pocket brochure. The subsequent survey clearly demonstrated this was a relatively unpopular mode. Rather, a cassette-narrated slide show was the overwhelming choice of Caltrans personnel.)

The second phase of the marketing research, currently underway, is designed to test the following questions: how Caltrans staff members form their initial expectations of the usefulness of information; how to measure the effects of individual biases, training and beliefs on the search for, and use of, available information; what impact exposure to different types of information will have on the policymaker's evaluation of additional information of this type, and on the policymaker's performance of the tasks assigned. In the second phase, selected policymakers in Caltrans will be exposed to relevant UPM information and questions via a computer-interactive data-collection and information-exposure system on video display terminals. This will allow controlled measurements of individual judgments of the usefulness of information for particular task assignments, both before and after exposure to the iceplant scale-related information, with a view toward identifying the type of information most useful and appropriate to particular work assignments.

In this study again, marketing principles such as audience segmentation and design of product and place have become integral and indispensable to the design of an active educational UPM program.

Cooperative Extension and Marketing

Extension education in and of itself refers to reaching out from some organization, institution, group, or individual to some audience. As such, the Cooperative Extension Service must obviously be included among those groups that practice extension (including the extension of UPM). The methods currently employed by cooperative extension illustrate traditional educational practices, that is, "education as selling." However, cooperative extension can also serve as a model for exploring the potential for pursuing the "education as marketing" approach to UPM.

Since 1914, with the Smith-Lever legislation, the Cooperative (or Agricultural) Extension Service has become the major extender of land grant university information throughout the nation and its territories. The educational philosophy of extension is based on the principle that educational activities should meet the problems, needs, and interests of those for whom they are planned (Seay, 1983; Harrington, 1977). Traditionally, extension personnel, particularly county-based agents, not only deliver information (an education as selling function) but also try to determine the needs and wants of the users (an education as marketing function). All too frequently, however, the former approaches are much better represented than the latter. Koehler (1983), in addressing the topic of information transfer in UPM from a perspective broader than cooperative extension, reinforces the reality that most extension work to date has been from an "education as selling" position.

Minnesota Program. Ascerno's work (1981 and pers. comm. [1984]) is moving in the direction of "education as marketing." Over a period of six years, Ascerno and his associates at the University of Minnesota Cooperative Extension Service have used computers and telephones for managing information on urban pest problems in that state. To date the system has been adapted to receive and respond to inquiries about pests from individual clients (homeowners and agency representatives) and store this information for future analysis. The system also allows for immediate public feedback on its information. For example, some urbanites may experience difficulty in locating a recommended product or may not be completely satisfied with the outcome of a particular control recommendation. Call-backs from enough urbanites provide the incentive for an immediate revision of the recommendations. Thus, in a sense extension information is being tested by the public. Currently, the information is used in a predictive manner to develop extension information releases (e.g., early warning pest alerts) and to plan special information sessions with pest control operators and arborists. In 1983 the computer-based system was modified for the purposes of predicting expected daily pest problems, which were then disseminated (or marketed) to specific audiences.

Michigan Program. Michigan State University's Project Pest is a second example of the use of marketing approaches to extend UPM information. Project Pest, viewed by its authors as a community development program in integrated pest management (IPM), was conducted during 1980-1982 in a suburban community in Michigan. Specialists in the fields of entomology, forestry, and community development joined with citizen representatives of a local township to increase awareness of alternatives to pesticides for management of pest problems in the yard (Fear et al., 1983). The first part of the program consisted of administering a questionnaire to assess the relevant needs of suburban homeowners in the township. The second phase involved community residents in the process of designing and implementing an education program for alternative approaches in UPM. It was assumed that resident involvement would eventually lead to a more realistic program based on existing community attitudes and practices. A pest management manual and a demonstration walking tour were developed to inform residents of their pest problems and ways to deal with them within an IPM framework. In the final phase, survey methods were used to evaluate the education effort. In general, most participating residents expressed overwhelming satisfaction with all project activities (Lambur et al., 1982; Lambur, 1983).

There is a basic compatability between the philosophy of extension (programs based on obvious user needs and wants, close proximity of the user and the educator) and marketing principles (audience segmentation, determination of what the audience needs and wants). Conceptually, it should be but a short step for extension professionals to efficiently move to better utilize marketing approaches for UPM. Practically, the step will be much larger. Tradition and organizational constraints oftentimes work against such gains. For example, the notion that extension, because of its limited resources, must maintain a low profile in disseminating urban information is a strong deterrent to innovative extension UPM efforts. We maintain that extension personnel must adopt a more aggressive attitude toward outreach as an initial step for more effective and widespread information transfer. The marketing approach offers extension an opportunity for taking this first step.

CLIENT AND PRACTITIONER EDUCATION

Among the potential audiences for UPM are the urban public in general and specific public and private audiences such as managers and employees of public hospitals, schools, parks, military bases, restaurants, and other businesses. Furthermore, the practitioners of UPM, particularly pest control operators (PCOs), constitute another significant audience. At the same time, practitioners may also serve as providers of UPM education to their clients. This dual role and responsibility creates particular challenges.

Client Education: Residential

PCOs are often, if not generally, perceived as chemical vendors. Clients, such as homeowners, merchants, and hospitals who call for pest control, usually expect the PCO to treat the problem with chemicals. This image developed after World War II as synthetic organic pesticides became the solution to most pest problems. Although tactics used by PCOs have evolved into a complex set of procedures, including the use of chemicals, the public has retained the antiquated image of the PCO (Levenson and Frankie, 1983). Few homedwellers, commercial clients or PCOs expect the PCO to be an educator.

The role of a PCO as an educator provides advantages for clients and the PCO. If the client comprehends the need for sanitation, mechanical exclusion, and habitat modification, then a pest management program has a better chance of success (Frankie, Granovsky, and Magowan, 1981; Todaro, 1984). Even the best conceived pest management program can be ineffective if a client does not understand that the client's role in a program makes a difference.

Client cooperation leading to more effective pest control translates into fewer complaints and call-backs for the PCO. Levenson and Frankie (1981) report that when Texas homedwellers received appropriate education about the habits of cockroaches, they accepted that occasional cockroaches in their homes did not necessarily constitute a breeding infestation. Instead of spraying immediately, homedwellers were advised to modify their personal habits related to food handling, storage, and disposal, make sanitation improvements, use boric acid selectively, and in some cases set out additional cockroach traps (Piper and Frankie, 1979). These homedwellers were also less likely to attribute such occasional intrusions to inadequate PCO service. An additional, but less obvious benefit to PCOs is the enhancement of their professional image associated with their educational role.

In general, client education can be accomplished directly through (1) individual personal contact (with sales or service personnel) or (2) organized seminars or workshops, and indirectly through (3) distribution of informational written material and (4) advertising. The first two items tend to be the most influential techniques; the third is a useful complement to the first two; and the fourth is important in introducing the company and its concepts to potential clients. Advertising can also be used to publicize the concept of the PCO as a professional pest manager selling expertise and informed service rather than solely chemical applications. The PCO should be portrayed as a consultant as well as a service person (Anon., 1981).

A new and relevant educational service to homeowners for the care and maintenance of ornamental and turf problems has been evolving on an experimental basis at the University of Maryland over the past several years. Under the direction of John Davidson, horticultural students are hired as scouts for detecting, diagnosing, and making corrective recommendations on yard pest problems. The service, which is conducted on a frequent basis during the growing season, is designed to provide homeowners with written and graphic materials (e.g., yard maps of vegetation) for ongoing detections and suggested remedial actions where necessary. The homeowner is obliged to take care of the actual treatments. A general overview and assessment of the information transfer and exchange program is provided in Davidson, Hellman, and Holmes (1981) and in Holmes and Davidson (1984).(See also *J. Arboriculture*, vol. 10, no. 3, 1984).

More recently, John Holmes has moved the research effort into the commercial arena by offering private practitioners software packages for computerized detection, diagnosis, and recommendations. Although this new commercial service offers the latest in technology for ongoing assessment of pest problems and exchange of UPM information, the question of wide acceptance among current practitioners remains an open question (Raupp and Noland, 1984).

The idea that PCOs must play some educational role is, of course, not new. However, education of clients has received insufficient emphasis. One particular difficulty concerns the recognition of clients who will most likely respond to education. Some clients care little about participating in a UPM program, and this kind of attitude does not encourage the PCO to attempt client education, but there are sufficient responsive clients to make education a useful aspect of PCO programs. The PCO must set the tone for this educational exchange as clients tend to be more receptive to service representatives who present a professional appearance and attitude and can intelligently answer questions and provide solutions.

Client Education: Commercial

In dealing with commercial accounts, the PCO has a double problem: he must contact both hourly workers and administrators. No matter how well informed, subordinate individuals usually follow a manager's lead. On the other hand, uninformed workers are unable to successfully implement the policies of knowledgeable managers. Therefore, the success of an effective UPM program can be impeded by either employees or management. Education must occur throughout the chain of command.

Some pest control companies organize seminars and workshops for their clients' employees. Unfortunately, these seminars may involve only lower-level employees and may not be given very high priority by clients. Ideally, workshops should involve all levels of management and should be held regularly. Indeed, it would be appropriate for cities, counties, or states to require employees in certain industries (such as those involving food handling) to participate in approved pest management courses. In this case, cooperative extension, health departments, and other public agencies should take an active role in providing or approving such instruction.

As an adjunct to personal contacts and informational workshops or as part of an advertising campaign, written material can be useful. Currently, a large body of brochures and pamphlets produced by the federal government, cooperative extension, PCO organizations, and manufacturers and distributors of pest control products already exists. Some of these materials are in the public domain and can be freely duplicated and others can be purchased at nominal cost. One PCO in the San Francisco Bay area (Grace, unpub.) has found his customers to be very receptive to a booklet distributed by Cooperative Extension titled "So, you've just had a structural pest control inspection" (Wilcox and Wood, 1980).

A classic example of effective education at the commercial (and residential) levels was recently reported by William Todaro (1984) for an IPM cockroach program in a Pennsylvania public housing project. In this program, Todaro placed considerable emphasis on educating PCOs, managers, and tenants of the project, in addition to testing new insecticides, modifying habitats (cockroach) within units, and developing a routine surveillance schedule. In one phase of the educational program, a booklet on cockroaches received testing (marketing) for effectiveness by the tenants. Todaro summarized the program by stating, "In public housing, pest management is a more realistic goal than pest extermination. But insecticides alone will not control cockroaches. The real key to pest management in these complexes is an involved manager and an aware, informed tenant who knows his responsibilities and who is obliged to be involved in the program." Gene Wood used a similar approach for managing cockroaches in urban housing developments in Maryland (National Research Council, 1980).

Client Education: Retail Sales

Retail distributors of pesticides are in a somewhat different position than PCOs. Usually, a one-time sale of goods is involved rather than a continuing service. However, retail distributors also benefit from consumer education. Informed consumers appear to discriminate in the short term among pest control products, choosing those best suited to certain pests or situations and thus purchasing a greater variety of products (Davidson, Hellman, and Holmes, 1981). More effective pest management and reduced toxicant exposure due to limited and specific pesticide applications are the concomitant benefits to the educated client and to the general public.

Practitioner Education: Initial Certification

Urban pest control has traditionally required little formal education of its practitioners. PCOs commonly entered the field with few skills and received their training on the job. Several texts are available to assist those in the industry with insect identification and choice of control methods (Cornwell, 1973; Ebeling, 1975; Mallis, 1982; Truman, Bennett, and Butts, 1976; Young, 1983), and new information is disseminated through journals, technical bulletins from trade associations, and annual conferences sponsored by the industry or universities.

Increased regulation of the pest control industry has created the need for more formalized approaches to education and training. PCOs in all 50 states are now required to pass written examinations to be certified to commercially apply pesticides. With two current exceptions, these certification programs are administered by state agencies and meet or exceed standards established by the U.S. Environmental Protection Agency (C.F.R., 1982*a*). In Nebraska and Colorado, the U.S. EPA currently examines and certifies PCOs and will continue to do so into 1985, although Colorado recently enacted appropriate regulatory legislation.

Minimum EPA examination standards require aspiring PCOs to demonstrate a broad knowledge of pests; label comprehension; pesticide safety; pest control chemicals, equipment, and application techniques; environmental protection; and applicable state and federal laws and regulations. Some state-administered examinations also require knowledge of business law. Obviously some amount of formal preparation is necessary to pass such a comprehensive examination. It is a rare individual who could acquire all the required knowledge from practical experience alone.

If practical experience alone is no longer sufficient, then where are the employees of today obtaining the training necessary to become the certified PCOs of tomorrow? In a more general sense, where is anyone interested in a career involving urban pest management—whether as a PCO, consultant, employee of a public agency, or even an architect or builder—able to obtain the necessary background?

We addressed this question in a recent study involving state pest control regulatory officials. Although other studies in recent years have queried industry representatives (Frankie, Granovsky, and Magowan, 1981) and the public at large (Frankie and Levenson, 1978; Robinson, 1980; Frankie et al., 1981; NPCA, 1982; Bennett, Runstrom, and Wieland, 1983; Levenson and Frankie, 1981; Byrne et al., 1984) about pest control practices and attitudes, none has collected comprehensive information on state policies and regulations since 1974 (Smythe and Williams, 1974; U.S. Environmental Protection Agency, 1974).

Using the membership list of the Association of Structural Pest Control Regulatory Officials (ASPCRO), a letter of inquiry was sent in February 1984 to the agency in each state charged with regulating structural pest control practices. This letter asked what education and experience were required for PCO licensing or certification, whether education could substitute for any required practical experience, whether continuing education was required for license renewal (recertification), and what sources of instruction were available to potential and established PCOs in that state. This initial inquiry was followed by additional letter and telephone contacts to achieve 100% response (50 states). Some results from this survey are presented in Tables 8-1 and 8-2; others are discussed elsewhere by Grace and Frankie (1985).

There appears to be an assumption among those who regulate the industry that training in UPM can be obtained at colleges and universities. The 23 states that require industry experience of applicants for PCO licensure will accept college-level courses in entomology or related fields in lieu of at least part of that experience (see Table 8-1). Unfortunately, there is currently little justification for this assumption of equivalence, given the lack of urban emphasis in most university pest management curricula. Of ten major U.S. universities (Arizona, Cornell, Louisiana State, Michigan State, North Carolina State, Purdue, Rutgers, Texas A&M, University of California at Berkeley, and University of California at Riverside) censused by us in 1984, only Purdue currently offers a complete pest management program with an urban, rather than agricultural emphasis. Several universities currently are developing curricula with an urban orientation; but most commonly, either a single course in urban pest problems is offered on a regular basis (Arizona, University of California, Riverside) or on an irregular basis (Louisiana State University) or no urban-oriented course is offered (Texas A&M University). However, some of these same universities (Texas A&M University) offer extensive and well-developed curricula in agricultural pest management. The University of California, Berkeley, has taken an intermediate position: although only one regular class in UPM is offered, relevant electives are suggested for interested students. It would appear that those regulating the pest control industry are unaware of the dearth of information on UPM currently offered at the university level, or they might not so readily equate a course in entomology with professional experience.

The educational background necessary for a career in UPM simply cannot be satisfied solely by exposure to courses in agricultural pest management. Although the pest management principles are the same, the ecosystems are different (c.f. Gill and Bonnett, 1973; Stearns and Montag, 1974; Frankie and Ehler, 1978; National Research Council, 1980; Sawyer and Casagrande, 1983). The UPM curriculum of Purdue University recognizes this difference and provides a background in technical and in business areas (accounting and economics). However, even this well-developed curriculum places little emphasis on the distinguishing feature of UPM: the human element. We maintain that UPM students should also be exposed to course work in urban planning and development, public policy, business and marketing, landscape

Table 8-1. Number of states (50 respondents) with specific requirements for licensing/certification of urban pest control operators.

Requirement	*Number of states*
Examination	50[a]
Completion of specific state-approved training	3[b]
Experience	23[c]
Continuing education/retraining required for license renewal or recertification	40[d]

[a]Currently administered by U.S. Environmental Protection Agency in Colorado and Nebraska.
[b]California and Rhode Island (all categories), New Mexico (termite only).
[c]Education accepted in place of at least some experience in all states.
[d]Alaska requires re-examination.

architecture, and communications. These social-oriented courses provide the necessary background for dealing professionally with at least some of the human aspects of pest problems. To be aware at the outset of the need for appropriate social input may also provide for substantial long-term savings in dollars and time on the part of all concerned parties (Merritt, Kennedy, and Gersabeck, 1983; Roberts and Dill, 1983).

Extension programs, considered bridges between the university and the public, are sources of training in UPM. Indeed, 13 of the states responding to our ASPCRO survey did specifically mention the Cooperative Extension Service as a source of training materials or courses on pest control practices for those entering the field (see Table 8-2). On the other hand, 5 of those responding specifically mentioned that the Extension Service does not offer any training in UPM in their state. Is this disparity another manifestation of confusion over the nature of the Extension Service commitment to the urban public?

Closer ties between experiment stations and extension personnel and the state agencies regulating urban pest control practices would be of mutual benefit. Effective training can only be provided if regulatory officials communicate their goals for the industry to educators. This relationship could readily be formalized by officially designating specific extension personnel as advisers to state agencies and contact persons for the urban pest control industry. Such an arrangement might necessitate the creation of new extension positions in some states, but perhaps expenses could be shared among the entities involved in this collaborative effort. Realistically, the pest control industry must be expected to assume part of the financial burden, perhaps through their state and national trade associations.

On a brighter note, it should be mentioned that the pest control industry itself has taken the initiative in creating training programs. Industry training efforts can be categorized as: (1) in-house training programs organized by

Table 8-2. Number of state pest control regulatory officials (50 respondents) mentioning specific sources of initial training and retraining materials/courses.

	Number of states	
Training source	*Initial licensing/ certification*	*License renewal/ recertification*
Cooperative extension	13[a]	16
Pest control industry	3	7
Purdue University correspondence course	2	1
Colleges	1[b]	1[c]
State regulatory agency	3	7

[a]Five regulatory officials specifically mentioned that the cooperative extension service does not offer training in their states.
[b]Community colleges in New York State.
[c]University of Kentucky.

pest control companies for their own employees; (2) training seminars or materials prepared by manufacturers and distributors of pest control chemicals and supplies; (3) training seminars or materials sponsored by trade associations; and (4) courses offered to the industry at large by private purveyors of educational materials.

In an industry consisting of many small businesses, only the larger pest control companies can justify the expense of organizing thorough in-house training programs. Many smaller companies, however, are able to take advantage of the programs prepared by distributors of pest control supplies. These educational programs, originally initiated as a service to their customers, appear to have grown into a minor industry of their own. For example, at least one western distributor (Van Waters and Rogers) now offers a separate catalogue of 35mm slides of urban pests and their biology. In several states, this "minor industry" has advanced a step further with the appearance of private businesses devoted exclusively to presenting courses and distributing educational materials.

New Mexico, Rhode Island, and California are the only states currently requiring that applicants for PCO licensure complete specific state-approved training courses regardless of their experience or educational background. In New Mexico eight hours of approved training are required for those wishing to become certified in termite control, while in Rhode Island all applicants must complete a one and a half day workshop in conjunction with the required examination. California PCO license applicants must complete state-approved courses in pesticides, pest identification and biology, contract law, rules and regulations, business practices, and (depending upon the type of license desired) fumigation safety or construction repair.

The intent of the strict educational requirements in California is to establish a common information base and to ensure that all PCO license applicants have a minimum amount of appropriate training (W. W. Wilcox, pers. comm.). Along with the materials available from private suppliers, self-paced extension courses have been developed to meet state criteria. However, since the Structural Pest Control Board of the state of California has no formal advisory relationship with professional educators, a quality control problem exists. It is difficult to ensure effectiveness and uniformity among the available training programs. In fact, there may be basic differences in the precepts of those providing training: the self-paced extension courses were created independently on a contract basis by university personnel, while private providers of educational materials are usually associated with the pest control industry. The former may be overly academic, and the latter excessively concerned with practical and business aspects. A formal coordinator (possibly an extension employee) is needed to provide structure to this rather chaotic situation, to match the needs of the industry and regulatory agency with the facilities of the university.

Practitioner Education: Recertification

In establishing criteria for federal approval of state regulatory programs, the U.S. EPA (C.F.R., 1982*b*) states that "the state plan should include ... provisions to ensure that certified applicators continue to meet the requirements of changing technology and to assure a continuing level of competency and ability to use pesticides safely and properly." Currently, 40 states require individuals applying pesticides commercially to attend periodic retraining sessions (variously designated as courses, seminars, or workshops) or to be re-examined to renew their certification. This general requirement conforms to EPA guidelines for the recertification of pesticide applicators at the federal level (on federal property). However, rigor and renewal periods vary substantially from state to state. Maryland, for example, requires attendance at one training seminar annually; Michigan requires attendance at one seminar every 3 years; Alabama requires attendance at two training sessions every 3 years; New Mexico requires completion of 8 hours of training every 5 years; and Maine requires 15 hours of training every 5 years.

Although continuing education (or re-examination) is required by most states, federal regulations do leave open the possibility of other options. For example, Tennessee offers PCOs three recertification choices—attendance at a workshop, re-examination, or presentation of an affidavit stating that the individual has received and is familiar with current informational material issued by the state Department of Agriculture. Since recertification procedures are suggested by the EPA rather than mandated, we would expect to see even more liberal alternatives in states traditionally wary of over-regulation.

Of course, the value of even the strictest educational regulations will depend entirely on the effectiveness of the educational programs offered.

Programs intended to meet recertification criteria are offered by the same variety of public and private sources as offer initial training (see Table 8-2). Once again, coordinators with a background in education could play a valuable role in maintaining the quality of these various educational efforts.

Practitioner Education: the Future

If stricter educational requirements and continuing education in UPM are to realize their potential as dynamic educational processes rather than simply as static bureaucratic requirements, university-based educators must take a leadership role. Cooperative extension (or other university) professionals should be specifically designated as advisers/consultants to the urban pest management industry, advisers/consultants to state (and federal) regulatory agencies, and coordinators on a state-wide basis of UPM educational efforts.

University UPM programs must be developed that focus on the diversity of urban ecosystems and recognition of UPM's unique socioeconomic and psychological elements. Universities have traditionally led the way in the development and promotion of new technologies. However, educational programs tailored to fit legislated regulations are inherently static, with the aim of maintaining the status quo at a certain acceptable level. If UPM is recognized as a dynamic and changing field, then educational programs should lead, rather than follow, industry trends and legislative regulatory efforts. This outcome can only be accomplished through marketing management approaches to analysis of industry and regulatory needs, implementation of appropriate educational "answers" to these needs, and continual reassessment of all factors impinging on the resultant educational exchange.

SUMMARY RECOMMENDATIONS

1. Research specialists developing urban pest management programs should make an active effort to include education and information exchange components into their management schemes. These components, which should be planned at the outset of the research program, should be built around an understanding of the needs, wants, expertise, limitations, and competing priorities of the relevant audiences. In the few urban IPM programs where this approach was adopted, the results were highly satisfactory (Fear et al., 1983; Todaro, 1984). In some cases where these components were not planned and developed at the outset, final implementation of research results was significantly impaired (Merritt, Kennedy, and Gersabeck, 1983).

2. New ways of aggressively reaching the public with UPM information should be researched and implemented by the Cooperative Extension Service. It has been clear for many years that current methods are not adequate for the task of effectively transferring and exchanging extension information (Kielbaso and Kennedy, 1983). Extension services interested in exploring the marketing approach for developing new UPM educational materials should seriously consider contracting with or hiring marketing specialists. Cooperative arrangements with marketing faculty at universities also offer a viable means for entering this field.

3. Appropriate curricula should be developed at land grant institutions offering UPM as one track of specialization in their pest management programs. Modern curricula in UPM are needed to better prepare university-trained practitioners for dealing with urban pests and the people-oriented environment the pests occupy. One additional benefit of such training may be the preparedness for dealing with pest problems before they occur, that is, by participation in the urban planning and development process. Excellent examples of how well-trained UPM practitioners may become involved as consultants to urban development projects are presented in Roberts and Dill (1983) with regard to mosquito and forest pest problems.

4. Professional urban pest managers should be encouraged to assume a greater role as educators as part of their general pest control service. Educational materials for distribution to clients and, more important, training in effective communication and application of educational principles should be provided to practitioners by both public institutions (e.g., land grant universities and cooperative extension) and industry associations. Agencies licensing PCOs should credit and, ideally, require such training as part of the continuing education needed for renewal of certifications and licenses.

5. In states where land grant university personnel do not already have formal and well-defined relationships with state regulatory agencies and the UPM industry, formal liaison positions should be developed. These should be permanent faculty or extension positions with specific responsibilities extending beyond occasional consultation services. Where new positions must be created, financing could be arranged jointly by the parties involved, possibly through allocation of a portion of professional licensing fees. Responsibilities would include: development and regular review of certification and recertification educational (and examination) criteria; development and critical review of curricula and educational materials intended to satisfy these criteria; coordination of university UPM curricula with industry (and public) needs by means such as the development of UPM student intern programs; and development of programs and materials to foster and facilitate client education by PCOs and other urban pest managers.

ACKNOWLEDGMENTS

L. H. Williams, C. E. Poindexter, J. P. LaFage, and N. Ehmann provided helpful background information for this chapter. State regulatory personnel from all states kindly cooperated with our national survey effort. C. S. Koehler, V. R. Lewis, and R. Smith reviewed an early draft of the paper.

REFERENCES

Anonymous. 1981. Purdue conference report. *Pest Control* **49**(4).

Ascerno, M. E. 1981. Diagnostic clinics: More than a public service. *Bull. Entomol. Soc. Am.* **27**:97-101.

Bennett, G. W., E. S. Runstrom, and J. A. Wieland. 1983. Pesticide use in homes. *Bull. Entomol. Soc. Am.* **29**(1):31-38.

Byrne, D. N., E. H. Carpenter, E. M. Thoms, and S. T. Cotty. 1984. Public attitudes toward urban arthropods. *Bull. Entomol. Soc. Am.* **30**(2):40-44.

Code of Federal Regulations. 1982*a*. Certification of Pesticide Applicators. Title 40, pt. 171.

Code of Federal Regulations. 1982*b*. Maintenance of State Plans. Title 40, pt. 171.8.

Cornwell, P. B. 1973. *Pest Control in Buildings.* Hutchinson, London.

Cremin, L. 1978. *Traditions of American Education.* Basic Books, New York.

Davidson, J., J. L. Hellman, and J. Holmes. 1981. Urban ornamentals and turf IPM. In *Proceedings of Urban Integrated Pest Management Workshop,* G. L. Worf (ed.). National Cooperative Extension Service, Dallas, Texas, pp. 68-72.

Ebeling, W. 1975. *Urban Entomology.* Division of Agricultural Sciences, University of California, Los Angeles.

Extension Committee on Organization and Policy. 1981. *Urban Integrated Pest Management.* Cooperative Extension Service, University of Georgia, Athens.

Farace, R. 1980. *Information Transfer in Urban Pest Management.* Prepared for Committee on Urban Pest Management, Environmental Studies Board, National Research Council. National Academy Press, Washington, D.C.

Fear, F. A., G. A. Simmons, M. T. Lambur, and B. O. Parks. 1983. A community development approach to IPM: Anatomy of a pilot effort to transfer IPM information on outdoor vegetation to suburban homeowners. In *Urban Entomology: An Interdisciplinary Approach,* G. W. Frankie and C. S. Koehler (eds.). Praeger, New York, pp. 127-150.

Frankie, G. W., and L. E. Ehler. 1978. Ecology of insects in urban environments. *Annu. Rev. Entomol.* **23**:367-387.

Frankie, G. W., and C. S. Koehler, eds. 1983. *Urban Entomology: Interdisciplinary Perspectives.* Praeger, New York, 496p.

Frankie, G. W., T. A. Granovsky, and C. Magowan. 1981. Study of attitudes and practices of pest control operators towards pests and pesticides in selected urban areas of California, Texas, and New Jersey. In *Proceedings of Urban Integrated Pest Management Workshop,* G. L. Worf (ed.). National Cooperative Extension Service, Dallas, Texas, pp. 9-32.

Frankie, G. W., R. M. Mandel, H. Levenson, and T. A. Granovsky. 1981. Survey of the arthropod pests and measures to control them in three metropolitan areas. In *Proceedings of Urban Integrated Pest Management Workshop,* G. L. Worf (ed.). National Cooperative Extension Service, Dallas, Texas, pp. 33-67.

Frankie, G. W., J. B. Fraser, and V. R. Lewis. 1982. Suggestions for implementing survey research results that deal with public attitudes and behavior towards arthropod pests in urban environments. In *Proceedings on Urban and Suburban Trees: Pest Problems, Needs, Prospects, and Solutions,* B. Parks, F. Fear, M. Lambur, and G. Simmons (eds.). Department of Resource Development and Department of Entomology, Michigan State University, East Lansing, pp. 174-179.

Gill, D., and P. Bonnett. 1973. *Nature in the Urban Landscape: A Study of City Ecosystems.* York Press, Baltimore.

Grace, J. K., and G. W. Frankie. 1985. Certification standards: A national overview. *Pest Control* **53**(5):46, 48, 50.

Harrington, F. H. 1977. Agricultural extension. In *The Future of Adult Education: New Responsibilities of Colleges and Universities.* Jossey-Boos, San Francisco.

Holmes, J. J., and J. A. Davidson. 1984. Integrated pest management for arborists: Implementation of a pilot program. *J. Arboric.* **10:**65-70.

Kielbaso, J. J., and M. K. Kennedy. 1983. Urban forestry and entomology: A current appraisal. In *Urban Entomology: Interdisciplinary Perspectives,* G. W. Frankie and C. S. Koehler (eds.). Praeger, New York, pp. 423-440.

Koehler, C. S. 1983. Information transfer in urban pest management. In *Urban Entomology: Interdisciplinary Perspectives,* G. W. Frankie and C. S. Koehler (eds.). Praeger, New York, pp. 151-160.

Kotler, P. 1972. A generic concept of marketing. *J. Marketing* **36:**46-54.

Kotler, P. 1984. *Marketing Management: Analysis, Planning and Control,* 5th ed. Prentice-Hall, Englewood Cliffs, N.J.

Kotler, P., and G. Zaltman. 1971. Social marketing: An approach to planned change. *J. Marketing* **35:**3-12.

Lambur, M. T. 1983. An evaluation of Project Pest's pest management manual as a means for transferring pest management information to urban homeowners. Ph.D. diss., Michigan State University, East Lansing.

Lambur, M. T., B. O. Parks, F. A. Fear, and G. A. Simmons. 1982. Taking public attitudes and practices into account: Pest management in a Michigan suburban community. In *Proceedings on Urban and Suburban Trees: Pest Problems, Needs, Prospects, and Solutions,* B. Parks, F. Fear, M. Lambur, and G. Simmons (eds.). Department of Resource Development and Department of Entomology, Michigan State University, East Lansing, pp. 151-173.

Levenson, H., and G. W. Frankie. 1981. Pest control in the urban environment. In *Progress in Resource Management and Environmental Planning,* vol. 3, T. O'Riordan and R. K. Turner (eds.). Wiley, New York, pp. 251-272.

Levenson, H., and G. W. Frankie. 1983. A study of homeowner attitudes and practices toward arthropod pests and pesticides in three U.S. metropolitan areas. In *Urban Entomology: Interdisciplinary Perspectives,* G. W. Frankie and C. S. Koehler (eds.). Praeger, New York, pp. 67-106.

Mallis, A., ed. 1982. *Handbook of Pest Control,* 6th ed. Franzak and Foster, Cleveland, 1,104p.

Merritt, R. W., M. K. Kennedy, and E. F. Gersabeck. 1983. Integrated pest management of nuisance and biting flies in a Michigan resort: Dealing with secondary pest outbreaks. In *Urban Entomology: Interdisciplinary Perspectives,* G. W. Frankie and C. S. Koehler (eds.). Praeger, New York, pp. 277-299.

National Pest Control Association. 1982. Results of NPCA consumer affairs committee's national opinion survey: The public appraises the pest control industry. *Pest Manage.* **1:**8-12.

National Research Council. 1980. *Urban Pest Management.* Report of the Committee on Urban Pest Management, Environmental Studies Board, Commission on Natural Resources. National Academy Press, Washington, D.C., 272p.

Peters, T. J., and R. H. Waterman, Jr. 1982. *In Search of Excellence.* Warner Books, New York.

Piper, G. L., and G. W. Frankie. 1979. *Integrated Management of Urban Cockroach Populations,* Final Report EPA Grant No. R805556010. Environmental Protection Agency, Washington, D.C.

Raupp, M. J., and R. M. Noland. 1984. Implementing landscape management programs in institutional and residential settings. *J. Aboric.* **10:**161-169.

Roberts, F. C., and C. H. Dill. 1983. Urban planning and insect pest management. In *Urban Entomology: Interdisciplinary Perspectives,* G. W. Frankie and C. S. Koehler (eds.). Praeger, New York, pp. 41-66.

Robinson, W. H. 1980. Homeowner knowledge of wood-infesting insects. *Melsheimer Entomol. Series* **29:**48-52.

Sawyer, A. J., and R. A. Casagrande. 1983. Urban pest management: A conceptual framework. *Urban Ecol.* **7:**145-157.

Schramm, W. 1973. *Men, Messages, and Media: A Look at Human Communication.* Harper and Row, New York.

Seay, M., ed. 1983. *Adult Education: A Part of a Total Educational Program: A Description of the Educational and Training Program of the TVA,* Bulletin of the Bureau of School Service, vol. 10, no. 4. College of Education, University of Kentucky, Lexington.

Smythe, R. V., and L. H. Williams. 1974. *Structural Pest Control Regulations,* Forest Research Service Paper SO-93. U.S. Department of Agriculture, Washington, D.C.

Stearns, F. W., and T. Montag, eds. 1974. *The Urban Ecosystem.* Dowden, Hutchinson & Ross, Stroudsburg, Pa.

Tassan, R. L., K. S. Hagen, and D. V. Cassidy. 1983. Imported natural enemies established against iceplant scales in California. *Calif. Agric.* **36**(9-10):16-17.

Todaro, W. 1984. Public housing: The pest control industry's ultimate challenge. *Pest Control Technol.* **12**(8):61, 62, 64, 76.

Truman, L. C., G. W. Bennett, and W. L. Butts. 1976. *Scientific Guide to Pest Control Operations,* 3rd ed. Purdue University/Harvest, Cleveland, 276p.

U.S. Environmental Protection Agency. 1974. *Digest of State Structural Pest Control Laws.* Washington, D.C.

Washburn, J. O., and G. W. Frankie. 1985. The biology of iceplant scales, *Pulvinariella mesembryanthemi* and *Pulvinaria delottoi* (Homoptera: Coccidae) in California. *Hilgardia* (in press).

Wilcox, W. W., and D. L. Wood. 1980. *So, You've Just Had a Structural Pest Control Inspection,* Leaflet No. 2999. Division of Agricultural Sciences, University of California, Berkeley, 15p.

Worf, G. L., ed. 1981. *Proceedings of Urban Integrated Pest Management Workshop.* National Cooperative Extension Service, Dallas, Texas, 176p.

Yankelovich, D. 1964. New criteria for market segmentation. *Harvard Bus. Rev.* **64213**(Mar.-Apr.):143-150.

Young, E. D. 1983. *Pest Control Training Manual.* Educational Services, Garden Grove, Calif.

9

Pest Management in Vegetables and Fruit

Alan C. York

Purdue University
West Lafayette, Indiana

HISTORY AND EVOLUTION OF GARDEN PEST CONTROL

When mankind first began using plants is unknown. At some point in history, before the cultivation of grain, perhaps even before the collection of grain seeds, humans were collecting berries and other small fruits, digging roots and other storage organs, and eating succulent stems of plants. The presence of insects or other pests in or on the selected food item probably made little difference. In fact people probably were also collecting and eating insects. Just as humans were not very choosy about the condition of the meat they were able to kill or scavenge, neither were they particular about the condition of edible plant material. With the exception of a few plant disease organisms and a few insect species most plant pests, short of being unsightly, will not harm the person eating the affected fruit or vegetable. Some pests, however, produced great discomfort and even death. It must also have been recognized very early that certain pests reduced the yield, lowered the quality, or shortened the storage life of edible plants. It was also recognized (Harpaz, 1973) that certain pests were unavoidable and that buyers must be prepared to accept some level of contamination. The Talmud says that a buyer must be prepared to accept 10% wormy figs. It goes on to say "if a man finds a hair and a fly in the food cooked by his wife, the fly albeit disgusting is excused since it is not her fault, but the hair is inexcusable and may become grounds for divorce."

We know also that certain pests were serious competitors for produce, and many examples of remedies are available from the writings of early scholars (Jones, 1973; Mayer, 1959). As reliance on certain kinds of produce increased, so did the seriousness of their pests. Perhaps the greatest example of this is the massive economic and sociological effects of the potato late blight epidemic of Ireland in the late 1840s. For the peasants of the middle ages, most of their caloric intake was from the produce they gathered or grew. Meat was a scarce item; poachers were hung. Vegetables of antiquity such as lettuce (Herodotus mentions it being consumed as early as 500 B.C.) (Sanders, 1919), onions (note that in the King James Version of the Bible in Numbers 11.5, onions, cucumber, melons, leeks, and garlic were among the foods the Israelites wandering in the wilderness longed for), cabbages, turnips, carrots, parsnips, and the more recently adopted potatoes, beans, tomatoes, and maize all have their serious pests, as do apples, peaches, and other fruits. Pests competed with humans for their cultivated, and often less hardy than wild, produce.

To the new world, immigrants brought their favorite staples, and encountered an abundance of pests to colonize them. Contend with pests, they did, by all manner and means available. To the present day considerable folklore and hand labor has been utilized in an attempt to protect crops from pests. My parents began our family in the early 1920s. Raising our rather large family in a rural area, my mother speaks of it being necessary to keep several of my older brothers and sisters busy for an entire day picking Colorado potato beetles and "potato bugs" (blister beetles) from potato plants.

Garden practices changed, however, as family size dropped from 5.6 persons per family in 1850 to 3.27 in 1981 (Anonymous, 1950). Affluency and technology brought changes in food procurement and eating habits. Vegetable consumption per capita dropped from a high of 308 pounds (265 fresh, 43 canned) in 1945, to a low of 156 pounds (106 fresh, 43 canned, 7 frozen) in 1960. Likewise, potato consumption dropped from 197 pounds per person in 1910 to 110 pounds in 1960. Since then, emphasis on nutrition and healthy bodies and, in the opinion of some commodity groups, e.g., the American Potato Council, advertising has resulted in an increase in vegetable and potato consumption. Vegetable consumption in 1981 amounted to 164 pounds per person, and potato consumption, 131 pounds (Anonymous, 1983). The recent introduction of baked potatoes in fast food restaurants will likely increase this figure significantly.

The result of this affluence and technology was apparently less dependence on home gardening. Time could be spent more enjoyably in other activities. Motivation for gardening, an ancient pasttime (Janick, 1982), was changing. Relf (1982) quotes an ancient Chinese proverb: "if you would be happy for a day, kill a pig. If you would be happy for a week, take a wife. If you would be happy for a lifetime, plant a garden." Households with gardens climbed from 39% in 1971 to a high of 49% in 1975 (high inflation, food costs, and fuel shortages). By 1978 this had dropped to 41%. Approximately equal percentages of people garden for recreation and relaxation and for budgetary or money-saving reasons (Davies, 1979). Those having gardened for longer periods of time placed less emphasis on the tangible benefits of harvesting (Kaplan, 1973). Almost half the gardens are under 550 sq ft, with another 25% between 550 sq ft and 2,400 sq ft (Davies, 1979).

History of Pesticide Use

People who garden are the audience toward which urban pest management (UPM) education is directed. Two generations of gardeners have been raised on produce mandated by law to be insect- and disease-free. They have learned to protect the "fruits of their labor" with a powerful arsenal of chemicals, largely developed for the commercial vegetable and fruit production industries. They have little knowledge of the labor-intensive gardens of a few decades ago and of the moderate to ineffective chemical remedies. Bordeaux implies a wine-growing region in France, not a 2:3 mixture of copper sulfate and hydrated lime in water. First used in France in 1882 for control of theft of grapes because of its distasteful appearance, Bordeaux mixture was serendipitously discovered to also control downy mildew on grapes. For a century it has been widely used to prevent many vegetable diseases and, when used in combination with other materials, as a repellent or control for such insect pests as flea beetles, potato leafhopper, cucumber beetles, and squash vine borer. Another common fungicide was Burgundy mixture: powdered copper sulfate, washing soda (sodium carbonate), and water. While sulphur has been known as a pesticide since about 2500 BC (Mayer, 1959), lime-sulphur, appearing in the 1880s, did not gain prominence until the early 1900s. Similarly, most gardeners of today are unaware that arsenic, in the forms of arsenious oxide or white arsenic (As_2O_3), arsenic oxide (As_2O_5), lead arsenate, calcium arsenate, and zinc arsenite were the mainstays of fruit and vegetable insect control for nearly a century. Paris green, a mixture of white arsenic, copper oxide, and acetic acid, although very phytotoxic, was used extensively for control of Colorado potato beetle, cabbage caterpillars, and other serious vegetable insects (Herrick, 1925). Consumers were emphatically cautioned to remove the outer wrapper leaves of cabbage in order to avoid the arsenic and lead accumulated after repeated applications to the growing crop. The first half of the twentieth century saw emulsions of carbolic acid (phenol) and bichloride of mercury poured over young radish, turnip, and cabbage plants in attempt to prevent destruction by root maggots. As many as seven applications of bichloride of mercury for control of onion maggot were recommended as late as 1937 (Gilbert and Caffrey, 1937). Another inorganic compound highly toxic to humans and widely used in the early decades of this century was sodium fluoride.

Pest control compounds, effective yet safe for humans, were rare. Several products of botanical origin were common. Nicotine, in the form of pulverized plant parts or more recently as nicotine sulphate or as a free nicotine extract, has had wide usage against sucking insects for centuries. Pyrethrum and rotenone (from several different plants) became popular for control of garden insects in the late 1920s and early 1930s. Neither of these three insecticides, however, provided any residual activity on crops, so that repeated applications were necessary and expensive.

Chemicals for weed control followed a similar pattern. In the 1890s chemicals such as copper salts, iron sulfate, sodium chlorate, sulfuric acid, sodium arsenite were used in attempt to reduce competition from weeds (Ahlgren, Klingman, and Wolf, 1951). Equipment corrosion, fire and explosion hazard, high per acre costs, and only moderate weed control temporarily reduced the emphasis on chemical weed control. The discovery in 1938 of Sinox (4,6-dinitro-o-cresol, sodium salt), a selective broad-leaved plant killer, renewed the interest in chemical weed control. Research began anew on the uses of sulfuric acid, sodium chloride, and oils. In the midst of World War II, the chemicals 2,4-D and DDT provided great promise for control of both weeds and insects. Wartime and postwar research indicated near miraculous powers for newly developed pest-mitigating materials. Although stated decades earlier, Lodeman (1903) generally expressed the feelings of the late 1940s and early 1950s, with his statement:

> The best is generally the most profitable commodity, and the poorest is the least so; and the grower of today has it in his power to produce the best. It rests entirely with him whether his apples shall be wormy or not, whether his trees shall retain their foliage or lose it from disease. There are few evils that affect his crops which he cannot control, in many cases almost absolutely. Only a few diseases remain which still refuse to submit to treatment, but the number is rapidly decreasing, and the time will come when these also will disclose some vulnerable point which will allow for their destruction. Foremost among the operations by means of which cultivated plants are protected from their enemies is spraying.

In just a few years, a revolution had occurred in insect control benefiting commercial agriculture and was beginning in disease control (Table 9-1). Furthermore, new emphasis was being placed on plant breeding for insect and disease control. Weed control was moving more slowly, but weed scientists and farmers were hopeful. A major textbook that appeared in 1951 (Ahlgren, Klingman, and Wolf, 1951) devoted much of its contents to 2.4-D and its proper role in weed management and also discussed new weed

Table 9-1. Materials recommended for insect and disease control in vegetables in 1944 and 1951.

1944[a]		*1951*[b]	
Insects	*Diseases*	*Insects*	*Diseases*
Rotenone	Bordeaux mixture	Rotenone	Bordeaux mixture
Pyrethrum	Copper-lime dust	Pyrethrum	Fixed-copper dusts
Nicotine	Fixed-copper dusts	Nicotine	Ziram
Calcium arsenate		DDT	Ferbam
Cryolite		DDD	Zineb
		Methoxychlor	Nabam
		Lindane	Thiram
		Chlordane	Spergon
		Aldrin	Orthocide 406
		Dieldrin	
		Dilan	
		Toxaphene	
		Parathion	
		TEPP	

[a]Data from Pratt, Blauvelt, and Dimock, 1944.
[b]Data from Chupp and Leiby, 1951.

control materials. It emphasized the role of weeds in reducing yields through competition for light, moisture, and nutrients and mentioned other losses such as cost of cultivation, lowered quality of crops, lowered quality of animal products, and animal losses due to poisonous weeds, protection of insects, aids to diseases, and the high costs of herbicides. Appropriately, it also emphasized the role of mechanical control (hoeing, tillage), burning, crop competition, mulching, crop rotation, and biological control, Today, while chemical weed control products are experiencing the rapid growth that insecticides did in the 1960s, nonchemical measures continue to play a major role in weed control.

Mechanical, biological, cultural, and genetic intervention are no less a part of disease and insect management in the urban garden. Plant disease management means interfering in the pathogen/host relationship in one of three ways: reducing the amount of inoculum available to initiate new infection (by attacking the pathogen or preventing its introduction to the garden); reducing the host susceptibility to either or both infection and severe damage from diseases (by careful selection of resistant, i.e., less susceptible, hosts); and by imposing barriers, e.g., fungicides, to the successful establishment of the pathogen on the host (Aldwinckle and Beer, 1979). Unlike many insect and weed pests that may be eliminated once present, most disease management relies on prevention of establishment or slowing down the spread of diseases. Traditionally, most plant diseases from pathogenic organisms were considered incurable. However, chemicals have recently been developed that show promise in doing so. It is hoped that these will not place less reliance on nonchemical methods as happened with insects and weeds.

PEST SEVERITY

In order to properly manage any pest, its biology and environmental requirements (food, hosts, overwintering sites, weather conditions favorable for growth and survival, movement, phenological occurrence) must be thoroughly understood. Many pests are present on a more or less predictable schedule. In addition, their potential for damage is known in most cases, and hence, the necessity for control is well established. Frequency of occurrence as well as severity will differ from region to region throughout the United States. What is appropriate and correct for the Midwest and Northeast (Table 9-2) will not, in all cases be correct for the rest of the country. In addition, normal fluctuations in pest populations will sometimes result in extreme differences in pest occurrence and severity.

Many pests that fall into the annual category actually have several generations per year; thus the potential for serious damage may occur almost continually throughout the growing season. It can be debated at length into which category certain pests, such as Colorado potato beetle, twospotted spider mite, or cabbage looper are placed. To some degree, certain choices reflect the regional bias of the author, and others reflect recent discovery and judgments on damage thresholds. Potato and tomato are both seriously defoliated in some areas of the United States by the potato beetle, particularly in the northeastern United States, where insecticide resistance is prevalent; however, both plants can lose a significant amount of foliage, up to 50% in many cases, at various growth stages without significant effect on yield (Cantelo and Cantwell, 1982; Cranshaw and Radcliffe, 1980; Ferro, Morzuch, and Margolies, 1983; Schalk and Stoner, 1979; Tamaki and Butt, 1978). The same is true for cabbage looper. Cabbage, broccoli, cauliflower, and so on need protection only just prior to harvest (Chalfant et al., 1979; Sears, Jaques, and Laing, 1983; Shelton et al., 1983). Conversely, late blight of potato, early blight of tomato, and bacterial wilt of cucurbits may cause serious damage or death to plants at any time. In part, these somewhat arbitrary rankings reflect both likely and potential damage or loss of yield.

However, yield loss is not always the appropriate criterion by which to designate pest status. For many gardeners the primary reason for gardening is not production, and loss of yield is not of primary concern. Those who garden for relaxation or recreation or for the aesthetics of a fine garden may find the presence of pests at any level offensive. For this group, pest manage-

Table 9-2. The nature of occurrence and the degree of severity of damage of some garden and fruit tree pests in Indiana.

	Frequency of occurrence	
Seldom	*Occasional*	*Annual*
Greatest severity	European corn borer	Striped cucumber beetle
Late blight (potato and tomato)	Lesser peach tree borer	Apple maggot
	Onion maggot	Cabbage maggot
	Peach tree borer	Codling moth
	Radish maggot	European red mite
	Squash vine borer	Plum curculio
	Fire blight	Potato leafhopper
	Downy mildew cucurbits	Spotted cucumber beetle
	Early blight (tomato)	Fusarium wilts
	Blossom end rot (tomato)	Verticillium wilt
		Apple scab
		Bacterial wilt (cucurbits)
		Powdery mildew (cucurbits)
		Brown rot of stone fruits
Moderate severity	Squash bug	Cabbage looper
Twospotted spider mite	Diamondback moth	Colorado potato beetle
Variegated cutworm	Mexican bean beetle	Corn earworm
Wireworms	Oystershell scale	Imported cabbageworm
Tobacco mosaic	Green peach aphid	Peach leaf spot
	Seed corn maggot	Anthracnose (tomato)
	Powdery mildew apple	
Least severity	Potato aphid	Apple aphid
Tomato hornworm	Corn rootworm adults	Asparagus beetle
Cabbage aphid	Rhizoctonia (potato)	Bean leaf beetle
Greenhouse whitefly	Smut (sweet corn)	Flea beetles
Redbanded leafroller		Oriental fruit moth
San Jose scale		
White grubs		
Tobacco hornworm		

ment strategies too often include, first and foremost, chemicals. It would be desirable to educate this group of gardeners to adopt the philosophy of pest management as a challenge, utilizing chemicals in gardening only where absolutely necessary. For many gardeners, it is likely that the rewards obtained from managing pests appropriately as an integral part of the gardening activity can be as great as those obtained from the vegetable production itself.

Sanitation, crop rotation, resistant and regionally adapted varieties, disease and insect-free planting stock, well-chosen horticulturally desirable planting sites, appropriate planting and harvesting schedules, trapping, and other cultural and mechanical techniques are primary tools in pest management. Pesticides are necessary for proper management of more fruit and vegetable pests than is desirable. However, in any program choice of chemical, timing of application, and method of application must be appropriate to the pest and crop involved.

PRINCIPLES OF INTEGRATED PEST MANAGEMENT

Five commonly accepted principles must be recognized if the philosophy of pest management in the true sense of the definition is to exist.

1. *Identification of the pest.* A seemingly self-evident statement, many gardeners treat gardens with little knowledge of what insect or disease is involved on the crop (Gardner and York, 1982).

2. *Identification of the management unit.* What are the boundaries of the pest? What crops are likely to be involved? What areas need to be treated or protected? From where and how might a particular pest come to invade a crop?

3. *Establishment of an economic (action) or aesthetic threshold.* Recognizing that economics in the true sense of the word do not play a major role in gardening, some will have difficulty with this item. There is for any pest, a density above which the pest is too numerous, too visible, or too damaging for the gardener. Furthermore, for most pest/crop relationships the economics has not been sufficiently researched to provide an "economic" threshold. For many of these relationships, sufficient research has been conducted to establish an "action" threshold. The action threshold is that level of insect, disease, or weed density in a crop shown to result in significantly reduced yields or quality. Realistically, there are pests so destructive that their mere presence is intolerable, e.g., late blight, fire blight, striped cucumber beetle (with bacterial wilt). But for most pests there is some level that is tolerable. Each apple need not be free of apple scab nor codling moth. Furthermore, it is impossible to set thresholds for most pests that will satisfy all gardeners. The thresholds are those established for vegetables and fruit under commercial production. Such standards are in some instances unrealistically high, for some home gardeners unaesthetically low, i.e., they may permit more injury than is desirable though not reducing yield or quality.

4. *Development of a management strategy.* It is not sufficient to wait until a pest is present and troublesome before taking action. For many of the most serious pests, action begins with the choice of planting site, variety chosen, source of planting stock, and even irrigation scheduling. Develop your arsenal and modify it as needed. Most important, keep good records of what is done and how well it worked.

5. *Utilization of appropriate monitoring or sampling techniques.* While some might include this as part of the management strategy, it is important enough to set apart for emphasis. With many of the insects involved, chemical control is necessary. Untimely control may be useless and may cause more harm than good. Prediction of occurrence or actual monitoring of occurrence is of critical importance in the management of many serious pests. In the case of less serious pests (moderate or light, see Table 9-2), simple occasional observation may provide sufficient information and lead time for implementation of desired management activity.

Monitoring should not be construed to be only a pre-pest occurrence activity. The degree of control obtained by a particular method and the need for additional activity are only ascertained by proper monitoring of pest populations. In some cases, a particularly appropriate strategy may not be possible until the following growing season.

It is best perhaps to recognize that monitoring or sampling consists of three types of activity: information sampling; problem diagnosis; population estimation or damage estimation.

Information Sampling

Information sampling and problem diagnosis are the two activities a pest management practitioner is likely to be involved with. Information sampling, often classified as either extensive or intensive depending upon whether a small number of samples are taken from a large area (extensive) or a large number of samples taken from a small area (intensive). Extensive samples tend to be those most frequently used by pest managers because of the necessity for compromise between information needed and cost of sampling. After assessing the presence of a pest through extensive sampling, a pest manager may sample intensively to determine population or damage levels. For example, when it is determined through pheromone traps, phenological indicators, or light traps that a migrant pest is in a general locale, it may be necessary to sample a garden intensively to determine if indeed it is present or if control is needed.

Weather information (current and predicted) is very important in pest management activities. While most pest populations develop more rapidly at warmer temperatures, some plant pathogens require cool temperatures, high relative humidity, frequent rainfall, or free water on plant surfaces for rapid growth. Other pathogens and arthropod pests are favored by hot dry weather.

Problem Diagnosis

Problem diagnosis is often necessary because pests are present, yet unidentified. Symptoms of pest presence (holes in leaves, spots) are the stimulus most often promoting control procedures in home gardens (Gardner and York, 1982) rather than the definite identification of a pest and its population density. While diagnostic keys of the pests commonly invading fruits and vegetables and of damage expected could be prepared and distributed, most problem diagnosis comes about by familiarity with the crops, pests, and by common sense.

Thorough knowledge of the biology of garden pests is necessary in order to adequately diagnose problems or sample pests. Lesions (spots) caused by so-called soil fungi such as anthracnose (on fruit) and Septoria (on leaves) of tomato show up first on lower leaves or fruit in contact with the soil. Likewise, green peach aphid will most likely be inhabiting the lower leaves of tomato and potato plants, while potato aphid will be feeding on leaves higher on the plant. Holes in the leaves on the upper portions of potato plants are likely to be caused by Colorado potato beetle, while those on lower leaves are very often from variegated cutworm. On peach trees, peach tree borer will be found feeding under the bark always within 12-18 in of the soil surface, while lesser peach tree borer feeds under the bark up around the first branches. In more difficult situations affecting plant growth or plant injury with less readily identified symptoms and no signs (pests themselves or products thereof), the pest manager should consider several possibilities: injury related to specific sites (rows, edges of lawns), cultural practices (mulching with herbicide-contaminated grass), or varieties (nematode-susceptible tomato); general injury to taxonomically related plants susceptible to the same pests (cucumber, cantaloupe, or honeydew melon) fed upon by cucumber beetle or affected by bacterial wilt, beans fed upon by potato leafhopper; damage to taxonomically unrelated plants (fertility effects, herbicide drift or carryover, water stress).

No substitute exists for an extensive knowledge of the effects of common and occasional pests as well as the interaction between pests and their management techniques. However careful observation, proper interrogation of parties involved and good judgment will suffice in many cases.

Population Estimation

The extent to which population estimation or damage estimation plays a role in home garden fruit and vegetable pest management is questionable. Sampling criteria normally observed in commercial agriculture such as randomness, freedom from bias, and representativeness (Southwood, 1978) are often impractical or impossible in residential settings. Regardless, the pest manager should understand these sampling considerations and restrictions in order to selectively violate them when necessary.

Randomness is the assurance that each unit in the population (plant, fruit, tree) has an equal chance to be sampled. Only if samples are randomly selected is the estimate of the pest population variation valid. Sampling only those units easily accessible violates randomness, and sampling only those showing damage violates representativeness.

Sampling only those plants showing damage may give an accurate estimation of the average (mean) pest density. However, it provides a very inaccurate estimate of the variation of the population density within the crop, a parameter necessary for determining the accuracy of the population mean.

Moreover, bias is often introduced in both population sampling and damage assessment. Defoliation of plants is usually estimated as greater than it actually is. However, such bias, if recognized, can be adjusted for in decision-making. It is best to avoid bias in sampling wherever possible.

Hence, sampling is of tremendous importance in pest management, whether rigidly followed or casually performed. Its objectives are to provide a basis for the principles mentioned earlier. As a simplification of these principles, a pest manager needs to ask these questions (Tette, 1972):

1. How do I make observations for a particular pest?
2. Where do I look for a pest?
3. When is the best time to look?
4. What will it look like if I find it?
5. What does it mean if I find it?
6. What can I do about it?
7. What will the short and long term effects of weather and crop growth have on a particular pest?

Least Manageable Pest

With many of the pests of a crop for which the period of occurrence is the same, e.g., June, July, and August, it may make little sense to a gardener to limit the choice of horticultural characteristics (flavor, color, storage ability) in favor of resistance to a particular pest if his only effective management practice for a second pest is one or more applications of a pesticide. This is particularly true with some of the vegetable foliar diseases, where the available and recommended fungicides must be applied in a preventative manner and are broad spectrum. A fungicide or insecticide applied to control the least manageable pest will likely suppress many of the other common and manageable pests of the target crop and of adjacent crops. This least manageable pest—be it disease, insect, weed, nematode, or vertebrate—must be kept in mind in examining the following case studies. In the amount of space available it is impossible to discuss strategies for all the pests of gardens and fruit. Several of the most common crops and their respective pests will be used to identify significant strategies often applicable to many pests. It is up to the practitioner to implement, as needs and resources allow, those appropriate to crops grown and pests present.

PEST MANAGEMENT STRATEGIES

It is not this book's purpose to exhaustively review all aspects of vegetable garden pest management. Table 9-3 will allow readers an entry point into the literature on all the major vegetable garden pests listed, while subsequent discussion of vegetable garden pest management will be confined to certain case studies.

Case Study I. Weed Control in Gardens

In a recent article, William (1981) reviews the many interactions between weeds, weed control systems, and insects and diseases in horticultural systems. He stresses the necessity of understanding a system and its many ramifications before proper pest management can occur. Weed management can be accomplished in many crops through the practices that benefit the crop in other ways.

Mulching. A major benefit to almost all vegetable crops is the practice of mulching (Hopen and Oebker, 1976). In addition to the prevention of weed growth, mulches increase yield and the earliness of yield, improve quality of yield, reduce certain diseases on plants, stabilize soil moisture, prevent leaching of fertilizers or plant nutrients, and modify soil temperatures. Because crop response to mulching differs considerably, the choice of mulch depends on the crop and the desired objective. Ashworth and Harrison (1983) found opaque plastic mulch to provide the most effective weed control for the longest time. Shredded hardwood bark (5 cm deep) was next best but also resulted in the highest soil pH, organic matter content, and potassium levels, although not significantly so. Straw (5 cm deep) provided the poorest weed control and maintained soil temperatures at significantly lower temperatures (at 10 cm depth) than all other mulches used. Tomato yield was equally good on all mulches, but snap beans did significantly poorer on straw, and cabbage best on opaque plastics. Clear plastics are not effective for weed control but result in significantly higher soil temperatures than do opaque plastics. Increased soil temperatures often translate into increased yields, particularly when soil temperatures are naturally cool (Trudel and Gosselin, 1982). An added benefit from mulches on many crops is reduced disease incidence as soil-borne disease organisms, e.g., tomato anthracnose, are not carried onto plant parts by splashing water. Winged

Table 9-3. Pests of vegetables and fruit trees and pest management strategies available.

Pest	*Strategy*	*References*
Apple scab	Resistant varieties (Prima, Priscilla, Sir Prize, Blairmont, Liberty, Redfree, Jonafree, Viking)	Brooks and Olmo, 1983; Janick, 1982
	Sanitation	Roberts and Boothroyd, 1972
	Chemicals (prevention)	Roberts and Boothroyd, 1972
Fire blight	Resistant varieties (good = Prima, Priscilla Liberty, Nova Easygro, Macfree Blairmont, Hazen; moderate = Red and golden delicious, Winesap, Stayman Carroll, Northwestern Greening, Britemac, Viking Quinte, Hawaili, Primegold, Splendor) Avoid: Jonathan, Ida Red, Rome Beauty	Aldwinckle and Beer, 1979 Aldwinckle and Beer, 1979 Cantelo and Cantwell, 1982 Aldwinckle and Beer, 1979
	Sanitation (pruning and sterilization)	Aldwinckle and Beer, 1979
	Chemicals (prevention)	Aldwinckle and Beer, 1979
European red mite	Chemicals (dormant oil)	
	Biological control (predators)	Asquith and Hull, 1979; Leink, Watve, and Weires, 1980
Apple maggot	Sticky spheres (adult control)	Ressig, Fein, and Roelofs, 1982
	Sanitation (pick up dropped fruit)	
	Chemicals (adult control)	Prokopy and Hauschild, 1979
San Jose Scale (apple)	Chemicals (dormant oils)	

(continued)

Table 9-3. *(continued)*

Pest	*Strategy*	*References*
Oystershell scale (apple)	Chemicals (dormant oils)	
Codling moth	Pheromones (monitoring and mating disruption)	Rock, Childers, and Kirk, 1978
	Chemicals (larval control)	Croft and Whalon, 1982; Embree and Whitman, 1978
	Bands around trees (larval capture)	Slingerland and Crosby, 1914
Plum curculio (apple and peach)	Trap fruit (monitoring)	
	Sanitation (dropped fruit)	Slingerland and Crosby, 1914
	Shake adults from trees and destroy	Slingerland and Crosby, 1914
	Biological agents (many)	Croft and Whalon, 1982
	Chemicals (adult control)	
Bacterial leaf spot (peach)	Resistant varieties (good = Idlewild	Boudreaux et al., 1983
	Bicentennial, Casellaqueen, Clayton, Corell	Brooks and Olmo, 1983
	Ellerbee, Harson; moderate = Derby)	
	(good = Harland, Jerseyglo, Stark Encore)	Brooks and Olmo, 1982
	Avoid: Blake, Suncrest, Suncling, Sunhigh	
	Chemicals (prevention very early)	
Brown rot of stone fruit (peach)	Resistant varieties (Harland)	Brooks and Olmo, 1982
	Harson, Bicentennial	Boudreaux et al., 1983
	Sanitation (dropped fruit)	Jones, 1971
	Chemicals (prevention)	
Peach tree borer and Lesser peach tree borer	Pheromones (monitoring)	Brunner and Howitt, 1981
	Chemicals (larval control)	Weiner and Norris, 1983; Yonce, 1980
	Hand removal of larvae	

Striped cucumber beetle Spotted cucumber beetle, and Bacterial wilt	Chemicals (prevention of beetle feeding only management on cantaloupe; on cucumber resistant varieties available within a year or two)	York, Reed, and Kinney, 1982
Powdery mildew (cucurbits)	Resistant varieties (cucumber—many, Poinsett, Early triumph, Slicemaster, Victory Hybrid, Bounty, Carolina) (Cantaloupe = Georgia 47, Honeyball 306, Homegarden, Cinco, Delicious 51, Early Dawn, Saticoy Hybrid, Summet)	Latin et al., 1983 Stevenson, 1979*a* Thomas and Webb, 1982
Seed corn maggot	Warm soil condition Shallow planting No fresh organic matter in soil Chemicals (seed treatment)	
Cabbage maggot	Time of planting (after oviposition of first generation) Phenological indicators for timing of planting and chemicals (adult control) Protective disks on soil around plants	Pedersen and Eckenrode, 1981 Pedersen and Eckenrode, 1981
Radish maggot	Cage to prevent oviposition Chemicals (adult control)	
Colorado potato beetle	Threshold observance	Cantwell and Cantelo, 1981; Ferro, Morzuch, and Margolies, 1983; Schalk and Stoner, 1979; Tamaki and Butt, 1978
	Hand picking *Bacillus thuringiensis* (adult control)	
Verticillium wilt and	Resistant varieties (potato—most are susceptible)	Hodson, Pond, and Munro, 1974;

(continued

Table 9-3. *(continued)*

Pest	*Strategy*	*References*
Fusarium wilt (potato)		O'Brien and Rich, 1976
	Seed piece selection	
	Rotation	
	Highly susceptible = Kennebec, Fundy, Irish Cobbler, Norchip (Fusarium, Shurchip)	
(tomato)	Resistant varieties = Better Girl, Big Girl, Better Boy, Beefmaster, Burpee's VF Hybrid, Campbell 1327, Centennial, Floramerica, Ace 55, Jet Star, Red Pak, Small Fry, Supersonic, Ultra Boy, Ultra Girl, Wonder Boy, Tropic	Stevenson, 1979*c*; Walker, 1971
Late blight potato	Resistant varieties (moderate to good = Alamo, Boone, Catoosa, Cherokee, Chieftain, Fundy Kennebec, Keswick, Merrimack, Onaway, Pennchip, Plymouth, Pungo, Reliance, Saco, Tawa, Wauseon)	O'Brien and Rich, 1976
tomato	Plant selection	
	Resistant varieties (good = New Yorker, Early cascade New Hampshire, Surecrop, New Yorker, Manalucie, West Virginia 63, Nova)	
	Chemicals (prevention and curative, Ridomil)	
Early blight (tomato)	Resistant varieties (Floramerica); early maturing varieties most susceptible	Stevenson, 1979*b*
	Avoid injury (pruning) to wet plants	
	Avoid wetting plants in evening	
	Chemicals (prevention)	
	Plant selection (disease-free)	
	Sanitation (clean up diseased materials)	

Blossom end rot (tomato)	Mulching	
	Regular water supply to plants	
	Calcium chloride	
Cabbage looper	Threshold observance	
Imported cabbageworm	*Bacillus thuringiensis*	
Diamondback moth and other caterpillars (cole crops)	Hand picking	
Mexican bean beetle	Hand picking	
	Biological control (introduced)	Barrows and Hooker, 1982
	Traumatic irrigation (knock larvae from plants to help control young potato leafhopper)	
Potato leafhopper	Pyrethrum insecticides (repeated applications)	
Flea beetles (many crops)	Rotenone (repeated applications necessary)	Gardner, 1981
Grasses (weeds)	Mulch (black plastic or organic)	Hopen and Oebker, 1976
Broad-leaved weeds	Hand removal	Harrison, 1983
	Chemical (DCPA = Dacthal™; chloramben = Amiben™; trifluralin = Treflan™, Preen™)	Binning, 1975

aphids and the mosaic diseases they transmit are reduced on many crops by reflective (aluminized) mulches (Black, 1975; Nawrocka et al., 1975). In addition, some researchers have found significant yield increases from the use of reflective mulch in the absence of insects and disease (Dufault and Wiggans, 1981; Porter and Etzel, 1982). When reflective mulch is used with snap beans growth is accelerated and yield significantly increased, and significantly fewer potato leafhopper are found on young plants (Wells, Dively, and Schalk, 1984).

Recent work with row covers, spun bonded polyester (Reemay) or slitted polyethylene, placed over the top of an entire row of growing plants, with and without opaque plastic mulch, also increase early yield and total yield by increasing air and soil temperatures (Loy and Wells, 1982).

There will be times when it is necessary to hand weed. Even this has been improved with the use of hand-held wick wipers and the herbicide glyphosate. Harrison (1983) found vegetable garden weed control with a hand-held wiper to be quicker, cheaper, and more effective in controlling some weeds. Yields were increased in the wiping treatment probably as a result of increased soil moisture and decreased root disturbance, which often occurs with deep hoeing.

Case Study II. Pest Management of Tomato and Potato

Insects. Gardner and York (1982) found that 29% of all garden insecticide uses were on tomato (19.1%) and potato (10.1%). Of households using insecticides on gardens, 24.5% treated potato and 46.6% treated tomato. In the upper midwest this is disproportionate to the number of insect pests occurring on these two crops. Both face serious damage from Colorado potato beetle, but such defoliation is only of importance to yield at certain times during the growing season, most importantly beginning at full bloom continuing into fruit (tuber) set and enlargement (Cantelo and Cantwell, 1982; Cranshaw and Radcliffe, 1980; Ferro, Morzuch, and Margolis, 1983; Tamaki and Butt, 1978). When plants are growing well, both crops can stand substantial insect injury. Research with Superior and Russet Burbank potatoes has found that plants in less than full bloom can tolerate up to 75% defoliation with only a 15-20% yield loss (Shields and Wyman, 1984). Related research with pole (snap) beans indicates that 50% defoliation per week results in a 34% yield loss (Waddill et al., 1984). Young potato and tomato plants may need to be protected from flea beetles. Later, when plants are growing well and flea beetles are no longer a problem, hand picking can reduce most insect pests to insignificant numbers.

Tobacco and tomato hornworms are common and consume much tomato foliage but are easily found by their damage and large droppings. They are seldom numerous and are often heavily parasitized. Green peach aphid and potato aphid seldom become a problem in gardens until after insecticide use for other insects has begun. The same is true for greenhouse whitefly and twospotted spider mite. It is particularly important in this respect to delay until absolutely necessary or, if possible, to eliminate applications of insecticides harmful to beneficial insects, including ladybird beetles, lacewings, syrphid flies, mantids, and others. Early insecticide applications of broad-spectrum insecticides often put one on an insect pest/insecticide treadmill from which it is often difficult to escape. There is no question that much of the leafminer (*Liriomyza* spp.) problem in tomatoes in California, and in other crops in other parts of the United States is caused by the reduction of natural enemies by certain insecticides, notably, methomyl (Trumble and Toscano, 1983). Resurgence of the target pest, because of the elimination of its natural enemies or the increase in numbers of a previously insignificant organism (*Liriomyza* spp.) to the point where it is now damaging to crops requires additional pesticide applications, which further exacerbate the situation. Furthermore, authors (McKee and Knowles, 1984) speculate that increased respiration in twospotted spider mite caused by application of pyrethroid insecticides for other pests may be responsible for the outbreaks of mites that often occur after use of these materials. Unfortunately, variability in susceptibility of natural enemies to insecticides occurs not only from family to family but also within a family from one crop ecosystem to another. Generally, the green lacewing has been shown to be somewhat tolerant to synthetic pyrethroid insecticides, while the ladybird beetles (Coccinellidae) and predacious mites are quite susceptible. An extremely good summary of synthetic pyrethroid susceptibilities has recently been published (Croft and Whalon, 1982).

Diseases. Disease management is of more consequence on both tomato and potato. Both crops should be grown on well-drained soils and rotated frequently with nonsolanaceous crops. Seed or seed pieces should be obtained from reputable dealers and not utilized from a previous year's harvest. Where Verticillium and Fusarium wilt fungus are a problem, solutions are limited to rotation, resistant varieties (see Table 9-3) or soil fumigation. Potatoes can be grown with an organic mulch. Tomato does particularly well on opaque plastic and Septoria leaf spot, anthracnose, and blossom-end rot can all be reduced in this way. Early blight and late blight are two diseases for which chemicals may be necessary management tools, but crop rotation, clean seed pieces, and destruction of foliage, fruit, and tubers on which inoculum can overwinter and serve as a source of reinoculation are very important. Resistant varieties are available, but under heavy inoculum pressure they frequently do not provide sufficient protection. Late blight on potato and tomato is favored by cool and wet weather in late summer and fall, whereas early blight is more likely to occur in warm weather with frequent heavy dews. Spores of both fungi are wind transmitted, but the fungi overwinter on diseased foliage (early blight) and tubers (late blight) from the previous year, so sanitation is an absolute must. Late blight occurs in many areas only rarely, but it can be a devastating disease if not controlled with fungicides. Early blight progresses much less rapidly on either plant, and disease progress can be slowed with several fungicides. There is evidence that two fungicides, triphenyltin hydroxide (Duter™) and $Cu(OH)_2$ (Kocide™ 101) reduce Colorado potato beetle feeding as much as 95% (Hare, Logan, and Wright, 1983). Potato treated with $Cu(OH)_2$ had 44-100% fewer Colorado potato beetle larvae than foliage treated with mancozeb (Dithane™ M-45; Hare, 1984).

Gardeners in urban areas with air pollution should be aware that sulfur dioxide and ozone can adversely affect tomato and potato. Synergistic effects severely affecting yields may occur when the early blight fungus and sulfur dioxide occur together (Lotstein and Davis, 1983). Tomato varieties not sensitive to sulfur dioxide include Heinz™ 1350, Ace, and Bonanza (Howe and Soltz, 1982). Potato varieties somewhat tolerant to air pollution include Alma, Belleisle, Hudson, Kennebec, Monona, Nampa, Russet Burbank, Sebago, Superior, and Targhee. Varieties severely affected are Centennial, Hi Plains, Norland, and Wischip (Hooker, Yang, and Potter, 1972, 1973).

Case Study III. Pest Management on Cole Crops

Insects. Cabbage, broccoli, cauliflower and other cole crops have the same spectrum of insect pests and diseases that require management. Minor differences exist in the management approach and timing as the consumed parts of the plants differ. All these crops need to be protected from cabbage maggot. The adult flies are active three times a year, ovipositing in the soil about the base of the plants. During these more or less discrete periods of egglaying, commercial growers are finding it beneficial to control adults with foliar-applied insecticides. Accurate timing of insecticide applications for peak adult flight is obtained either by keeping track of heat unit accumulations or by monitoring common phenological indicators (Pedersen and Eckenrode, 1981). The first adult peak occurs in late May at 375 day degrees over a threshold of 43°, coinciding in New York and Indiana with full bloom of the common weed yellow rocket (*Barbareavulgaris*). Second generation peak emergence follows in about five weeks depending upon temperature (an additional 1,175 day degrees) when day lily (*Hemerocallis* spp.) has been in full bloom for about one week. The third generation of adults follows after an additional 1,175 day degrees, coinciding with the bloom of Canada thistle (*Cirsium arvense*) and early goldenrod (*Sildago juncea*). Plants set after the peak flight period will likely escape much oviposition. Tar paper circles set closely around the stems of the young plants to prevent adult flies from reaching the soil and laying eggs is an old remedy that many gardeners report works well if the holes for the stem are close to the size of the stem. Stoner (1982) reports that cabbage maggot infestation is reduced by the use of aluminum foil mulch over unmulched plots. Soil drenches of insecticide poured around the base of plants shortly after the peak flight period will kill young larvae in the soil sufficiently to allow suitable crop growth. Several species of flea beetle will feed without discretion on any of the cole crops and can seriously damage young seedlings. They are easily controlled with rotenone and most other insecticides (Gardener, 1981). Companionate plants may reduce flea beetle numbers on collards and other cole crops but will not provide adequate protection and have been found to reduce yields significantly (Latheef and Ortiz, 1984; Latheef, Ortiz, and Sheikh, 1984). Interplanting tomato with collards increased damage from imported cabbageworm (Maguire, 1984).

When properly fertilized and growing strongly (side-dressed fertilizer of an organic or inorganic nature is almost a requirement as cole crops require a great deal of nitrogen), plants can withstand considerable feeding from caterpillars without harm (Sears, Jaques, and Laing, 1983; Shelton et al., 1983). Of the three species (imported cabbageworm, cabbage looper, diamondback moth) commonly found on cole crops, the cabbage looper is usually most difficult to control and consumes approximately 1.5 times the foliage of the imported cabbageworm and several times the consumption of the diamondback moth larva.

In Texas, cabbage can easily tolerate up to 0.3 total caterpillars per plant and maintain 80% marketable heads (Kirby and Slosser, 1984). Others report an action threshold of 0.5 larvae per head (Leibee et al., 1984). When heading begins, or if necessary before, gardeners may wish to protect cabbage, broccoli, and cauliflower with an insecticide. *Bacillus thuringiensis*, a bacterial insecticide, is a good choice for gardens because it is toxic only to caterpillars and unlikely to cause pest outbreaks. It is important with Bt to shorten up the interval between treatments from ten days to about five days for a week or so before harvest to ensure clean produce. Regardless of what insecticides are used, good coverage of plants with spray and the inclusion of a wetting agent or a spreader/sticker will improve control.

In most gardens fungicides will not be needed if disease-free planting stock is used and if gardeners rotate crops. Mulch will reduce the amount of inoculum splashed onto plants, reduce the disease incidence, and result in significantly higher yields (Hopen and Oebker, 1976; Stoner, 1982).

Case Study IV. Cucurbit Pest Management

The various cucurbit crops require widely differing management programs if they are to be grown successfully. Cantaloupe, most cucumber, honeydew melon, and hubbard and banana squash are susceptible to bacterial wilt, a disease transmitted only by striped and spotted cucumber beetles. Watermelon and other winter squash, butternut, acorn, Boston marrow, and buttercup, as well as zucchini and summer squash and pumpkin are not susceptible. Plants of all susceptible varieties become somewhat more tolerant of the disease as they become older.

Control of cucumber beetles is necessary to prevent the disease. Carbaryl should be used early in the season. When plants begin to bloom methoxychlor is the material of choice as it is almost nontoxic to honeybees. In areas with high populations of cucumber beetles, insecticides may need to be applied at approximately weekly intervals until harvest is underway. Unfortunately, after multiple applications of carbaryl on long-season crops such as cantaloupe, watermelon, and winter squash, twospotted spider mite infestations are common. Hence insecticide applications should be kept to the minimum necessary to protect plants from cucumber beetles. Before the development of currently effective insecticides, gardeners often caged plants in cheesecloth or screening to prevent beetle access until plants were older and well established. This is only marginally effective. Resistant varieties of cucumber and cantaloupe are greatly needed. The cucumber Saladin is reportedly tolerant to bacterial wilt (Stevenson, 1979*a*), but no cantaloupe varieties currently available have any useful tolerance to the disease (G. Reed, pers. comm.) New varieties of cucumber have resistance to multiple diseases, making the job of the gardener easier (Peterson et al., 1982). Considerable disease resistance, with the exception of bacterial wilt, has been incorporated into new varieties of cantaloupe. Cinco cantaloupe, for example, is resistant to watermelon mosaic virus-1, downy mildew, powdery mildew, and Alternaria leaf blight (Thomas and Webb, 1982).

The cucurbits are warm-season crops and respond very positively to plastic mulches and row covers, which increase soil temperature and air temperature around the plants early in the growing season and stimulate early plant growth. This enables plants to more effectively withstand diseases and other pests. Significant increases in earliness of yield and total yield are also obtained (Loy and Wells, 1982; Warren and Gerber, 1983).

Powdery mildew, downy mildew, anthracnose, and angular leaf spot can be problems at times and may require multiple fungicide applications if resistant varieties are not being grown.

Case Study V. Apple and Peach Pest Management

Diseases. Disease management for the gardener is being made considerably more manageable by the introduction of varieties resistant to several defoliating diseases. Redfree, an apple variety developed by a cooperative program of Purdue University, Rutgers University, and the University of Illinois is immune to apple scab and cedar apple rust and resistant to powdery mildew, fireblight, and European red mite (Williams et al., 1981). This tree will need practically no sprays on young, nonbearing trees (Janick, 1982). Novole, in addition to resistance to apple scab, fire blight, and powdery mildew, is also resistant to the bark feeding of pine vole and meadow voles (Cummins, Aldwinckle, and Beyers, 1983). This tree is only for use as parent in breeding stock but indicates the kinds of pest resistance being incorporated into new varieties.

Trees without apple scab resistance require multiple applications of fungicide to prevent defoliation. Disease incidence may be reduced by raking up and destroying leaves and leaf litter on which fungus overwinters, but windborne spores are often sufficiently numerous to cause infection.

Fireblight will kill susceptible trees. Pruning out of infected branches is critical. Branches should be cut 8-10 in below infected areas with pruning tools sterilized with a 10% solution of laundry bleach between each cut. This solution is corrosive to pruning tools, so tools must be rinsed with clean water after pruning is completed. Late winter pruning should be done wherever possible, when there is less chance of spreading the disease. Dormant sprays of copper sulfate (1 dry ounce per 1 gallon of water) or Bordeaux mixture, plus Superior oil, applied two weeks before bud break and repeated 7-10 days later will help prevent season buildup of inoculum of diseases and mites (Aldwinckle and Beer, 1979). Resistance to streptomycin, an antibiotic sometimes used by commercial growers, has occurred and treatment is usually too expensive and not recommended for homeowners.

Fortunately, peach is not susceptible to fire blight bacterium. New varieties being released are resistant to the two common and serious diseases: bacterial leaf spot and brown rot of stone fruit (Brooks and Olmo, 1982, 1983). Bacterial leaf spot control is extremely difficult even with chemical applications; materials must be applied before the growing season starts if disease is to be prevented. Controlling brown rot with fungicides will require four to five applications early in the growing season. Any fruit dropping to the ground or remaining on the tree (mummies) with brown rot should be collected and destroyed. This will help reduce the level of inoculum.

Insects. If mites have been a problem the preceding season, dormant season sprays with Superior oil are necessary. Otherwise mite control should not be necessary; predators are usually sufficiently numerous to maintain control. Likewise, small aphid colonies on branch tips are not serious and should be ignored or destroyed by pruning.

Plum curculio, on the other hand, will need control on both peach and apple. First adults appear in trees when temperatures average 60-65° for 3 days, usually when apple trees are in pink-bud stage or are beginning to bloom. Adults will begin laying eggs in small fruit as soon as they form. As a monitoring tool, Granny Smith apples can be hung by a wire through the core in contact with a branch in the upper half of the tree at a rate of three to four apples per tree (Jean-Pierre et al., 1984). If placed before any other fruit has begun to form, they are very attractive to females for oviposition. Success of the plum curculio depends upon the female's ability to lay an egg in a fruit, and nearly all successfully attacked fruit will fall from the trees as the larvae begin to develop. Hence, all drops should be picked up and destroyed soon after falling to prevent larvae from completing development and crawling into the soil to pupate. The second generation adults emerge from the soil in Indiana in late July. Peaches again require protection but rapidly ripening apples are usually not damaged. Depending on the severity of damage the previous year and the individual gardener's tolerance, trees may need insecticides to control the adults before oviposition. Because of a habit of falling from the tree when disturbed, shaking trees will dislodge adults onto sheets or other materials for later destruction. This may not provide adequate control, however, for if temperatures are above 70° adults may fly when dislodged rather than fall to the ground.

Ripening apples are attacked by other severe pests. In unsprayed orchards of dwarf apples in New York the most serious insect pests were in decreasing order: apple maggot, plum curculio, codling moth, oriental fruit moth, and lesser appleworm. Disease-resistant trees were no less susceptible to insect injury than nonresistant trees (Reissig et al., 1984*b*).

Apple maggot flies will emerge from overwintering in the soil usually in early to late June in Indiana (about 1,137 day degrees over a threshold of 43°). After emergence, the adults seek food and a mate, in that order. Adults can be attracted to feeding traps for about 10 days after emergence. After this period, they will be attracted to sex pheromone traps or sticky sphere traps. Adult females, seeking oviposition sites, are attracted to red spheres about the size of apples, 2-3 in in diameter, and these sticky wooden spheres, with or without an attractant, are very effective for monitoring and controlling flies (Reissig, Fein, and Roelofs, 1982).

Four years' data from New York orchards have shown apple maggot infested apples reduced to an average of 5.4% (1 -12.8%) compared to 1.9% infested fruit in the insecticide treatments (Reissig et al., 1984*a*). One to three spheres per tree (one per 25 apples) are placed equidistant around the circumference of the canopy. Trapping will be most attractive to flies and therefore most effective if no apples are within 9-18 in around the spheres (Drummond, Groden, and Prokopy, 1984). At this rate of spheres, cost is $4-$11 per tree.

If infestations have been severe in previous years and dropped apples not picked up or if flies are coming in from untreated trees nearby, chemical control may be necessary. The first application should occur soon after the first fly is caught on a sticky red sphere. The second treatment should occur as desired but no sooner than 10 days later or when the next fly is caught, whichever is later. Care must be taken to recoat spheres 4-6 weeks or sooner after placement, as stickiness will decline with time.

Codling moth is now an apple pest throughout most of the world. Easily controlled by arsenicals, it rapidly developed resistance to these chemicals. DDT was effective until resistance occurred after 10-15 years of use. Currently satisfactory control is obtained with several insecticides, but five to six applications may be needed to control larvae on enlarging and ripening fruit. Larvae pupate and overwinter in sheltered locations such as under loose bark or in litter at the base of trees. Burlap wrapped around trees provides an excellent hiding place for larvae as they crawl down trees seeking shelter. This can be removed occasionally and larvae destroyed. These harborages may also serve as shelters for natural enemies as well. When introducing earwigs (five to six per tree) for successful control of apple aphid, researchers found it necessary to provide shelter on the trunks of the trees (Carroll and Hoyt, 1984).

Peak emergence of adults in the spring occurs 10-14 days after apple blossoms begin to show pink color. Eggs laid on leaves soon hatch, and larvae bore into fruit. Larvae feed for 3-4 weeks, pupate, and emerge as adults for a second generation about mid-July to early August. If previous infestations of fruit are higher than desired, sex pheromone traps can be used to monitor appearance of adults, thus ensuring timely application of insecticides. In the northwest United States it has been found that when population levels were low three to five applications per week of female sex pheromone was successful in preventing males from finding females resulting a 90% reduction of damaged pears (Moffitt and Westigard, 1984). Furthermore, since insecticides were not used in these orchards, pear psylla, a normally severe pest, was controlled by natural enemies (Westigard and Moffitt, 1984).

Recently the description of a cheap, simple pheromone for codling moth was described that costs only $0.16 to $0.20 compared to $1.20 for the commercial traps (Howell, 1984).

It should be obvious that pest management in vegetable gardens and fruit trees is complicated by the wide range of pests and crops occurring therein. The problems are further complicated by the complex biology of some of the pests and the difficulty in identification, particularly of the diseases. Good tools exist for identifying pests of vegetables and fruit (Brunner and Howitt, 1981; Jones, 1971; Macnab, Sherf, and Springer, 1983; Wescott, 1973). Local extension service personnel should be able to assist with timing and history of disease incidence.

As mentioned earlier, the least identifiable or the least manageable pest may dictate the program to be followed for the crop. Many gardeners will find it easier to apply an "all-purpose pesticide" (containing, e.g., malathion and methoxychlor insecticides and captan fungicide) to fruit trees as the urge or inspiration strikes (although usually not at the proper time nor with the proper frequency to be really effective) than to try to monitor pest populations and make intelligent pest management decisions.

Another area of difficulty is that of "organic" gardening. While much that is recommended by well-meaning groups is excellent (Slingerland and Crosby, 1914) and fits well with traditional pest management practices, a great deal more is not supported by any scientifically valid research. Many of the remedies are folk tales and more is simply misinterpretation of pest conditions and solutions. Many of the home-concocted spray mixtures may have value, particularly those which have long history of use, e.g., soap as an insecticide or miticide. Others are worthless and may be phytotoxic to plants. The best advice to be given is to try it. One should try out recommendations on difficult to control insects and diseases. It makes no sense to try them on easily managed or on sporadically occurring nondamaging pests. Observe the results, and watch carefully for any plant damage resulting from the treatments.

FUTURE DIRECTIONS

Much exciting research promises to help the vegetable and fruit gardener. Because of the high monetary return of fruits and vegetables growers felt that high dollar inputs for insecticides and fungicides were justified. This attitude, coupled with the high cosmetic standards required of fruits and vegetables, resulted in less research on host plant resistance for fruits and vegetables than for traditionally low-value crops, e.g., wheat. Exceptions were those pests that could not reasonably be controlled by other traditional methods, e.g., Fusarium and Verticillium wilt of tomato. Today, concern over the expected effective life of pesticides, environmental contamination, and increasing cost efficiency has stimulated increased research in host plant resistance on fruits and vegetables. Increasingly, resistance to multiple pests is being incorporated so that one is not faced with a "least manageable pest" that must be treated with pesticide. To be certain that resistant varieties are adopted to their geographical region, gardeners should consult local extension publications (Stevenson, Harrison, and Heimann, 1984).

Timing of insect life cycles and schemes to use that timing are becoming more precise for many of our most serious pests (Liu, McEwen, and Ritcey, 1982). Companies such as Albany International (P.O. Box 537, Buckeye, Arizona 85326) are actively involved in selling pheromone and other sampling traps for insect pests. Research into anti-feedant compounds to be applied to crops shows promise for serious insect pests, including striped cucumber beetle and Colorado potato beetle (Reed, Freedman, and Ladd, 1982). Research into improved chemical pest management continues. Increased target specificity for insects and diseases is not likely to be profitable for industry, but new herbicides that control annual and perennial grasses have become best sellers. New modes of action, including systemic fungicides, are becoming a reality. A new class of herbicides developed at the University of Illinois kills weeds upon their exposure to sunlight after application. The insecticide diflubenzuron, a chitin inhibitor with very low toxicity to nontarget organisms, provides sufficient control of many insects, including the Mexican bean beetle (Zungoli, Steinhauer, and Linduska, 1983) to make it of value in vegetable production. Renewed research on the fungi *Beauveria bassiana* and *Metarrhizium anisopliae* may provide improved management of pests entering the soil, such as plum curculio (Tedders et al., 1982), Colorado potato beetle, cabbage maggot, and apple maggot. There is as yet inconclusive evidence on the overall pest management suitability of the pyrethroid insecticides, but it is likely that they will be available to the gardener and that they may be compatible with integrated pest management programs (Bashir and Crowder, 1983; Hull and Starner, 1983).

REFERENCES

Ahlgren, G. H., G. C. Klingman, and D. E. Wolf. 1951. *Principles of Weed Control.* Wiley, New York.

Aldwinckle, H. S., and S. V. Beer. 1979. Fire blight and its control. *Hortic. Rev.* **1**:423-474.

Anonymous. 1950. *U.S. Bureau of the Census 1949 Statistical Abstracts of the United States, 1789-1945.* Government Printing Office, Washington, D.C.

Anonymous. 1981. *Letter from New England Insect Traps.* P. O. Box 938, Amherst, Mass., 01004.

Anonymous. 1983. *U.S. Bureau of the Census 1983 Statistical Abstracts of the United States, 1789-1945.* Government Printing Office, Washington, D.C.

Ashworth, S., and H. Harrison. 1983. Evaluation of mulches for use in the home garden. *HortScience* **18:**180-182.

Asquith, D., and L. A. Hull. 1979. Integrated pest management systems in Pennsylvania apple orchards. In *Pest Management Programs for Deciduous Tree Fruits and Nuts,* D. J. Boethel and R. D. Eikenbary (eds.). Plenum Press, New York, pp. 203-222.

Barrows, E. M., and M. E. Hooker. 1982. Parasitization of the Mexican bean beetle (Coleoptera: Coccinellidae) by *Pediobius foveolatus* (Hymenoptera: Eulophidae) in urban community vegetable gardens. In *Research for Small Farms,* H. W. Kerr, Jr., and L. Knutson (eds.), Miscellaneous Publication 1422. Agricultural Research Service, Washington, D. C., pp. 272-274.

Bashir, N. H., and L. A. Crowder. 1983. Mechanisms of permethrin tolerance in the common green lacewing (Neuroptera: Chysopidae). *J. Econ. Entomol.* **76:**407-409.

Binning, L. K. 1975. Vegetable gardens without weeds. *Brooklyn Botanic Garden Record: Plants and Gardens* **31:**36-39.

Black, L. L. 1975. "Aluminum" mulch-less disease, higher vegetable yields. *Louisiana Agric.* **23:**16-18.

Boudreaux, J. E., C. E. Johnson, P. L. Hawthorne, W. A. Young, R. L. Cunningham, R. J. Raiford, F. J. Peterson, N. L. Horn, P. W. Wilson, and D. W. Newsom. 1983. Idlewild peach. *HortScience* **18:**375.

Brooks, R. M., and H. P. Olmo. 1982. Register of new fruit and nut varieties. *HortScience* **17:**17-23.

Brooks, R. M., and H. P. Olmo. 1983. Register of new fruit and nut varieties. *HortScience* **18:**155-161.

Brunner, J. R., and A. J. Howitt. 1981. *Tree Fruit Insects,* North Central Regional Extension Publication No. 63. Michigan State University, East Lansing, Mich.

Cantelo, W. W., and G. E. Cantwell. 1982. Colorado potato beetle on tomatoes: Economic damage thresholds and control with *Bacillus thuringiensis.* In *Research for Small Farms,* H. W. Kerr, Jr., and L. Knutson (eds.). Miscellaneous Publication 1422. Agricultural Research Service, Washington, D.C., pp. 56-61.

Cantwell, G. E., and W. W. Cantelo. 1981. *Bacillus thuringiensis:* A potential control agent for the Colorado potato beetle. *Am. Potato J.* **58:**457-468.

Carroll, D. P., and S. C. Hoyt. 1984. Augmentation of European earwigs (Dermaptera: Forficulidae) for biological control of apple aphid (Homoptera: Aphididae) in an apple orchard. *J. Econ. Entomol.* **77:**738-740.

Chalfant, R. B., W. H. Denton, D. J. Schuster, and R. B. Workman. 1979. Management of cabbage caterpillars in Florida and Georgia by using visual thresholds. *J. Econ. Entomol.* **72:**411-413.

Chupp, C., and R. W. Leiby. 1951. *The Control of Diseases and Insects Affecting Vegetable Crops,* Cornell Extension Bulletin 206. Cornell University, Ithaca, N.Y.

Cranshaw, W. S., and E. B. Radcliffe. 1980. Effect of defoliation on yield of potatoes. *J. Econ. Entomol.* **73:**131-134.

Croft, B. A. 1982. Tree fruit pest management. In *Introduction to Insect Pest*

Management, R. L. Metcalf and W. H. Luckmann (eds.). Wiley, New York, pp. 465-498.

Croft, B. A., and M. E. Whalon. 1982. Selective toxicity of pyrethroid insecticides to arthropod natural enemies and pests of agricultural crops. *Entomophaga* **27**:3-21.

Cummins, J. N., H. S. Aldwinckle, and R. E. Beyers. 1983. Novole apple. *HortScience* **18**:772-774.

Davies, J. O. 1979. *Gardening in America, 1978.* National Association for Gardening, Burlington, Vt.

Drummond, F., E. Groden, and R. J. Prokopy. 1984. Comparative efficacy and optimal positioning of traps for monitoring apple maggot flies (Diptera: Tephritidae). *Environ. Entomol.* **13**:232-235.

Dufault, R. J., and C. Wiggans. 1981. Response of sweet peppers to solar reflectors and reflective mulches. *HortScience* **16**:65-67.

Embree, C. G., and R. J. Whitman. 1978. *Codling Moth,* Orchard Outlook No. 12. Nova Scotia Department of Agriculture, Halifax.

Ferro, D. D., B. J. Morzuch, and D. Margolies. 1983. Crop loss assessment of the Colorado potato beetle (Coleoptera: Chrysomelidae) on potatoes in western Massachusetts. *J. Econ. Entomol.* **76**:349-356.

Gardner, R. D. 1981. Home vegetable garden insecticide use in Indiana and rotenone efficacy on vegetable insect pests. M.S. thesis, Purdue University, West Lafayette, Ind.

Gardner, R. D., and A. C. York. 1982. *Home Garden Insecticide Use in Indiana,* Agricultural Experiment Station Research Bulletin 92. Purdue University, West Lafayette, Ind.

Gilbert, W. W., and D. J. Caffrey. 1937. *Diseases and Insects of Garden Vegetables,* Farmer's Bulletin No. 1371. U.S. Department of Agriculture, Washington, D.C.

Hare, D. 1984. Suppression of the Colorado potato beetle, *Leptinotarsa decemlineata* (Say) (Coleoptera: Chrysomelidae), on solanaceous crops with a copper-based fungicide. *Environ. Entomol.* **13**:1010-1014.

Hare, J. D., P. A. Logan, and R. J. Wright. 1983. Suppression of Colorado potato beetle, *Leptinotarsa decemlineata* (Say), (Coleoptera: Chrysomelidae) populations with antifeedant fungicides. *Environ. Entomol.* **12**:1470-1477.

Harpaz, I. 1973. Early entomology in the Middle East. In *History of Entomology,* R. F. Smith, T. E. Mittler, and C. N. Smith (eds.). Annual Reviews, Palo Alto, Calif., pp. 21-36.

Harrison, H. F., Jr. 1983. Hoeing and hand-held wiper application of glyphosate for weed control in vegetables. *HortScience* **18**:333-334.

Herrick, G. W. 1925. *Manual of Injurious Insects.* Holt, New York.

Hodson, W. A., D. D. Pond, and J. Munro. 1974. *Diseases and Pests of Potatoes,* Canada Department of Agriculture Publication 1492. Agriculture Canada, Ottawa.

Hooker, H. J., T. C. Yang, and H. S. Potter. 1972. *Air Pollution Effects on Potato and Bean in Southern Michigan,* Report 167. Michigan State University, East Lansing, Mich.

Hooker, H. J., T. C. Yang, and H. S. Potter. 1973. Air pollution of potato in Michigan. *Am. Potato J.* **50**:151-161.

Hopen, H. J., and N. F. Oebker. 1976. *Vegetable Drop Responses to Synthetic Mulches: An Annotated Bibliography,* Illinois Agricultural Experiment Station Special Publication 42. University of Illinois, Urbana, Ill.

Hough-Goldstein, J. A., and K. A. Hess. 1984. Seedcorn maggot (Diptera: Anthomyiidae) infestation levels and effects on five crops. *Environ. Entomol.* **13**:962-965.

Howe, T. K., and S. S. Soltz. 1982. Sensitivity of tomato cultivars to sulfur dioxide. *HortScience* **17**:249-250.

Howell, F. J. 1984. New pheromone trap for monitoring codling moth (Lepidoptera: Olethreutidae) populations. *J. Econ. Entomol.* **77**:1612-1614.

Hull, L. A., and V. R. Starner. 1983. Impact of four synthetic pyrethroids on major natural enemies and pests of apple in Pennsylvania. *J. Econ. Entomol.* **76**:122-130.

Janick, J. 1979. Horticulture's ancient roots. *HortScience* **14**:299-313.

Janick, J. 1982. Pest resistance in horticultural crops: Tomato and apple. In *Research for Small Farms,* H. W. Kerr, Jr., and L. Knutson (eds.), Miscellaneous Publication 1422. Agricultural Research Service, Washington, D.C., pp. 66-71.

Jean-Pierre, R. LeBlanc, S. B. Hill, and R. O. Paradis. 1984. Oviposition in scout-apples by plum curculio, *Contrachelus nenuphar* (Herbst) (Coleoptera: Curculionidae), and its relationship to subsequent damage. *Environ. Entomol.* **13**:286-291.

Jones, A. L. 1971. *Diseases of Tree Fruits in Michigan,* Extension Bulletin E-174. Michigan State University, East Lansing, Mich.

Jones, D. P. 1973. Agricultural entomology. In *History of Entomology,* R. F. Smith, T. E. Mittler, and C. N. Smith (eds.). Annual Reviews, Palo Alto, Calif., pp. 307-331.

Kaplan, R. 1973. Some psychological benefits of gardening. *Environ. Behav.* **5**:145-161.

Kirby, R. D., and J. E. Slosser. 1984. Composite economic threshold for three lepidopterous pests of cabbage. *J. Econ. Entomol.* **77**:725-733.

Latheef, M. A., and J. H. Ortiz. 1984. Influence of companion herbs on *Phyllotreta cruciferae* (Coleoptera: Chrysomelidae) on collard plants. *J. Econ. Entomol.* **77**:80-82.

Latheef, M. A., J. H. Ortiz, and A. Q. Sheikh. 1984. Influence of intercropping on *Phyllotreta cruciferae* (Coleoptera: Chrysomelidae) populations on collard plants. *J. Econ. Entomol.* **77**:1180-1184.

Latin, R. X., D. L. Matthew, E. C. Tigchelaar, S. E. Weller, G. E. Wilcox, and R. J. Barman. 1983. *Vegetable Production Guide,* Publication ID-56. Extension Service, Purdue University, West Lafayette, Ind.

Leibee, G. L., R. B. Chalfant, D. J. Schuster, and R. B. Workman. 1984. Evaluation of visual damage thresholds for management of cabbage caterpillars in Florida and Georgia. *J. Econ. Entomol.* **77**:1008-1011.

Leink, S. E., C. M. Watve, and R. W. Weires. 1980. Phytophagous and predacious mites on apple in New York. *Search* **6**:14.

Liu, H. J., F. L. McEwen, and G. Ritcey. 1982. Forecasting events in the life cycle of the onion maggot, *Hylemya antiqua* (Diptera: Anthomyiidae): Application to control schemes. *Environ. Entomol.* **11**:751-755.

Lodeman, E.G. 1903. *The Spraying of Plants.* Macmillan, New York.

Lotstein, R. J., and D. D. Davis. 1983. Influence of chronic sulfur dioxide exposure on early blight of tomato. *Plant Dis.* **67**:797-800.

Loy, J. B., and O. S. Wells. 1982. A comparison of slitted polyethylene and spunbonded polyester for plant row covers. *HortScience* **17**:405-407.

McKee, M. J., and C. O. Knowles. 1984. Effects of pyrethroids on respiration in the twospotted spider mite (Acari: Tetranychidae). *J. Econ. Entomol.* **77**:1376-1380.

Macnab, A. A., A. F. Sherf, and J. K. Springer. 1983. *Identifying Diseases of Vegetables.* Pennsylvania State University, University Park, Pa.

Maguire, L. A. 1984. Influence of surrounding plants on densities of *Pieris rapae* (L.) eggs and larvae (Lepidoptera: Pieridae) on collards. *Environ. Entomol.* **13**:464-468.

Mayer, K. 1959. *4500 Jahre Pflanzenshutz Zeittafel zur Geschichte des Pflanzenshutzes und der Shadlingsbekampfung unter besonderer Berucksichtigung der Verhaltnisse in Deutschland.* Eugen Ulmer, Stuttgart.

Moffitt, H. R., and P. H. Westigard. 1984. Suppression of the codling moth (Lepidoptera: Tortricidae) population on pears in southern Oregon through mating disruption with sex pheromone. *J. Econ. Entomol.* **77**:1513-1519.

Morge, G. 1973. Entomology in the western world in antiquity and in medieval times. In *History of Entomology,* R. F. Smith, T. E. Mittler, and C. N. Smith (eds.). Annual Reviews, Palo Alto, Calif., pp.37-80.

Nawrocka, B. Z., C. J. Eckenrode, J. K. Uyemoto, and D. H. Young. 1975. Reflective mulches and foliar sprays for suppression of aphid-borne viruses in lettuce. *J. Econ. Entomol.* **68**:694-698.

O'Brien, M. J., and A. E. Rich. 1976. *Potato Diseases,* Handbook No. 474. U.S. Department of Agriculture, Washington, D.C.

Pederson, L. H., and C. J. Echenrode. 1981. Predicting cabbage maggot flights in New York using common wild plants, *New York's Food and Life Sciences Bulletin 87.* New York Agricultural Experiment Station, Geneva, N.Y.

Peterson, C. E., P. H. Williams, M. Palmer, and P. Loriward. 1982. Wisconsin 2757 cucumber. *HortScience* **17**:268.

Porter, W. E., and W. W. Etzel. 1982. Effects of aluminum-painted and black polyethylene mulches on bell pepper. *HortScience* **7**:942-943.

Pratt, A. J., W. E. Blauvelt, and A. W. Dimock. 1944. Victory gardening, *Cornell Extension Bulletin 631.* Cornell University, Ithaca, N. Y.

Prokopy, R. J., and K. I. Hauschild. 1979. Comparative effectiveness of sticky red spheres and pherocon am standard traps for monitoring apple maggot flies in commercial orchards. *Environ. Entomol.* **8**:696-700.

Reed, K. K., B. Freedman, and T. L. Ladd, Jr. 1982. Insecticidal and antifeedant activity of neriifolin against codling moth, striped cucumber beetle, and Japanese beetle. *J. Econ. Entomol.* **75**:1093-1097.

Reissig, W. H., B. L. Fein, and W. L. Roelofs. 1982. Field tests of synthetic apple volatiles as apple maggot (Diptera: Tephritidae) attractants. *J. Econ. Entomol.* **76**:1294-1298.

Reissig, W. H., R. W. Weires, C. G. Forshey, W. L. Roelofs, R. C. Lamb, H. S. Altwinckle, and S. R. Alm. 1984*a*. Management of the apple maggot, *Rhagoletis pomonella* (Walsh) (Diptera: Tephritidae), in disease-resistant dwarf and semi-dwarf apple trees. *Environ. Entomol.* **13**:684-690.

Reissig, W. H., R. W. Weires, G. C. Forshey, W. L. Roelofs, R. C. Lamb, and H. S.

Aldwinckle. 1984*b*. Insect management in disease-resistant dwarf and semi-dwarf apple trees. *Environ. Entomol.* **13:**1201-1207.

Relf, P. D. 1982. Consumer horticulture: A psychological perspective. *HortScience* **17:**317-319.

Roberts, D. A., and C. W. Boothroyd. 1972. *Fundamentals of Plant Pathology.* W. H. Freeman, San Francisco.

Rock, G. C., C. C. Childers, and H. J. Kirk. 1978. Insecticide applications based on codlemone trap catches vs. automatic schedule treatments for codling moth control in North Carolina apple. *J. Econ. Entomol.* **71:**650-653.

Sanders, T. W. 1919. *Vegetables and Their Cultivation.* W. H. & L. Collingridge, London.

Schalk, J. M., and A. K. Stoner. 1979. Tomato production in Maryland: Effect of different densities of larvae and adults of the Colorado potato beetle. *J. Econ. Entomol.* **72:**826-829.

Sears, M. K., R. P. Jaques, and J. E. Laing. 1983. Utilization of action thresholds for microbial and chemical control of lepidopterous pests (Lepidoptera: Noctuidae, Pieridae) on cabbage. *J. Econ. Entomol.* **76:**368-374.

Shelton, A. M., M. K. Sears, J. A. Wyman, and T. C. Quick. 1983. Comparison of action thresholds for lepidopterous larvae on fresh-market cabbage. *J. Econ. Entomol.* **76:**196-199.

Shields, E. J., and J. A. Wyman. 1984. Effect of defoliation at specific growth stages on potato yields. *J. Econ. Entomol.* **77:**1194-1199.

Slingerland, V., and C. R. Crosby. 1914. *Manual of Fruit Insects.* Macmillan, New York.

Southwood, T. R. E. 1978. *Ecological Methods: With Special Reference to the Study of Insect Populations,* 2nd ed. Halsted Press, London, 524p.

Stevenson, W. R. 1979*a*. Cucumber, bacterial wilt profile. *Plant Disease Profiles: Vegetables.* FACTS Profile Series, Purdue University, West Lafayette, Ind.

Stevenson, W. R. 1979*b*. Muskmelon, powdery mildew profile. *Plant Disease Profiles: Vegetables.* FACTS Profile Series, Purdue University, West Lafayette, Ind.

Stevenson, W. R. 1979*c*. Tomato, early blight profile. *Plant Disease Profile: Vegetables.* FACTS Profile Series, Purdue University, West Lafayette, Ind.

Stevenson, W. R. 1979*d*. Tomato, fusarium wilt profile. *Plant Disease Profiles: Vegetables.* FACTS Profile Series, Purdue University, West Lafayette, Ind.

Stevenson, W. R., H. C. Harrison, and M. F. Heimann. 1984. *Disease Resistant Vegetables for the Home Garden,* Cooperative Extension Publication A3110. University of Wisconsin, Madison.

Stoner, A. K. 1982. Combining sequential cropping of vegetables and modern cultural practices to maximize land use on small farms. In *Research for Small Farms,* H. W. Kerr, Jr., and L. Knutson (eds.), Miscellaneous Publication 1422. Agricultural Research Service, Washington, D.C., pp. 49-55.

Tamaki, G., and B. A. Butt. 1978. *Impact of* Perillus bioculatus *on the Colorado Potato Beetle and Plant Damage,* Technical Bulletin 1581. U.S. Department of Agriculture, Washington, D.C.

Tedders, W. L., D. J. Weaver, E. J. Wehunt, and C. R. Gentry. 1982. Bioassay of *Metarhizium anisopliae, Beauveria bassiana,* and *Neoaplectana carpocapsae* against larvae of the plum curculio, *Conotrachelus nenuphar* (Herbst). *Environ. Entomol.* **11:**901-904.

Tette, J. P. 1972. Researching methods for implementing IPM on small farms in the northeast. In *Research for Small Farms,* H. W. Kerr, Jr., and L. Knutson (eds.), Miscellaneous Publication 1422. Agricultural Research Service, Washington, D.C., pp. 72-75.

Thomas, C. E., and R. E. Webb. 1982. Cinco muskmelon. *HortScience* **17:**684-685.

Trudel, M. J., and A. Gosselin. 1982. Influence of soil temperature in greenhouse tomato production. *HortScience* **17:**628-629.

Trumble, J. T., and N. C. Toscano. 1983. Impact of methamidophos and methomyl on populations of *Liriomyza* species (Diptera: Agromyzidae) and associated parasites in celery. *Can. Entomol.* **115:**1415-1420.

Waddill, V., K. Pohronezny, R. McSorley, and H. H. Bryan. 1984. Effect of manual defoliation on pole bean yield. *J. Econ. Entomol.* **77:**1019-1023.

Walgenbach, J. F., and J. A. Wyman. 1984. Dynamic action threshold levels for the potato leafhopper (Homoptera: Cicadellidae) on potatoes in Wisconsin. *J. Econ. Entomol.* **77:**1335-1340.

Walker, F. C. 1971. *Fusarium Wilt of Tomato,* Monograph No. 6. American Phytopathological Society, St. Paul, Minn.

Warren, A. W., and J. M. Gerber. 1983. *Economic Feasibility of Slitted Row Covers on Muskmelon to Promote Earliness and Yield,* Proceedings of the Illinois Vegetable Growers Schools, Horticultural Series 40. University of Illinois, Urbana.

Weiner, L. F., and D. M. Norris. 1983. Evaluation of sampling and control methods for lesser peachtree borer (Lepidoptera: Sesiidae) and American plum (Lepidoptera: Pyralidae) in sour cherry orchards. *J. Econ. Entomol.* **76:**1118-1120.

Wells, P. W., G. P. Dively, and J. M. Schalk. 1984. Resistance and reflective foil mulch as control measures for the potato leafhopper (Homoptera: Cicadellidae) on Phaseolus species. *J. Econ. Entomol.* **77:**1046-1051.

Wescott, C. 1973. *The Gardener's Bug Book.* Doubleday, New York.

Westigard, P. H., and H. R. Moffitt. 1984. Natural control of the pear psylla (Homoptera: Psyllidae): Impact of mating disruption with the sex pheromone for control of the codling moth (Lepidoptera: Tortricidae). *J. Econ. Entomol.* **77:**1520-1523.

William, R. D. 1981. Complementary interactions between weeds, weed control, practices, and pests in horticultural cropping systems. *HortScience* **16:**508-513.

Williams, E. B., J. Janick, F. H. Emerson, D. F. Dayton, L. F. Hough, and C. Bailey. 1981. "Redfree" apple. *HortScience* **16:**798-799.

Yepsen, R. B., Jr., ed. 1976. *Organic Plant Protection.* Rodale, Emmaus, Pa.

Yonce, C. E. 1980. Effectiveness of chlorpyrifos for control of *Syanthedon pictipes* and *S. exitiosa* in peach orchard tests of young trees with emphasis on timing applications. *J. Econ. Entomol.* **73:**827-828.

York, A. C., G. L. Reed, and K. K. Kinney. 1982. Effects of soil insecticides on cantaloupe yield. *Insectic. Acaricide Tests* **8:**103-104.

Zungoli, P. A., A. L. Steinhauer, and J. J. Linduska. 1983. Evaluation of diflubenzuron for Mexican bean beetle (Coleoptera: Coccinellidae) control and impact on *Pediobius foveolatus. J. Econ. Entomol.* **76:**188-191.

10

Urban Landscape Pest Management

Daniel A. Potter
University of Kentucky, Lexington

Research in entomology and the applied plant sciences has historically been concerned mostly with agricultural, silvicultural, and medical/veterinary pest problems, while horticultural areas have been relatively neglected. This lack of research is unfortunate because there is a growing need for improved pest control technology in the urban and suburban environment. During the last 40 years, large areas of land throughout the United States have been and continue to be developed into subdivisions, shopping centers, schools, or industrial sites. As cities and suburbs have grown, so have public awareness of and appreciation for green lawns, tree-lined streets, well-manicured golf courses, and other recreational areas. With continued urbanization seemingly inevitable, the demand for turfgrass, shade trees, and other ornamental plants will continue to increase.

THE VALUE OF HORTICULTURAL PLANTS

The value of trees, shrubs, and turfgrass is both functional and aesthetic. Urban vegetation supplies oxygen, reduces glare, soil erosion, water runoff, and noise pollution, and filters dust from the air. Tree rows can provide privacy or serve as an effective windbreak. Green plants also are important for temperature modification. On a summer day, one acre of turfgrass can lose an estimated 2,400 gallons of water through transpiration, providing a cooling effect equivalent to a 70 ton air conditioner (Kageyama, 1982).

Surface temperature on living turf may be much lower than on other surfaces. For example, in one study (Mecklenburg, 1971) the surface temperature of artificial turf was found to be 153°F, compared to 83°F on adjacent living grass. Shade trees can also reduce the temperatures of city streets by more than 10° (Federer, 1976). Quality lawns or landscapes also provide important economic and commercial benefits. Green grass and plants around a factory or business convey a favorable impression to employees and to the general public. An attractive lawn adds an estimated average of more than 3% to the resale value of a modest home (Kageyama, 1982), while trees alone increase the value of residential property by an average of 6-9% (Morales, 1980).

Perhaps most important, trees, shrubs, and turfgrasses help to make urban areas a pleasant place to live and work. Turfgrasses enhance the safety and enjoyment of many sports and leisure activities, while trees, shrubs, and flowers provide shade, beauty, and recreation for millions of people. These aesthetic benefits are increasingly important to the physical health and psychological well-being of the urban population (Starkey, 1979).

Turfgrasses cover an estimated 25 million acres in the United States (Kageyama, 1982). The value of the national turfgrass industry is not precisely known, but in 1982 annual expenditures for turfgrass maintenance were estimated at more than $1 billion in New York State alone (Anonymous, 1983). This figure does not include the value of unpaid labor or the cost of equipment replacement. Gross sales of lawn care and landscape management companies in 1983 were more than $2.2 billion (Anonymous, 1984*b*). Much of the expense of turf maintenance can be attributed to the large acreage, the high cost of specialty products and the small amounts purchased by each individual.

The value of the nursery industry in the United States tripled during the 1970s. In 1981 total retail sales were estimated at $18.4 billion and involved more than 46,000 retail outlets (Hensley and Siebert, 1982). Recent surveys indicate that there may be as many as 342 million urban trees in the United States (Kielbaso and Kennedy, 1983). This number includes about 60 million street trees, with an estimated value (as of 1985) of nearly $25 billion.

THE URBAN ENVIRONMENT AND PEST PROBLEMS

Urban environments can be severe. Hot temperatures, restricted air movement, and severely disturbed or highly compacted soils frequently provide a harsh environment for landscape plantings. Landfill and grading activities accompanying the development of new home lots and subdivisions often result in the removal of natural topsoil, making the tasks of establishing and maintaining turfgrass and woody plants very difficult. Additional stresses result from drought, improper cultivation, and exposure to road salt and air pollution. Trees and shrubs may be subjected to trauma from trunk injuries inflicted by lawn mowers, vandalism, road vehicles, or construction equipment, which can provide invasion sites for pathogens and insects.

Although the biological reasons are not fully understood, stressed plants are especially prone to attack by arthropod pests such as mites, borers, and scales. For example, native understory trees such as rhododendron and dogwood may become more susceptible to borers when transplanted into the urban landscape (Neal, 1982; Potter and Timmons, 1981). It is known that exposure to abiotic or biotic stress can induce biochemical changes in plants, including altered levels of foliar nitrogen, amino acids, and sugars. Even relatively small biochemical changes in a plant can affect the feeding preferences, survival, and population growth of insects using it as a food source (Kimmins, 1971; Lewis, 1979; White, 1969, 1974, 1984). Environmental stress is also a predisposing factor for many diseases of woody plants (Schoeneweiss, 1975). For example, European white birch may become more susceptible to *Botrysphaeria dothidea* following exposure to water stress (Crist and Schoeneweiss, 1975). Kentucky bluegrass, the predominant cool season turfgrass in the United States, undergoes severe stress during the hot summer months, particularly in the transitional climatic zone (Beard, 1973). This predisposes it to many insect, disease, and weed problems.

PESTICIDE USE PATTERNS

Enormous cost and time are required for the chemical industry to test and market a new pesticide (Kageyama, 1982). Because of the relatively small size of the urban landscape market, few pesticides have been developed specifically for use on trees, shrubs, and turf. The availability of new pesticides generally depends upon their success in the much larger agricultural market. For example, isofenphos, a widely used turfgrass insecticide, was originally developed for control of soil insects in corn. Bendiocarb, a carbamate insecticide, is one example of a pesticide that was developed initially for use against landscape and structural pests.

Formulation technology is providing a number of new products to better serve the needs of the urban market. For example, combination fertilizer plus insecticide or herbicide products are available that enable the homeowner to perform two tasks in a single time-saving operation. Improved packaging and delivery systems are being developed that allow a pesticide to be applied safely and uniformly to the target site. These include pre-measured hose-end applicators, insecticidal tree spikes, tree injection systems, and other methods that reduce the hazards of pesticide use.

An important change in pesticide use patterns in the urban landscape has been the rapid growth of the commercial tree, shrub, and lawn care industry since about 1975. Gross annual sales of the lawn care industry increased at an average annual rate of 22% between 1977 and 1984 (Anonymous, 1984*b*). Approximately 6% of the 45 million owner-occupied single family homes with lawns and about 13% of those with incomes over $20,000 contracted for professional lawn care in 1984 (Anonymous, 1984*b*). Most of these services involve an annual schedule of fertilizer and pesticide applications (Table 10-1). Many of these companies are also involved in tree and shrub protection and other landscape management activities.

Insect Control in Turf

The highly persistent chlorinated cyclodiene insecticides aldrin, dieldrin, chlordane, and heptachlor came into general use about 1945 and were the mainstay of turfgrass insect control until shortly after 1970. These chemicals provided one of the most effective soil insect controls ever devised. Spray timing was not critical because a single application generally provided control of soil-inhabiting insect pests for several years (Tashiro, 1981). However, widespread pest resistance and concern over environmental persistence resulted in cancellation of cyclodiene registrations on turf.

Subsequently, organophosphate and carbamate insecticides such as chlorpyrifos, diazinon, trichlorfon, and bendiocarb came into general use against turf insects. Proper timing of applications is essential with these insecticides because they quickly degrade (Kuhr and Tashiro, 1978). Several applications may be necessary to control multivoltine insects such as sod webworms, greenbug aphids, and chinch bugs. Moreover, these chemicals tend to become bound in the upper soil and organic thatch layer, and thus their effectiveness is limited against soil-inhabiting pests (Niemczyk, Krueger and Lawrence, 1977; Sears and Chapman, 1979).

Isofenphos, an organophosphate, became widely used on turfgrass in the 1980s. Residues of isofenphos remain effective in the soil for at least 200 days (Tashiro, 1981), so spray timing is not as critical as with other turf insecticides. The residual characteristic of isofenphos is particularly advantageous to lawn care operators because it obviates the need for critical spray timing and immediate post-treatment irrigation.

Insect Control on Trees and Shrubs

During the past 20 years, DDT and other highly persistent broad spectrum insecticides have been largely replaced by short residual organophosphates and carbamates such as malathion, acephate, diazinon, and carbaryl for controlling insects on trees and shrubs. However, certain of the less persist-

Table 10-1. A turfgrass maintenance program typical of those commonly applied by professional lawn care companies in the midwestern United States.

Date of application	*Treatment*	*Target pest*
Mid-February–mid-April	2,4-D, dicamba, and/or MCPP DCPA, bensulide, or benefin N, P_2O_5, K_2O	Broadleaf weeds Preemergent crabgrass
Mid-April–mid-June	2,4-D, dicamba, and/or MCPP chlorpyrifos or aspon N	Broadleaf weeds Surface-feeding insects
Mid-June–mid-August	diazinon or isofenphos N	White grubs
Mid-August–mid-October	2,4-D, dicamba, and/or MCPP N, P_2O_5, K_2O	Broadleaf weeds
Mid-October–January	N, P_2O_5, K_2O	

Source: A. J. Powell (pers. comm.)

ent organochlorides such as lindane and methoxychlor are still available. These insecticides are generally of low to moderate toxicity to humans and of moderate to high toxicity to insects.

There is currently a trend toward the use of relatively nonhazardous and environmentally safe insecticides or delivery systems on trees and shrubs. *Bacillus thuringiensis,* a bacterial insecticide that is nontoxic to higher animals, is widely used against lepidopteran larvae, especially defoliators of shade trees (Olkowski et al., 1978). Research on entomophagous nematodes, fungi, and viruses may soon offer additional microbial insecticides for use on trees and shrubs. Horticultural spray oils are effective against scales, mites, and other small arthropods (Johnson, 1982). Tree implantation or injection of systemic insecticides can sometimes be an effective alternative to conventional sprays, especially in situations where spray drift is a concern. However, the effect of the wounds made for insertion can be harmful in some applications (Kielbaso, 1978). Insect growth regulators (IGRs), a group of chemicals that interfere with the molting process of insects, may offer another alternative to conventional insecticides in the near future.

Disease Control on Trees, Shrubs, and Turf

Prior to World War II, control of diseases on turfgrass and woody ornamentals depended almost entirely on Bordeaux mixture (lime and copper sulfate) and, later, on heavy metal fungicides containing mercury or cadmium (Couch,

1971). These chemicals tended to be phytotoxic and, like DDT, some accumulated in the environment. The latter property led to cancellation of mercury in some states. Development of antibiotic fungicides such as cycloheximide in the 1950s, new organic fungicides (dithiocarbamates) such as thiram, zineb, and maneb in the 1960s, and systemic fungicides such as benomyl, metalaxyl, and triademefon in the 1970s and 1980s has provided a degree of efficacy never before realized. However, resistance to certain fungicides such as benomyl and metalaxyl is becoming more widespread each year (Couch, 1971). As with insecticides, there is increased interest in tree implantation or injection with systemic fungicides (Kielbaso, 1978; Phair and Ellmore, 1982).

Weed Control

Before the development of 2,4-D in the 1940s, sodium arsenite, potassium cyanate, and ammonium sulfate were widely used for weed control in turfgrass. These early herbicides were sometimes hazardous to use and often phytotoxic to desirable grasses (Engle and Ilnicki, 1969). The phenoxy herbicides, including 2,4-D, 2,4,5-T, mecoprop, and silvex were generally safer and less expensive, and because they are translocated, they could be used in low volume. Phenoxy herbicides, with the exception of 2,4,5-T and silvex, are still the mainstay of broadleaf weed control on urban lawns.

PROBLEMS WITH PESTICIDE USE

The potential problems and hazards of pesticide use in urban areas have been discussed in an earlier chapter. This section summarizes certain detrimental side-effects of pesticides that pertain specifically to horticultural plants.

Thatch Accumulation in Turfgrass

Thatch is a tightly bound layer of dead and living roots and stems that accumulates between the soil surface and green vegetation in turfgrass (Beard, 1973). Excessive thatch results from an imbalance between production and decomposition of organic matter. Among the problems associated with thatch buildup are restricted penetration of fertilizer and binding of insecticides (Cornman, 1952; Niemczyk, Krueger, and Lawrence, 1977), reduced water infiltration (Taylor and Blake, 1982), and shallow root growth, with increased vulnerability to heat and drought stress.

Kentucky bluegrass lawns maintained with multiple applications of pesticides and fertilizer often accumulate a thatch problem within four to five years. Pesticides or fertilizer can affect thatch accumulation by directly or indirectly (e.g., through soil acidification) reducing populations of microorganisms (Smiley and Craven, 1978) or earthworms (Potter, Bridges, and Gordon, 1985; Randall, Butler, and Hughes, 1972; Turgeon, Freeborg, and Bruce, 1975), which are important to plant litter decomposition (Edwards and Heath, 1963).

Pest Resurgences or Secondary Pest Outbreaks

Exposure to some herbicides and fungicides can increase the susceptibility of turfgrasses to certain diseases (Madison, 1961). For example, leaf spot caused by species of *Drechslera* on Kentucky bluegrass may be more severe following treatment with post-emergent herbicides (Hodges, 1978) or benomyl (Hartman and Powell, 1980). Herbicides may encourage plant disease by increasing pathogen growth or virulence, by inhibiting microflora that compete with a pathogen, by decreasing the host's structural defense, or by inducing physiological changes in the host that increase disease expression (Altman and Campbell, 1977; Katan and Eshel, 1973).

Outbreaks of certain scale insects, whiteflies, mites, and other pests sometimes occur after application of insecticides (Debach and Rose, 1977; Luck and Dahlsten, 1975; McClure, 1977; Merritt, Kennedy, and Gersabeck, 1983). Such outbreaks are most often attributed to elimination of natural enemies that normally help to regulate the pest population. Applications of common turfgrass insecticides may reduce predatory arthropod populations in lawns for at least six weeks (Cockfield and Potter, 1983). This may result in short-term reductions in predator-induced mortality of certain pests, such as sod webworms (Cockfield and Potter, 1984). Reinert (1978*b*) has reported that natural enemies prevented rapid buildup of chinch bug populations in unsprayed lawns but that outbreaks occurred on lawns that received repeated insecticide treatments. Similarly, populations of winter grain mites may increase following application of carbaryl to lawns (Streu and Gingrich, 1972).

Pest Resistance

Acquired resistance to insecticides has been documented for a number of epigeal and soil-inhabiting turfgrass insects, including webworms, chinch bugs, bluegrass billbugs, greenbugs, and several species of white grubs (Reinert, 1982*a;* Tashiro, 1982). In recent years, resistant strains of pathogens have been reported for many important diseases of turf and woody ornamentals (Smith, 1982). These problems have been attributed to the exclusive and intensive use of the newer, systemic fungicides, most of which have relatively specific sites of action (Dekker, 1976; Clay, Hardy, and Hammond, 1985).

Phytotoxicity and Drift

Phytotoxicity can occur when pesticides are used on or around certain ornamental plants. Some plant species are particularly sensitive to certain fungicides or insecticides and may be disfigured or defoliated if sprayed (Partyka et al., 1980; Patel and Gammel, 1978). This is often a problem in the home landscape where plant species or varieties may not be known. Liquid pesticides that contain xylene or a petroleum-based carrier often cause foliage injury if applied at temperatures above 80°F. Wettable powders are safer to use above these temperatures. Horticultural spray oils used for control of scales and other insects may be phytotoxic to certain plant species. Use of some granular herbicide formulations, such as dicamba, may be followed by downward or lateral movement of herbicide into the root zone. Uptake of the herbicide may be harmful to trees and shrubs.

Herbicide spray drift can injure shrubs and flower or vegetable gardens when broadleaf weed controls are applied to lawns. However, drift tends to be less of a problem for lawn applicators, who generally apply large droplets, than for sprayers of tall trees. Pesticide drift has become a political issue in crowded urban areas and has resulted in an increase in local ordinances restricting the use of pesticides (Kenney, 1983).

NONCHEMICAL PEST MANAGEMENT PROCEDURES

Promoting plant health through the use of sound cultural practices is probably the most important management strategy for minimizing pest problems on ornamental plants. A dense, actively growing turf is better able to compete with weeds for light, water, and nutrients and to resist or recuperate from fungal diseases and insect injury. Similarly, trees and shrubs should be watered, fertilized, and pruned to promote vigor. Plant species and cultivars should be selected not only for their aesthetic qualities but also for their compatibility with local growing conditions and relative resistance to pests.

This section provides an overview of nonchemical methods for combating pests of outdoor ornamental plants. General guidelines for cultivation of turfgrass and woody ornamentals are covered in several excellent texts (e.g., Beard, 1973; Harris, 1983; Madison, 1971; Pirone, 1978*b*). References for examples given in this section and additional pest management methods are summarized in Tables 10-2 through 10-4.

Cultural Methods: Turfgrass

Uniformity of growth is an important component of turfgrass quality (Beard, 1973). Weeds often disrupt uniformity by having a noticeably different color, texture, leaf width, or growth habit. Insects and diseases cause discoloration, reduced shoot and/or root density, and unsightly dead areas. Routine cul-

Table 10-2. Pest management strategies for some important diseases and insects infesting turfgrass.

Pest	*Pest management strategies*	*References*
Diseases		
Drechslera leaf spot and melting out (*Drechslera* spp.)	Avoid excessive N, overwatering, or drought stress; control thatch; raise mowing height; remove diseased clippings; resistant cultivars	Beard, 1973; Couch, 1980; Madison, 1971
Fusarium blight (*Fusarium* spp.)	Avoid excessive N or drought stress; control thatch; sharpen mower blade; resistant cultivars	Beard, 1973; Couch, 1980; Madison, 1971
Dollar spot (*Sclerotinia homoeocarpa*)	Avoid N deficiency or drought stress; control thatch; raise mowing height; resistant cultivars	Beard, 1973; Couch, 1980; Madison, 1971
Red thread (*Corticium fuciforme*)	Avoid N deficiency; maintain high K level; resistant cultivars	Beard, 1973; Couch, 1980; Madison, 1971
Brown patch (*Rhizoctonia solani*)	Avoid excessive N or overwatering; control thatch; raise mowing height; sharpen mower; remove diseased clippings; maintain high K level; avoid excessive shoot growth entering winter; resistant cultivars	Beard, 1973; Couch, 1980; Madison, 1971
Insects		
Japanese beetles (*Popillia japonica*)	Milky spore powder; attractant traps	Klein, 1981*a* Gordon and Potter, 1985; Klein, 1981*b*
Sod webworms (*Crambus, Pediasia Chrysoteuchia* spp.)	Pheromone traps; resistant cultivars	Kamm, 1982
Chinch bugs (*Blissus* spp.)	Resistant cultivars	Reinert, 1982*b*
Mole crickets (*Scapteriscus* spp.)	Resistant cultivars	Reinert, 1982*b*
Bluegrass billbugs (*Sphenophorus parvulus*)	Resistant cultivars	Reinert, 1982*b*
Armyworms	Pheromone traps; resistant cultivars	Kamm, 1982 Reinert, 1982*b*
Cutworms	Pheromone traps; resistant cultivars	Kamm, 1982 Reinert, 1982*b*
Rhodesgrass mealybugs (*Antonina graminis*)	Parasitic wasps	Dean et al., 1979

Table 10-3. Pest management strategies for some important diseases of ornamentals and shade trees.

Disease and causal agent	*Pest management strategies*	*References*
Dutch elm disease (*Ceratocystis ulmi*)	Remove and destroy diseased, weak, and dead trees and branches; enhance tree vigor; prevent root grafts by trenching between diseased and healthy trees; resistant genotypes	Schreiber and Peacock, 1974
Pinewood nematode (*Bursaphelenchus xylophilis*)	Remove and destroy diseased, weak, and dead trees; avoid drought stress	Spencer and Jones, 1982
Fire blight (*Erwinia amylovora*)	Sanitation pruning; remove diseased trees; remove suckers; avoid cultural practices that promote excessive succulent growth; resistant genotypes	Hartman, 1982; Tattar, 1978
Crown gall (*Agrobacterium tumefaciens*)	Biocontrol with *Agrobacterium radiobacter* strain 84; sanitation pruning; avoid unnecessary wounding	Kerr, 1980
Verticillium wilt (*Verticillium dahliae*)	Avoid drought stress; fertilize early in spring; resistant genotypes	Tattar, 1978
Scab (*Venturia* spp.)	Resistant genotypes	Hartman, 1982
Powdery mildew (many species)	Improve air movement and increase sunlight around plants; resistant genotypes	Tattar, 1978; Jones, 1982
Black spot (*Diplocarpon rosae*)	Prune out infected rose canes in late winter	McCoy and Philley, 1982
Black knot (*Dibotryon morbosum*)	Sanitation pruning	Tattar, 1978
Phytophthora root rot (*Phytophthora* spp.)	Select well-drained planting sites; use drain tiles to remove excess water; maintain balanced fertilization; resistant genotypes	Benson, 1982; Tattar, 1978

tural practices such as fertilization, irrigation, and mowing can have a pronounced influence on turf vigor as well as the severity of common pest problems (Beard, 1973; Couch, 1980).

Fertility. Maintaining balanced fertility is important for good growth of turfgrasses. Insufficient nitrogen results in thin turf, which favors weed encroachment and encourages certain diseases such as dollar spot and red

Table 10-4. Pest management strategies for some important insect pests of ornamentals and shade trees.

Pest	*Pest management strategies*[a]	*References*
Japanese beetle *(Popillia japonica)*	Attractant traps; milky spore powder; protective mesh over plants	Gordon and Potter, 1985; Klein, 1981*b*, Klein, 1981*a*
Gypsy moth *(Porthetria dispar)*	Destroy egg masses; burlap or sticky tree bands; BT	Westcott, 1973
Bagworms *(Thyridopteryx ephemeraeformis)*	Handpick larvae or overwintering egg masses in bags; BT	Westcott, 1973
Web or tent-making caterpillars (several genera)	Prune out egg masses; prune, or mechanically destroy nests; BT	Westcott, 1973
Cankerworms *(Alsophila, Paleocrita)*	Sticky tree bands; BT	Westcott, 1973
Bronze birch borer *(Agrilus anxius)*	Maintain tree vigor; prune out and destroy dead or infested branches in early spring	Nielsen, 1979*b*
Clearwing borers (Sesiidae)	Maintain tree vigor; pheromone traps; avoid bark injuries	Nielsen, 1979*b*; Potter and Timmons, 1981
Spider mites (Tetranychidae)	Syringe foliage with water hose	Westcott, 1973
Aphids (Aphididae)	Syringe foliage with hose; release lady beetles or lacewings; parasites	Olkowski et al.; 1978; Westcott, 1973
Lacebugs (Tingidae)	No effective methods	
Scale insects (Coccoidea)	Horticultural spray oils	Johnson, 1982; Westcott, 1973

[a]*BT = Bacillus thuringiensis*

thread. However, excessive fertilization may enhance seed germination of crabgrass and certain other weeds, and encourage outbreaks of Drechslera leaf spot and melting out, Fusarium patch and blight, and *Rhizoctonia* brown patch. Heavy spring fertilization of Kentucky bluegrass growing in the transitional zone can reduce summer drought tolerance, resulting in thin, weak turf and more pest problems. Supplemental nitrogen often elicits enhanced growth and reproduction of herbivorous insects (Mattson, 1980), but the effect of fertilization on greenbugs, chinch bugs, and other herbivores in turfgrass has not been evaluated.

Irrigation. Almost all fungal pathogens require free water or a high relative humidity to germinate and grow. Some *Pythium* spp. and *Rhizoctonia* spp. tend to be more damaging on turf that is irrigated excessively or that receives frequent, shallow waterings. Turf should be irrigated in morning or early afternoon to encourage rapid evaporation of water droplets from the foliage but never at night. Some diseases, such as Drechslera and Bipolaris leaf spot and Fusarium blight, are often more severe on drought-stressed turf. Adequate irrigation helps turfgrass recover from root-feeding insect injury (Potter, 1982).

Mowing. Excessively low mowing depletes carbohydrate reserves and can weaken grass plants, making them more susceptible to diseases. Wounds caused by mowing are a principal site of invasion by many turfgrass pathogens. *Drechslera* and *Bipolaris* spp., *Sclerotinia homoeocarpa,* and other pathogens can be discouraged by avoiding close mowing of turf. Weeds such as goosegrass and crabgrass, which require light for seed germination, grow better in closely mowed than in dense, high-cut turf.

Thatch and Clipping Removal. Excessive thatch reduces turf vigor and provides a favorable microenvironment for fungal pathogens (Couch, 1980). Thatch also provides a refuge for insect pests and can shield them from insecticides. Removal of thatch by vertical mowing or cultural management programs generally helps reduce pest problems. However, vertical mowing or other cultivation should be done only during months when the potential for turf recovery and growth is high. Removal of diseased clippings may help reduce the severity of leaf spot and certain other diseases.

Liming. Field observations in the 1950s suggested that Japanese beetle larvae were adversely affected by high soil pH. However, research has shown that larval growth and survival does not differ within the pH range that is optimal for turfgrass growth (Vittum and Tashiro, 1980). Therefore, application of lime is probably of little value for reducing grub populations. Lime has been shown to affect the severity of certain diseases, such as Ophiobolus patch and Fusarium patch, in turfgrasses (Couch, 1980).

Cultural Methods: Trees and Shrubs

The importance of promoting the vigor of woody plants and avoiding plant stress has been emphasized by Schoeneweiss (1975, 1981), Nielsen (1979*b,* 1981), and many others. Proper watering, pruning, and fertilization will usually keep trees and shrubs in a vigorous condition and help them avoid or recover from infestation by insects or infection by pathogens. Use of vigorous, well-adapted plant species and selection of proper planting sites are also important.

Mulching can reduce or eliminate weeds from landscape plantings and can help conserve soil moisture. Mulches or lawnmower guards help prevent trunk injuries, which can provide entry to borers (Potter and Timmons, 1981) and certain pathogens. Some leaf diseases can be minimized by planting and pruning, which permits greater penetration of light and movement of air through the leaf canopy (Harris, 1983). In home plantings, severe disease problems often occur following transplanting (Schoeneweiss, 1975). Proper handling and planting techniques are necessary to avoid "transplant shock" and ensuing pest problems.

Sanitation

Sanitary practices are very important for prevention of weeds and certain insects and diseases in turfgrass. Seed should be free of objectionable weed species and sod, sprigs, or plugs should be purchased from a reputable dealer and inspected for freedom from pests. Soil that is purchased for establishment or used for topdressing should be free of weed seeds and pathogenic fungi or be sterilized before use. Weed seeds, fungal conidia, and small insects such greenbugs can be transferred on wet shoes, mowers, or cultivation equipment. Maintenance equipment should be thoroughly cleaned before moving from infested to uninfested areas. Infected grass clippings are an important source of inoculum for certain fungal diseases such as Drechslera leaf spot and Rhizoctonia brown patch.

Sanitary measures for woody ornamentals and flowering plants include use of disease-free seeds, bulbs, and cuttings; sterilization of soil or compost growing media; pruning infested plant parts; and removal of dead, dying, or infested plants (Pirone, 1978*a*). When pruning diseased plant tissue, cutting tools should be surface sterilized with a disinfectant solution between each cut. Many pathogenic fungi produce fruiting bodies which produce large numbers of infectious spores, especially during wet weather. Pruning and shearing operations generally should not be conducted under damp conditions to reduce disease spread. Felling and immediate removal of infected trees can help reduce the spread of diseases such as pine wilt and dutch elm disease by eliminating the breeding sites of their insect vectors.

Resistant Genotypes

Historically, breeding programs for woody ornamentals have been more concerned with desirable horticultural or aesthetic characteristics than with resistance to pests. Cultivar resistance evaluations have been made for very few plant species. Nonetheless, cultivars do exist that are resistant to certain insects and diseases, and selection of these genotypes can substantially reduce future pest problems. Morgan, Frankie, and Gaylor (1978) provide an overview of the use of insect-resistant plant materials in the landscape. A number of disease-resistant species or cultivars of woody ornamentals are listed by Walker (1982).

Little emphasis has been placed on development of insect resistant turfgrasses; however, cultivars with varying degrees of resistance to chinch bugs, bluegrass billbugs, mole crickets, and other pests have been selected (Reinert, 1982*b*). Unfortunately, little effort has been made by turfgrass breeders to combine resistance factors with other desirable traits. Only a few resistant genotypes have been released as new cultivars (Reinert, 1982*b*; see also Case History III), and these are resistant only to specific pests.

A close association between levels of an endophytic fungus, *Epichloe typhina,* and resistance of perennial ryegrass and tall fescue to certain stem- and leaf-feeding insects has been recently demonstrated (Barker, Pottinger, and Addison, 1983; Clay, Hardy, and Hammond, 1985; Funk et al., 1983; Johnson et al., 1985; Prestige, Pottinger, and Barker, 1982). The fungus is carried as intracellular hyphae in infected plants and is transmitted by seed via the maternal parent. The discovery of endophyte-enhanced resistance to insects may encourage the development of turfgrass cultivars having high levels of endophyte. Recently, seedlings of perennial ryegrass and tall fescue were artificially infected with endophyte (Latch and Christensen, 1985). The results suggest that endophyte levels could be augmented in cultivars with other desirable characteristics or possibly transferred between species. The possibility of endophyte-enhanced resistance to insects in Kentucky bluegrass and other turfgrasses is under investigation. This research may eventually produce turfgrass cultivars resistant to a number of insect pests.

Resistance to major diseases should be an important consideration when selecting a turfgrass for lawn establishment. Although turfgrass species and cultivars differ widely in their resistance to various diseases, the degree of resistance may vary from region to region or may change over time because of development of new strains of pathogens (Beard, 1973; Vargas, 1981). No one grass species or cultivar is resistant to all major turf diseases. Development of disease resistant cultivars is a primary objective of most turfgrass breeding programs (Vargas et al., 1980).

Biological Control

Biological control research on urban horticultural pests has received relatively little attention (National Research Council, 1980). Consequently, the natural enemy complexes of most landscape pests have been neither identified nor studied.

The most successful application of classical biological control (importation and release of a pest's natural enemies) for urban landscape pests has involved the use of parasitic wasps against aphids on shade trees in California (Olkowski et al., 1976; Olkowski et al., 1978). This approach may be applicable to other insect pests on street trees or other large, single species plantings. However, its success requires the cooperation of trained professional tree managers within municipal governments (Olkowski et al., 1976).

With the diversity and patchy distribution of ornamental plants and their associated pest complexes, it seems improbable that classical biological control will be widely implemented at the level of individual homeowners. Only a few species of beneficial arthropods (e.g., lacewings, lady bird beetles, preying mantis eggs) are commercially available. These insects are generalist predators, and their purchase and subsequent release will not control specific pest problems. It is difficult to envision a practical, cost-effective way that natural enemies could be made available for inundative release by the general public. Also, unless collective action were taken over entire neighborhoods (an unlikely event, given the diversity of user attitudes), biological controls implemented by one homeowner might be disrupted by insecticide applications on neighboring properties. Area-wide inoculative releases of natural enemies may eventually be effective against well-established, introduced species such as the elm leaf beetle, Japanese beetle, and gypsy moth, but additional research is needed.

Bacillus popilliae, which causes milky disease in Japanese beetle grubs, is an effective biological control agent on turfgrass. Once the bacterial spores become well-distributed through the soil, the treatment will provide control of grubs for many years thereafter. However, formulations of *B. popilliae* are relatively expensive, and several years may elapse before good control is attained. Another microbial insecticide, *Bacillus thuringiensis,* is effective and widely used against many kinds of caterpillars on ornamental plants. A highly successful biological control method for crown gall disease using a bacterium has recently been developed. This method is discussed more completely in Case History II.

Traps and Attractants

Chemical sex attractants, or pheromones, have been identified and synthesized for a number of important horticultural pests (Mayer and McLaughlin, 1975). Pheromone traps are useful to detect infestations and to monitor pest activity so that insecticides can be applied at the proper time (see Case History I). Light traps are effective for monitoring the night-flying adults of many turf pests. Unfortunately, the relationship between adult captures in pheromone or light traps and subsequent larval density has received little attention. Commercial Japanese beetle traps containing a floral lure and sex attractant are widely available. However, recent experiments (Gordon and Potter, 1985) indicate that use of traps will not prevent or reduce adult feeding damage to nearby landscape plants and may result in even greater damage than would have occurred without a trap.

Physical and Mechanical Controls

Physical and mechanical control methods can be effective on a limited scale. Sticky tree bands afford some protection from cankerworms, gypsy moth and elm leaf beetle larvae, and other climbing defoliators. Galls, egg masses, and the nests of web-making caterpillars can be pruned out. Large, relatively sessile insects such as bagworms can be handpicked on small trees and shrubs. Populations of mites or aphids can be reduced by syringing the foliage of infested plants with a water hose. Additional examples are listed in Table 10-4.

CASE HISTORIES

The following case histories illustrate how pest management principles can be successfully implemented against landscape pests. Case Histories I, II, and III concern the application of sex attractants, biological control, and host plant resistance, respectively, to particular pest problems. Case History IV outlines a comprehensive integrated pest management (IPM) program for landscape plantings.

I. Pheromones of Clearwing Borers

Clearwing borers (Lepidoptera: Sesiidae) are common destructive pests of woody plants (Nielsen, 1978; Potter and Timmons, 1983*c*). The tunneling of borer larvae in living wood destroys vascular tissues and causes loss of vigor, structural weakness, or complete girdling and tree death. Clearwing moths lay their eggs near trunk wounds or in crevices in the bark. The young borers hatch in about two weeks and quickly tunnel into the tree. Once beneath the bark, the borers are protected from insecticides and are seldom detected until serious damage has been done.

Homeowners, arborists, and nurserymen once relied on persistent chemicals such as DDT and dieldrin for borer control. These pesticides were applied early in the spring before adult emergence, and residues persisted on the bark throughout the larval hatching period. Critical spray timing was therefore unnecessary.

Shorter-lived insecticides such as chlorpyrifos (Nielsen, 1979*b*) must now be used because of federal restrictions on DDT and dieldrin. These sprays must be applied just before the larvae hatch, so that the borers are killed before or as they enter the host plant. However, hatching periods differ among species, and between years and geographic areas. Until recently, little was known about the seasonal activity of borers, and there was no way to accurately determine when they were vulnerable to sprays.

When a female clearwing moth emerges, a volatile chemical sex attractant, or pheromone, is emitted from her abdomen. Males detect this airborne pheromone with their antennae and follow it upwind to the receptive female. Egg-laying begins soon after mating.

Research by Nielsen and Balderston (1973) in the early 1970s revealed that some species of clearwing moth are cross-attractive. About the same time, research chemists of the U.S. Department of Agriculture identified and synthesized the main components of the sex pheromones of the peachtree and lesser peachtree borers (Tumlinson et al., 1974). Soon afterward it was discovered that other species of clearwing moth were attracted to pure isomers or isomeric blends of these chemicals (Nielsen, 1979*a*). Field testing in several states led to development of attractants for borers that attack lilac, ash, dogwood, oak, rhododendron, and other woody plants (Nielsen, Purrington and Shambaugh, 1979). This research led to the first commercial marketing of clearwing borer traps in 1978.

Clearwing moth pheromone trap kits are available at a modest cost. The trap is hung in the landscape or nursery in early spring and checked regularly for male clearwing moths. Capture of males indicates that mating and egg laying by a particular species has begun. A protective spray should be applied to the host trees or shrubs about 10-14 days after the first male is caught. This ensures that the insecticidal treatment will coincide with the beginning of the larval hatching period. The duration of the flight period is measured by weekly removal of captured males, and a second spray is applied if flight lasts for more than 8-10 weeks (Nielsen, 1978).

Use of clearwing moth pheromone traps helps to take the guesswork out of spray timing and reduces the probability of unnecessary applications. For example, before the traps were available it was common practice for preventive sprays for lilac, oak, rhododendron, and dogwood borers to be applied monthly from May to October. However, by timing the insecticidal treatment with pheromone traps, these borers can be controlled with a single application (Neal, 1981; Nielsen and Purrington, 1978; Potter and Timmons, 1983*b*). The traps can also reveal if a borer species is very abundant or nearly absent, and control decisions can be modified accordingly.

Forecasting systems based on accumulated thermal units or degree days can be developed by analyzing pheromone trap data over several seasons

(Potter and Timmons, 1983*a*). Mass trapping to reduce borer populations in shelter belts and nurseries may provide another application for pheromones (Nielsen, 1979*a*). As a research tool, synthetic pheromones have helped clarify the biology of a number of pest species (e.g., Neal, 1984; Potter and Timmons, 1983*b;* Purrington and Nielsen, 1977) and have provided a basis for more effective management.

II. Biological Control of Crown Gall Disease

A means of reducing losses caused by crown gall disease became available to nurserymen and horticulturalists in the United States in 1978. This method is the first commercial use of a bacterium to control any plant disease (Kerr, 1980).

Crown gall is caused by the soil-inhabiting bacterium *Agrobacterium tumefaciens.* The genes controlling pathogenicity of *Agrobacterium* are located on a plasmid, which is itself parasitic within strains of the bacterium (Kerr, 1980). The bacterium enters the plant through wounds and induces abnormally rapid cell division resulting in massive tumors or galls on the roots and stems. The galls interfere with water and nutrient transport and eventually result in wilting, stunting, or plant death. Crown gall affects many ornamental plants, including euonymus, rose, forsythia, viburnum, flowering peach, and willow. Nursery plants infected with crown gall disease cannot be sold. Crop losses in only 10 states in 1976 were estimated at $23 million (Lacy and Reed, 1982).

Biological control of crown gall was first reported in Australia (New and Kerr, 1972). A nonpathogenic strain of bacteria, *Agrobacterium radiobacter* strain 84, was found to inhibit *A. tumefaciens* by competing for colonization sites around wounds. Strain 84 also produces a specific antibiotic (agrocin 84) that selectively inhibits the pathogenic bacteria (Kerr, 1980; Moore, 1979). The beneficial bacteria provide residual protection while the wound heals and becomes nonsusceptible to the pathogenic bacteria.

A. radiobacter strain 84 is available commercially as Galltrol A for control of crown gall on ornamentals and nonbearing fruit trees. It is distributed as a bacterial culture on agar, which is viable for up to 45 weeks when refrigerated (Moore, 1979). The bacteria are scraped from the plate and mixed with water in a clean bucket or container. Cuttings, seeds, or roots of young plants are dipped in the protective bacterial suspension prior to planting. The treatment costs only a few cents per plant and usually provides nearly 100% control (Kerr, 1980). However, to be effective, *A. radiobacter* must be used as a preventive rather than a curative means of control. It should not be considered a substitute for good sanitation and sound nursery management (Moore, 1979).

Before the development and registration of *A. radiobacter*, sanitation and pruning of galled stems were the only available control measures for crown gall disease. It is probably the most successful example of biological disease control in plant pathology (Moore, 1979). Biological control of other plant diseases by inoculation with antagonistic microorganisms is being investigated (Gidrat, 1979).

III. Chinch Bug and Virus Resistance in St. Augustinegrass

St. Augustinegrass, *Stenotaphrum secundatum* (Walt.) Kunze, is the most widely planted ornamental turfgrass in the southern coastal states, comprising more than 46% of the home lawns in Florida, 50% of the lawns in Texas, and 96% of the lawns throughout the Gulf coast region (Reinert, 1978*a*). The southern chinch bug, *Blissus insularis* Barber, is a major impediment to the successful culture of St. Augustinegrass in this region. Expenditures to control this insect were estimated in 1971 to exceed $25 million in Florida alone (Strobel, 1971). Commercial lawn care companies in Florida often apply 6–12 insecticide applications annually for chinch bug control (Reinert and Niemczyk, 1982). In south Florida, southern chinch bug populations have become resistant to several chlorinated hydrocarbon insecticides and to the organophosphate insecticides diazinon and chlorpyrifos (Reinert and Niemczyk, 1982).

Experiments were initiated in Florida in 1971 to find varieties of St. Augustinegrass resistant to *B. insularis.* Reinert (1972) evaluated 78 accessions of St. Augustinegrass under field conditions, and 16 accessions were further screened for resistance under heavy population densities in the laboratory (Reinert and Dudeck, 1974). Several accessions exhibited significant resistance to chinch bugs, involving antibiosis, or death of insects confined on the grasses, along with reduced oviposition and delayed developmental rates. Resistance of one of the accessions, FA-110, to southern chinch bug was later confirmed in Texas (Carter and Duble, 1976).

In other evaluations, accession FA-110 was found to also exhibit resistance to the St. Augustine decline strain of Panicum mosaic virus (SAD-PMV). This virus is the most severe disease of St. Augustinegrass in Texas, causing an estimated loss of $100 million during 1970 (Bruton, Reinert and Toler, 1979; Reinert, Bruton and Toler, 1980). Accession FA-110 was subsequently renamed "Floratam" and was released jointly by the University of Florida and Texas A & M University.

Since its release to the turfgrass industry, "Floratam" has been widely planted throughout the Gulf coast states and in California. Its use has resulted in considerable reductions in insecticide usage and in replacement costs for damaged turf. "Floratam" compares favorably with other St.

Augustinegrass cultivars in soil adaptation and color, but it is relatively coarse-textured, having large stolons and long internodes. It also has problems with winter hardiness in the more northern parts of its range (Wilson, Reinert and Dudeck, 1977). Research is continuing to develop cultivars of St. Augustinegrass having the combined attributes of cold hardiness, finer texture, high density, low maintenance, and resistance to southern chinch bug and SAD-PMV (Reinert, 1982*b*; Reinert, Bruton and Toler, 1980).

IV. Urban Integrated Pest Management in Maryland

Recently, extension entomologists at the University of Maryland have developed and evaluated several comprehensive IPM programs for urban settings (Davidson, Hellman and Holmes, 1981; Hellman, Davidson and Holmes, 1982; Holmes and Davidson, 1984; Raupp and Noland, 1984). The following is an overview of the 1980-1981 Maryland urban IPM program for home landscapes. A more complete account of the program was provided by Hellman, Davidson and Holmes (1982).

The Maryland homeowner IPM program employed scouts who inspected individual home landscapes to detect and monitor pest populations and plant injury. Scouts were mostly upper level entomology and horticulture students with training in plant and pest identification. During a preliminary visit to each homesite, the scout developed a detailed landscape map showing the location of trees, shrubs, flower beds, and turfgrasses. The map was copied to provide a blank map for each week of the 14 week program.

The scout visited each homesite weekly to inspect all plants for pests and cultural problems. He then filled out a survey report form and indicated the location of developing pest problems on the landscape map. Pests or samples of plant injury that the scouts could not identify were taken to the extension plant diagnostic clinic or to University of Maryland extension specialists for identification. Each scout monitored about 60 home landscapes per week. Soil samples were taken from lawns in April or May to identify fertility and pH problems.

A scout supervisor used the weekly scouting reports to prepare an annotated landscape map for each homesite. The map pointed out pest infestations or cultural problems and advised the homeowner as to management options. In addition to the individualized maps, homeowners received a weekly newsletter that provided an update on local pest activity, along with supplementary fact sheets or extension leaflets concerning specific cultural and pest problems occurring within their landscape. By the end of the summer, each homeowner had accumulated an individualized notebook of pest management information.

Scouts in the Maryland program did not make control recommendations or apply pesticides. Implementation of control measures was left to the homeowner. As with preventive spray services, subscribers to the program were charged a fee proportional to the size of their property.

Post-season surveys (Hellman, Davidson and Holmes, 1982) showed that about 98% of the participants were satisfied with the homeowner IPM program and that most felt the general appearance of their landscape had improved. The program also succeeded in increasing public awareness and use of IPM in landscape situations. For example, homeowners completing the program were found to be less likely to knowingly purchase highly pest-susceptible plants and somewhat less willing to use pesticides except when plant health was threatened.

Programs such as the aforementioned could be offered commercially and may eventually provide an alternative to the preventive spray programs presently being marketed. However, it must be noted that the Maryland program was developed and supported by university personnel who had access to educational materials and a greater information base than would be available to commercial lawn and tree care companies. Furthermore, the resubscription rate to the Maryland program was much lower than generally occurs with preventive spray services (J. Davidson, pers. comm.). Apparently, many subscribers believed that once they had the information needed to manage their landscape more effectively there was no reason to continue the IPM service.

PROBLEMS OF IMPLEMENTING IPM IN THE URBAN LANDSCAPE

Unfortunately, a number of practical and socioeconomic factors may impede the implementation of pest management technology for urban landscape plants. These factors include difficulty in diagnosing pest problems, lack of basic information on pest biology, limited availability of practical, nonpesticidal control methods, absence of economic or aesthetic thresholds or other decision making guidelines for major pests, and cost.

Diagnosis and Timing of Control Measures

Diagnosing pest problems of ornamental plants requires familiarity with many disciplines relating to plant health, including entomology, plant pathology, agronomy, and horticulture. Diagnosis is complicated by the great diversity of plant species and pest problems within the urban landscape. For example, about 2,500 species of insects and mites attack ornamental plants in the United States (Johnson and Lyon, 1976). Often, pest problems are overlooked until severe damage becomes apparent. Insects such as armored scales and bagworms may be mistaken for plant structures. Others, such as wood borers, leafminers, and white grubs feed within plant tissues or underground. Physiological disorders, especially those related to environmental stress, are common in landscape plants and are often difficult to distinguish from disease and insect problems (Partyka et al., 1980). It is therefore not surprising that homeowners, garden center clerks, and pest control specialists often fail to recognize or misidentify pest problems.

Control of many horticultural pests requires that pesticide applications coincide with a relatively short-lived vulnerable life stage, such as the first instar, or crawler stage, of scale insects or the newly hatched larvae of tree borers. Timing of control measures is complicated by phenological variation among species and between different geographic areas.

Few nonspecialists have the expertise to identify a pest problem, determine the pest's life stage and the size of its population, estimate its potential for damage, and then evaluate the available pest management options. In 1981 almost 60% of the total sales of ornamental plants and related pesticides and application equipment were by hardware stores, home centers, lumber yards, mass merchandizers, and other outlets not specializing in such products (Hensley and Siebert, 1982). These sales suggest that an increasing number of control recommendations are being made by persons without formal training in entomology, plant pathology, or horticulture.

A number of diagnostic aids are available for troubleshooting pests and other disorders of landscape plants. Selected publications that may be of particular value to homeowners and plant care professionals are listed below.

Insects

Destructive Turf Insects. H. D. Niemczyk. Gray Printing, Fostoria, Ohio. 48p. 1981.

Insect and Other Pests Associated with Turf. J. R. Baker, ed. North Carolina Agricultural Extension Service, Raleigh. 108p. 1982.

Insects That Feed on Trees and Shrubs. W. T. Johnson and H. H. Lyon. Cornell University Press, Ithaca, New York. 463p. 1976.

The Gardener's Bug Book, 4th ed. C. Westcott. Doubleday, Garden City, New York. 689p. 1973.

Diseases

A Turf Manager's Guide to Microscopic Identification of Common Turfgrass Pathogens. P. O. Sanders. Pennsylvania Turfgrass Council, University Park, Pennsylvania. 28p. 1977.

Compendium of Turfgrass Disease. R. W. Smiley. American Phytopathological Society, St. Paul, Minn. 102p. 1983.

Diseases of Turfgrasses: Turfgrass Slide Monograph. L. M. Callahan, H. B. Couch, T. E. Freeman, C. J. Gould, and W. C. Steinstra. Crop Science Society of America, Madison, Wisc. 22p. 1978.
Diseases of Woody Ornamental Plants and Their Control in Nurseries. R. K. Jones and R. C. Lambe, eds. North Carolina Agricultural Experiment Station, Raleigh. 130p. 1982.
Westcott's Plant Disease Handbook, 4th ed. C. Westcott, rev. R. K. Horst. Van Nostrand Reinhold, New York. 803p. 1979.

Weeds

Identifying Seedling and Mature Weeds. J. M. Stucky. North Carolina Agricultural Research Service, Raleigh. 197p. 1980.
Scotts Guide to the Identification of Dicot Turf Weeds. J. Converse. O. M. Scott & Sons, Marysville, Ohio. 110p. 1979.
Scotts Guide to the Identification of Grasses. J. Converse. O. M. Scott & Sons, Marysville, Ohio. 82p. 1979.

General

Diseases and Pests of Ornamental Plants, 5th ed. P. P. Pirone. Wiley, New York. 566p. 1978.
The Ortho Problem Solver, 2nd ed. Ortho Consumer Products Division, Chevron Chemical, San Francisco. 1,040p. 1982.
Scott's Guide to the Identification of Turfgrass Diseases and Insects. J. Converse. O. M. Scott & Sons, Marysville, Ohio. 105p. 1982.
Turfgrass Pests. comp. W. R. Bowen. University of California Publication 4053, Berkeley. 53p. 1980.
Woody Ornamentals: Plants and Problems. R. E. Partyka, J. W. Rimmelspach, B. G. Joyner, and J. A. Carver. Hammer Graphics, Piqua, Ohio. 427p. 1980.

Aesthetic Injury Levels

One of the greatest problems confronting urban landscape pest management is the difficulty of determining the level at which a pest infestation becomes intolerable, that is, the level at which control measures are justified. A central concept of agricultural pest management is that pest population density should be maintained below the economic injury level or the level at which the expected monetary loss resulting from the infestation becomes greater than the cost of controlling the pest. With horticultural pests, economic loss may not be involved, may be impossible to measure, or may not be a factor in decision making.

Pest injury on ornamentals may be measured in terms of reduced aesthetic value or by the cost of replacing the infested plant. Even nuisance, embarrassment, or inconvenience associated with an infestation, such as caterpillar frass falling on a patio or picnic area, could be an important consideration. In some situations, the concept of an aesthetic injury level (the highest level of pests or damage that would be acceptable to most people; Olkowski, 1974), can be used as a basis for management decisions. However, aesthetic injury levels will vary according to the location of the plant and the purpose for which it is used. There has been almost no research on economic or aesthetic injury levels for horticultural pests.

Lack of Information on Pests and IPM

Very few researchers specialize in pest problems of horticultural plants (Niemczyk, 1982; Olkowski and Olkowski, 1978). Therefore, the informational base on urban landscape pests lags far behind that for agronomic systems. Often, such basic biological information as the number of generations per year or the preferred host cultivars of particular pests is lacking. Research efforts in the past have been largely oriented toward pesticides, and in many instances, the effectiveness of alternative pest management strategies has not been evaluated (National Research Council, 1980).

Only a small number of specific, nonpesticidal control methods has been developed for horticultural pests (see Tables 10-2-10-4). These methods are often effective in suppressing pest populations, but in some cases they require considerable expertise to use or are slower-acting, less thorough, or less reliable than conventional pesticides. In general, effective monitoring or sampling methods have not been developed for urban landscape pests. Furthermore, because there is so little information on aesthetic or economic injury levels, there is often no clear basis for distinguishing between tolerable and intolerable pest densities or for predicting the damage potential of a particular infestation.

Socioeconomic Considerations

A major problem in implementing horticultural IPM on a commercial scale is cost. It is currently faster, simpler, and more profitable for commercial lawn and tree care firms to apply routine preventive sprays than to hire or train qualified consultants to engage in sampling, monitoring, and other pest management activities. The time required to return several times to individual landscapes is prohibitive in the highly competitive lawn care business. Also, service might be cancelled if the client were informed that regular spraying was unnecessary. Pesticide-oriented programs provide reliable short-term control as well as the opportunity to accumulate repeat customers. In fact, a recent survey indicated that the great majority of those who use a professional lawn care company are satisfied with the service (Anonymous, 1984*a*). Consequently, there is little economic incentive to adopt nonpesticidal alternatives. Given the prevailing public demand for pest-free plants, it is questionable whether companies that emphasize pest management could complete with those that promise pest eradication.

CONCLUSION

Integrated pest management strategies have been developed and are employed far more in agriculture than in the urban landscape. One reason is that extension and research programs for agricultural crops have historically been given priority and financial support over those for horticultural commodities. Much of the basic research on agricultural pests and IPM has been subsidized by the U.S. Department of Agriculture or by the major crop industries. The turfgrass and ornamentals industry, however, has only recently begun to support basic research in any significant way. Very few university and Department of Agriculture research service personnel are assigned to research on horticultural pests. The rate of progress of urban horticultural IPM is likely to remain limited unless personnel and funding for such research are dramatically increased.

There is currently a serious lack of detailed information on the biology, population dynamics, and host plant relationships of major pests of ornamental plants. Unless basic research needs are met, the information base on which comprehensive urban IPM programs must be built will be so weak that such programs will fail.

Little research has been done to facilitate forecasting of insect emergence to optimize insecticidal applications in the urban landscape. Such information is especially critical for pests such as wood borers and scale insects that may be only briefly vulnerable to sprays. Reliable forecasting systems based on degree days or other parameters have been developed for only a small number of landscape pests (Akers and Nielsen, 1984; Potter and Timmons, 1983*a*). Development of additional synthetic pheromones, food-type lures, or other trapping methods would facilitate the problem of detecting and monitoring insect populations. These tools would allow extension personnel or lawn and tree care professionals to "fine-tune" spray schedules annually or to eliminate unnecessary applications.

The need to develop meaningful economic or aesthetic injury levels for the more important pests of ornamental plants is also great. The lawn and tree care industry cannot be expected to move from the current preventive spray programs to a more threshold-oriented approach if decisionmaking guidelines do not exist. Because standards and expectations vary greatly, this research will have to consider not only insect/plant relationships but also the public's perception of pest problems and how these attitudes can be modified.

Additional research is needed to identify or develop insect and disease resistant cultivars of ornamental plants. The research should also consider the relationship between environmental stress and plant susceptibility to pests. Possibly, many insect and disease problems could be avoided by minimizing stress through comprehensive plant health care programs.

Biological control technology may be applicable to certain urban pest problems, but additional research is needed. The natural enemy complexes of many important landscape pests have been neither identified nor studied. Such research may eventually allow landscape managers to use pesticides or other practices in a manner compatible with survival of parasites and predators (Olkowski et al., 1978). Deliberate importation of natural enemies offers the prospect of long-range regulation of exotic pest populations in shade trees and other relatively stable urban vegetation. However, work in this area has barely started (Olkowski et al., 1978).

Finally, improved delivery systems must be developed to disseminate pest management information and to educate the public and the landscape pest control industry about the benefits of IPM. These activities must include additional pilot projects to demonstrate the feasibility of urban IPM, cost/benefit comparisons between conventional and alternative pest control methods, IPM clinics and training sessions, newsletters, and other means of technology transfer. Conventional pesticides will undoubtedly continue to be a critical component of urban landscape IPM in the future. The challenge to researchers and extension personnel is to develop technology that will substantially reduce pesticide use while increasing the efficacy of those pesticide applications that are necessary.

ACKNOWLEDGMENTS

Many of the ideas expressed in this chapter were developed during discussion with Professors J. R. Hartman (plant pathology), R. E. McNiel (horticulture) and A. J. Powell (agronomy) of the University of Kentucky. I also thank P. F. Colbaugh, J. R. Hartman, N. Jackson, M. K. Kennedy, J. W. Neal, Jr., A. J. Powell, L. H. Townsend, and K. V. Yeargan for reviewing earlier drafts of the manuscript and making valuable suggestions for its improvement.

REFERENCES

Akers, R. C., and D. G. Nielsen. 1984. Predicting *Agrilus anxius* Gory (Coleoptera: Buprestidae) adult emergence by heat unit accumulation. *J. Econ. Entomol.* **77**:1459-1463.

Altman, J., and C. L. Campbell. 1977. Effect of herbicides on plant disease. *Annu. Rev. Phytopathol.* **15:**361-385.
Anonymous. 1983. *Lawn Institute Harvests, July 1983.* The Lawn Institute, Pleasant Hill, Tenn.
Anonymous. 1984*a*. How do customers perceive their lawn care service? *Am. Lawn Applic.* **5**(1):24.
Anonymous. 1984*b*. Lawn care/landscape receipts vault to more than $2 billion. *Lawn Care Ind.* **8**(6):1, 8.
Barker, G. M., R. P. Pottinger, and P. J. Addison. 1983. Effect of tall fescue and ryegrass endophytes on Argentine stem weevil. In *Proceedings of the 36th New Zealand Weed and Pest Control Conference,* pp. 216-219.
Beard, J. B. 1973. *Turfgrass: Science and Culture.* Prentice-Hall, Englewood Cliffs, N.J., 658p.
Benson, D. M. 1982. Phytophthora root rot. In *Diseases of Woody Plants and Their Control in Nurseries,* R. K. Jones and R. C. Lambe (eds.). North Carolina Agricultural Extension Service, pp. 14-17.
Bruton, B. D., J. A. Reinert, and R. W. Toler. 1979. Effects of southern chinch bug (*Blissus insularis*) and the St. Augustine decline strain of Panicum mosaic virus (PMV-SAD) on seventeen accessions and two cultivars of St. Augustinegrass. *Phytopathology* **69:**525-526.
Carter, R. P., and R. L. Duble. 1976. *Variety Evaluations in St. Augustinegrass for Resistance to the Southern Chinch Bug,* Progress Report PR-3374C. Texas Agricultural Experiment Station, 39p.
Clay, K., T. N. Hardy, and A. M. Hammond, Jr. 1985. Fungal endophytes of grasses and their effects on an insect herbivore. *Oecologia* **66:**1-5.
Cockfield, S. D., and D. A. Potter. 1983. Short-term effects of insecticidal applications on predaceous arthropods and oribatid mites in Kentucky bluegrass turf. *Environ. Entomol.* **12:**1260-1264.
Cockfield, S. D., and D. A. Potter. 1984. Predation on sod webworm eggs as affected by chlorpyrifos application to Kentucky bluegrass turf. *J. Econ. Entomol.* **77:**1542-1544.
Cornman, J. F. 1952. Mat formation on putting greens. *The Golf Course Rep.* **20**(4):8-14.
Couch, H. 1971. Turfgrass disease control in the twentieth century. *The Golf Superintendant,* July.
Couch, H. B. 1980. Relationship of management practices to incidence and severity of turfgrass diseases. In *Advances in Turfgrass Pathology,* P. O. Larsen and B. G. Hoyner (eds.). Harcourt Brace Jovanovich, Duluth, Minn., pp. 65-72.
Crist, C. R., and D. F. Schoeneweiss. 1975. The influence of controlled stresses on susceptibility of European white birch stems to attack by *Botryosphaeria dithidea. Phytopathology* **65:**369-373.
Davidson, J., J. L. Hellman, and J. Holmes. 1981. Urban ornamentals and turf IPM. In *Proceedings of Integrated Pest Management Workshop,* G. L. Worf (ed.). National Cooperative Extension Service, Dallas, Texas, pp. 68-72.
Dean, H. A., M. F. Schuster, J. C. Boling, and P. T. Richard. 1979. Complete biological control of *Antonia graminis* in Texas with *Neodusmetia sangwani* (a classic example). *Bull. Entomol. Soc. Am.* **25:**262-267.

Debach, P., and M. Rose. 1977. Environmental upsets caused by chemical eradication. *Calif. Agric.* **32:**8-10.

Dekker, J. 1976. Acquired resistance to fungicides. *Annu. Rev. Phytopathol.* **14:**405-428.

Edwards, C. A., and G. W. Heath. 1963. The role of soil animals in breakdown of leaf material. In *Soil Organisms,* J. Doesksen and J. van der Drift (eds.). North Holland, Amsterdam, pp. 76-85.

Engle, R. E., and R. D. Ilnicki. 1969. Turf weeds and their control. In *Turfgrass Science,* A. A. Hansen and F. V. Juska (eds.). American Society of Agronomists, Madison, Wisc., pp. 240-287.

Federer, C. A. 1976. Trees modify the urban microclimate. *J. Arboric.* **2:**121-127.

Funk, C. R., P. M. Halisky, M. C. Johnson, M. R. Siegel, A. V. Steward, S. Ahmad, R. H. Hurley, and I. C. Harvey. 1983. An endophytic fungus and resistance to sod webworms: Association in *Lolium perenne* L. *Bio/Technology* **1:**189-191.

Gidrat, D. 1979. Biocontrol of plant diseases by inoculation of fresh wounds, seeds, and soil with antagonists. In *Soil-borne Plant Pathogens,* B. Schippers and W. Gams (eds.). Academic Press, New York, pp. 538-551.

Gordon, F. C., and D. A. Potter. 1985. Efficiency of Japanese beetle traps in reducing plant defoliation in the landscape and effect on larval density in turf. *J. Econ. Entomol.* **78:**774-778.

Harris, R. W. 1983. *Arboriculture.* Prentice-Hall, Englewood Cliffs, N.J., 688p.

Hartman, J. 1982. Crab apple diseases. In *Diseases of Woody Plants and Their Control in Nurseries,* R. K. Jones and R. C. Lambe (eds.). North Carolina Agricultural Extension Service, pp. 47-50.

Hartman, J. R., and A. J. Powell. 1980. Application of fungicides in winter for control of diseases of Kentucky bluegrass. *Fungic. Nematic. Tests* **36:**146.

Hellman, J. L., J. A. Davidson, and J. Holmes. 1982. Urban ornamental and turfgrass integrated pest management in Maryland. In *Advances in Turfgrass Entomology,* H. D. Niemczyk and B. G. Joyner (eds.). Hammer Graphics, Piqua, Ohio, pp. 31-38.

Hensley, D. L., and A. Siebert. 1982. Nursery industry needs new selling strategies to grow and compete. *Am. Nurseryman* **155**(10):59-62.

Hodges, C. F. 1978. Postemergent herbicides and the biology of *Drechslera sorokiniana:* Influence on severity of leaf spot on *Poa pratensis. Phytopathology* **68:**1359-1363.

Holmes, J. J., and J. Davidson. 1984. Integrated pest management for arborists: Implementation of a pilot program in Maryland. *J. Arboric.* **9:**145-150.

Johnson, M. C., D. L. Dahlman, M. R. Siegel, L. P. Bush, G. C. M. Latch, and D. A. Potter. 1985. Insect feeding deterrents in endophyte-infected tall fescue. *Appl. Environ. Microbiol.* **49:**568-571.

Johnson, W. T. 1982. Horticultural spray oils for tree pest control. *Weeds, Trees and Turf,* May, pp. 36-40.

Johnson, W. T., and H. H. Lyon. 1976. *Insects That Feed on Trees and Shrubs.* Cornell University Press, Ithaca, N.Y., 464p.

Jones, R. K. 1982. Powdery mildew. In *Diseases of Woody Plants and Their Control in Nurseries,* R. K. Jones and R. C. Lambe (eds.). North Carolina Agricultural Extension Service, pp. 24-25.

Kageyama, M. E. 1982. Industry's contribution to turfgrass entomology. In *Advances in Turfgrass Entomology,* H. D. Niemczyk and B. G. Joyner (eds.). Hammer Graphics, Piqua, Ohio, pp. 133-138.

Kamm, J. A. 1982. Use of insect sex pheromones in turfgrass management. In *Advances in Turfgrass Entomology,* H. D. Niemczyk and B. G. Joyner (eds.). Hammer Graphics, Piqua, Ohio, pp. 39-41.

Katan, J., and Y. Eshel. 1973. Interactions between herbicides and plant pathogens. *Residue Rev.* **45:**145-177.

Kenney, J. 1983. Dangerous and deadly poisons have been applied by. . . . *Am. Lawn Appl.* **4**(5):26-27.

Kerr, A. 1980. Biological control of crown gall through production of agrocin 84. *Plant Dis.* **64:**24-30.

Kielbaso, J. J., ed. 1978. *Proceedings of the Symposium on Systemic Chemical Treatments in Tree Culture.* Michigan State University, East Lansing, 358p.

Kielbaso, J. J., and M. K. Kennedy. 1983. Urban forestry and entomology: A current appraisal. In *Urban Entomology: Interdisciplinary Perspectives,* G. W. Frankie and C. S. Koehler (eds.). Preager, New York, pp. 423-440.

Kimmins, J. P. 1971. Variations in foliar amino acid composition of flowering balsam fir (*Abies balsamea* (L.) Mill.) and white spruce (*Picea glauca* (Moench) Voss) in relation to outbreaks of the spruce budworm (*Choristoneura fumifera* (Clem.). *Can. J. Zool.* **49:**1005-1011.

Klein, M. G. 1981*a.* Advances in the use of *Bacillus popilliae* for pest control. In *Microbial Control of Pests and Plant Diseases, 1970-1980,* H. D. Burges (ed.). Academic Press, London, pp. 183-192.

Klein, M. G. 1981*b.* Mass trapping for suppression of Japanese beetles. In *Management of Insect Pests with Semiochemicals,* E. R. Mitchell (ed.). Plenum, New York, pp. 183-190.

Kuhr, R. J., and H. Tashiro. 1978. Distribution and persistence of chlorpyrifos and diazinon applied to turf. *Bull. Environ. Contam. Toxicol.* **20:**652-656.

Lacy, G. H., and H. E. Reed. 1982. Crown gall. In *Diseases of Woody Plants and Their Control in Nurseries,* R. K. Jones and R. C. Lambe (eds.). North Carolina Agricultural Extension Service, pp. 25-26.

Latch, G. C. M., and M. J. Christensen. 1985. Inoculation of grasses with the endophytic fungi. *N.Z. J. Agric. Res.* (in press).

Lewis, A. C. 1979. Feeding preference for diseased and wilted sunflower in the grasshopper, *Melanoplus differentialis. Entmol. Exp. Appl.* **26:**202-207.

Luck, R. F., and D. L. Dahlsten, 1975. Natural decline of a pine needle scale (*Chionaspis pinifoliae* (Fitch)) outbreak at South Lake Tahoe, California, following cessation of adult mosquito control with malathion. *Ecology* **56:**893-904.

McClure, M. S. 1977. Resurgence of the scale, *Fiorinia externa* (Homoptera: Diaspididae) on hemlock following insecticide application. *Environ. Entomol.* **6:**480-484.

McCoy, N. L., and G. Philley. 1982. Rose diseases. In *Diseases of Woody Plants and Their Control in Nurseries,* R. K. Jones and R. C. Lambe (eds.). North Carolina Agricultural Extension Service, pp. 72-74.

Madison, J. H. 1961. The effect of pesticides on turfgrass disease incidence. *Plant Dis. Rep.* **45:**892-893.

Madison, J. H. 1971. *Practical Turfgrass Management.* Van Nostrand Reinhold, New York, 466p.

Mattson, W. J. 1980. Herbivory in relation to plant nitrogen content. *Annu. Rev. Ecol. Syst.* **11:**119-161.

Mayer, M. S., and J. R. McLaughlin. 1975. *An Annotated Compendium of Insect*

Sex Pheromones, Florida Agricultural Experiment Station Monograph Series, No. 6, 88p.

Mecklenburg, R. A. 1971. *Artificial vs. Living Turf.* Department of Horticulture, Michigan State University, East Lansing, 4p.

Merritt, R. W., M. K. Kennedy, and E. F. Gersabeck. 1983. Integrated pest management of nuisance and biting flies in a Michigan resort: Dealing with secondary pest outbreaks. In *Urban Entomology: Interdisciplinary Perspectives,* G. W. Frankie and C. S. Koehler (eds.). Praeger, New York, pp. 277-299.

Moore, L. W. 1979. Practical use and success of *Agrobacterium radiobacter* strain 84 for crown gall control. In *Soil-borne Plant Pathogens,* B. Schippers and W. Gams (eds.). Academic Press, New York, pp. 553-568.

Morales, D. J. 1980. The contribution of trees to residential property value. *J. Arboric.* **6:**305-308.

Morgan, D. L., G. W. Frankie, and M. J. Gaylor. 1978. Potential for developing insect-resistant plant materials for use in urban environments. In *Perspectives in Urban Entomology,* G. W. Frankie and C. S. Koehler (eds.). Academic Press, New York, pp. 267-294.

National Research Council. 1980. *Urban Pest Management.* Report prepared by the Committee on Urban Pest Management, Environmental Studies Board, Commission on Natural Resources. National Academy Press, Washington, D.C., 272p.

Neal, J. W., Jr. 1981. Timing insecticide control of rhododendron borer with pheromone trap catches of males. *Environ. Entomol.* **10:**264-266.

Neal, J. W., Jr. 1982. Rhododendron borer: A worthy competitor. *J. Am. Rhododendron Soc.* **36:**57-60.

Neal, J. W., Jr. 1984. Bionomics and instar determination of *Synanthedon rhododendri* (Lepidoptera: Sesiidae) on rhododendron. *Ann. Entomol. Soc. Am.* **77:**552-560.

New, P. B., and A. Kerr. 1972. Biological control of crown gall: Field observations and glasshouse experiments. *J. Appl. Bacteriol.* **35:**279-287.

Nielsen, D. G. 1978. Sex pheromone traps: A breakthrough in controlling borers of ornamental trees and shrubs. *J. Arboric.* **4:**181-183.

Nielsen, D. G. 1979*a.* Clearwing moth pheromone research: A perspective. In *Pheromones of the Sesiidae,* SEA-AR, ARR-NE-G. U.S. Department of Agriculture, Washington, D.C., pp. 75-83.

Nielsen, D. G. 1979*b.* Integrated control strategies established for tree insects. *Weeds, Trees and Turf* **18**(5):20-26.

Nielsen, D. G. 1981. Alternative strategy for arborists: Treat the tree, not the customer. *Weeds, Trees and Turf* **20**(7):40-42.

Nielsen, D. G., and C. P. Balderston. 1973. Evidence for intergeneric sex attraction among aegeriids. *Ann. Entomol. Soc. Am.* **66:**227-228.

Nielsen, D. G., and F. F. Purrington. 1978. Sex attractants: A new warning system to time clearwing borer control practices. In *Ohio Agricultural Research Development Center Research Circular 236,* pp. 7-8.

Nielsen, D. G., F. F. Purrington, and G. F. Shambaugh. 1979. EAG and field responses of sesiid males to sex pheromones and related compounds. In *Pheromones of the Sesiidae,* SEA-AR, ARR-NE-G. U.S. Department of Agriculture, Washington, D.C., pp. 11-26.

Niemczyk, H. D. 1982. The status of USDA-SEA-AR and U.S. university input of professional personnel to turfgrass entomology, 1980. In *Advances in Turfgrass*

Entomology, H. D. Niemczyk and B. G. Joyner (eds.). Hammer Graphics, Piqua, Ohio, pp. 127-132.

Niemczyk, H. D., H. R. Krueger, and K. O. Lawrence. 1977. Thatch influences movement of soil insecticides. *Ohio Rep.* **62:**26-28.

Olkowski, W. 1974. A model ecosystem management program. *Proc. Tall Timbers Conf. Ecol., Anim. Control Hab. Manage.* **5:**103-117.

Olkowski, W., and H. Olkowski. 1978. Urban integrated pest management. *J. Arboric.* **4:**241-246.

Olkowski, W., H. Olkowski, R. van den Bosch, and R. Hom. 1976. Ecosystem management: A framework for urban pest control. *Bioscience* **26:**384-389.

Olkowski, W., H. Olkowski, A. I. Kaplan, and R. van den Bosch. 1978. The potential for biological control in urban areas: Shade tree insect pests. In *Perspectives in Urban Entomology,* G. W. Frankie and C. S. Koehler (eds.). Academic Press, New York, pp. 311-347.

Partyka, R. E., J. W. Rimespach, B. G. Joyner, and S. A. Carver. 1980. *Woody Ornamentals: Plants and Problems.* Chemlawn, Columbus, Ohio, 428p.

Patel, S. I., and W. A. Gammel. 1978. Insecticides and fungicides phytotoxicity, fumigants and precautions. *Proc. South. Nurseryman's Conf.* **23:**77-85.

Phair, W. E., and G. S. Ellmore. 1982. Improved trunk injection for control of Dutch elm disease. *J. Arboric.* **10:**273-278.

Pirone, P. P. 1978*a*. *Diseases and Pests of Ornamental Plants,* 5th ed. Wiley, New York, 566p.

Pirone, P. P. 1978*b*. *Tree Maintenance,* 5th ed. Oxford University Press, New York, 588p.

Potter, D. A. 1982. Influence of feeding by grubs of the southern masked chafer on quality and yield of Kentucky bluegrass. *J. Econ. Entomol.* **75:**21-24.

Potter, D. A., and G. M. Timmons. 1981. Factors affecting predisposition of flowering dogwoods to attack by the dogwood borer. *HortScience* **16:**677-679.

Potter, D. A., and G. M. Timmons. 1983*a*. Forecasting emergence and flight of the lilac borer (Lepidoptera: Sesiidae) based on pheromone trapping and degree day accumulations. *Environ. Entomol.* **12:**400-403.

Potter, D. A., and G. M. Timmons. 1983*b*. Flight phenology of the dogwood borer (Lepidoptera: Sesiidae) and implications for control in *Cornus florida* L. *J. Econ. Entomol.* **76:**1069-1074.

Potter, D. A., and G. M. Timmons. 1983*c*. Biology and management of clearwing borers in woody plants. *J. Arboric.* **9:**145-150.

Potter, D. A., B. L. Bridges, and F. C. Gordon. 1985. Effect of nitrogen fertilization on earthworm and microarthropod populations in Kentucky bluegrass turf. *Agron. J.* **77:**367-372.

Prestige, R. A., R. P. Pottinger, and G. M. Barker. 1982. An association of *Lolium* endophyte with ryegrass resistance to Argentine stem weevil. In *Proceedings of the 35th New Zealand Weed and Pest Control Conference,* pp. 119-122.

Purrington, F. F., and D. G. Nielsen. 1977. Biology of *Podosesia* (Lepidoptera: Sesiidae) with description of a new species from North America. *Ann. Entomol. Soc. Am.* **70:**906-910.

Randell, R., J. D. Butler, and T. D. Hughs. 1972. The effect of pesticides on thatch accumulation and earthworm populations in Kentucky bluegrass turf. *HortScience* **7:**64-65.

Raup, M. J., and R. M. Noland. 1984. Implementing landscape plant management programs in institutional and residential settings. *J. Arboric.* **10:**161-169.

Reinert, J. A. 1972. Turf-grass insect research. *Fla. Turfgrass Manage. Conf. Proc.* **20:**79-84.

Reinert, J. A. 1978*a*. Antibiosis to the southern chinch bug by St. Augustine accessions. *J. Econ. Entomol.* **71:**21-24.

Reinert, J. A. 1978*b*. Natural enemy complex of the southern chinch bug in Florida. *Ann. Entomol. Soc. Am.* **71:**728-731.

Reinert, J. A. 1982*a*. Insecticide resistance in epigeal insect pests of turfgrass: 1. A review. In *Advances in Turfgrass Entomology,* H. D. Niemczyk and B. G. Joyner (eds.). Hammer Graphics, Piqua, Ohio, pp. 71-76.

Reinert, J. A. 1982*b*. A review of host resistance in turfgrasses to insects and acarines with emphasis on the southern chinch bug. In *Advances in Turfgrass Entomology,* H. D. Niemczyk and B. G. Joyner (eds.). Hammer Graphics, Piqua, Ohio, pp. 3-12.

Reinert, J. A., and A. E. Dudeck. 1974. Southern chinch bug resistance in St. Augustinegrass. *J. Econ. Entomol.* **67:**275-277.

Reinert, J. A., and H. D. Niemczyk. 1982. Insecticide resistance in epigeal pests of turfgrass: II. Southern chinch bug resistance to organophosphates in Florida. In *Advances in Turfgrass Entomology,* H. D. Niemczyk and B. G. Joyner (eds.). Hammer Graphics, Piqua, Ohio, pp. 77-80.

Reinert, J. A., B. D. Bruton, and R. W. Toler. 1980. Resistance of St. Augustinegrass to southern chinch bug and St. Augustine decline strain of Panicum mosaic virus. *J. Econ. Entomol.* **73:**602-604.

Schoeneweiss, D. F. 1975. Predisposition, stress, and plant disease. *Annu. Rev. Phytopathol.* **13:**193-211.

Schoeneweiss, D. F. 1981. The role of environmental stress in diseases of woody plants. *Plant Dis.* **65:**308-314.

Schreiber, L. R., and J. W. Peacock. 1974. Dutch Elm Disease and Its Control, Forest Service Information Bulletin 193. U.S. Department of Agriculture, Washington, D.C.

Sears, M. K., and R. A. Chapman. 1979. Persistence and movement of four insecticides applied to turfgrass. *J. Econ. Entomol.* **72:**272-274.

Smiley, R. W., and M. M. Craven. 1978. Fungicides in Kentucky bluegrass turf. *Agron. J.* **70:**1013-1019.

Smith, L. D. 1982. Fungicides for nursery ornamentals. In *Diseases of Woody Plants and Their Control in Nurseries,* R. K. Jones and R. C. Lambe (eds.). North Carolina Agricultural Extension Service, pp. 101-106.

Spencer, S., and R. K. Jones. 1982. Pinewood nematode. In *Diseases of Woody Plants and Their Control in Nurseries,* R. K. Jones and R. C. Lambe (eds.). North Carolina Agricultural Extension Service, p. 22.

Starkey, D. G. 1979. Trees and their relationship to mental health. *J. Arboric.* **5:**153-154.

Streu, H. T., and J. B. Gingrich. 1972. Seasonal activity of the winter grain mite in turfgrass in New Jersey. *J. Econ. Entomol.* **65:**427-430.

Strobel, J. 1971. Turfgrass. *Fla. Turfgrass Manage. Conf. Proc.* **19:**19-28.

Tashiro, H. 1981. Limitations of organophosphate soil insecticides on turfgrass scarabaeid grubs and resolution with isofenphos. In *Proceedings of the 4th International Turfgrass Research Conference.* University of Guelph, Ontario, pp. 425-432.

Tashiro, H. 1982. The incidence of insecticide resistance in soil inhabiting turfgrass insects. In *Advances in Turfgrass Entomology,* H. D. Niemczyk and B. G. Joyner (eds.). Hammer Graphics, Piqua, Ohio, pp. 81-84.

Tattar, T. A. 1978. *Diseases of Shade Trees.* Academic Press, New York, 362p.

Taylor, D. H., and G. R. Blake. 1982. The effect of turfgrass thatch on water infiltration rates. *Soil. Sci. Soc. Am.* **46:**616-619.

Tumlinson, J. H., C. E. Yonce, R. E. Doolittle, R. R. Heath, C. R. Gentry, and E. R. Mitchell. 1974. Sex pheromones and reproductive isolation of the lesser peachtree borer and the peachtree borer. *Science* **185:**614-616.

Turgeon, A. J., R. P. Freeborg, and W. N. Bruce. 1975. Thatch development and other effects of preemergent herbicides in Kentucky bluegrass turf. *Agron. J.* **67:**563-565.

Vargas, J. M. 1981. *Management of Turfgrass Diseases.* Burgess, Minneapolis, 204p.

Vargas, J. M., K. T. Payne, A. J. Turgeon, and R. Deitweiler. 1980. Turfgrass disease resistance: Selection, development, and use. In *Advances in Turfgrass Pathology,* B. G. Joyner and P. O. Larsen (eds.). Harcourt Brace Jovanovich, Duluth, Minn., pp. 179-182.

Vittum, P. J., and H. Tashiro. 1980. Effect of soil pH on survival of Japanese beetle and European chafer larvae. *J. Econ. Entomol.* **73:**577-579.

Walker, J. T. 1982. Disease resistance among woody ornamentals. In *Diseases of Woody Ornamental Plants and Their Control in Nurseries,* R. K. Jones and R. C. Lambe (eds.). North Carolina Agricultural Extension Service, pp. 112-114.

Westcott, C. 1973. *The Gardener's Bug Book.* Doubleday, Garden City, N.Y., 688p.

White, T. C. R. 1969. An index to measure weather-induced stress of trees associated with outbreaks of psyllids in Australia. *Ecology* **40:**905-909.

White, T. C. R. 1974. A hypothesis to explain outbreaks of looper caterpillars with special reference to populations of *Selidosema suavis* in a plantation of *Pinus radiata* in New Zealand. *Oecologia* **16:**279-301.

White, T. C. R. 1984. The abundance of invertebrate herbivores in relation to the availability of nitrogen in stressed food plants. *Oecologia* **63:**90-105.

Wilson, C. A., J. A. Reinert, and A. E. Dudeck. 1977. Winter survival of St. Augustinegrass in north Mississippi. *Quar. News Bull. South. Turfgrass Assoc.* **12**(3), August.

11

Vertebrate Pest Management

Rex E. Marsh
University of California, Davis

Although interest in urban wildlife is not new, it has increased in the past 10 years. Many urban residents find great enjoyment in seeing birds and squirrels in their backyards or other animals in parks or open spaces. An indication of this interest is apparent from the numerous books available concerning landscaping to attract birds and mammals and about the identification and enjoyment of urban wildlife.

Urban wildlife takes on different meanings depending on whether you are a bird or mammal enthusiast who enjoys watching wildlife as a hobby, a wildlife biologist interested in studying wildlife, a wildlife manager engaged in managing wildlife for increased diversity or increased numbers, or a pest manager or vector control specialist who has the responsibility for preventing or resolving human-wildlife conflicts (Leedy, 1979; Leedy and Adams, 1984; Leedy, Maestro, and Franklin, 1978; Noyes and Progulske, 1974). Urban wildlife management includes both wildlife enhancement and pest wildlife control. Vertebrate control or animal damage control (vertebrate pest management) is considered a specialized area of the more traditional nonurban wildlife management; however, vertebrate pest management takes on significant importance in urban environments.

In this chapter, wildlife management per se is not being addressed but rather wildlife that may become troublesome in urban areas and take on the status of a pest, at least under certain circumstances. These are referred to as vertebrate pests. Vertebrate pests include any vertebrate, native or introduced, domestic or feral, that constantly or periodically has an adverse effect on human health or well-being or conflicts in some significant way with human activities.

Vertebrate pests make up a very diverse group of animals (including amphibians, snakes, birds, and mammals), and in an urban setting their habitat and ecology are often atypical because of the presence of structures of human origin, landscapes of exotic plants, and altered terrain, all a part of human-modified urban environments. For most wildlife species, except possibly the introduced rats and mice, it is unfair to categorically characterize them as pest species, since most are considered pests only in some situations and may be neutral or highly desirable in others. For example, deer may be a pest in some urban areas because they feed on gardens, yet they are highly prized for their aesthetic value in parks and cemeteries. Some find enjoyment in feeding pigeons, whereas others never view them as anything other than a messy pest.

HOW VERTEBRATES ADVERSELY AFFECT HUMANS

The human-wildlife interactions that become problematic are so varied in urban settings that it is impossible to adequately describe them all in a single chapter. However, vertebrate pests or their problems can be categorized in several ways. In Table 11-1, the problems are designated by general categories according to how they affect people's well-being or compete with their activities. Such descriptive categories make it easier to understand the problems, for example, than would a category based on the phylogenetic order of the species implicated.

The categories in Table 11-1 are not absolute or mutually exclusive, and some pest problems could logically be in two or more categories. For example, rats are responsible for causing a number of fires, and fire can be both an economic loss as well as a possible health or life threat. Bird droppings may be a nuisance on the sidewalks or on boat docks; they can also be a health threat because of the disease organisms they contain.

When wildlife in urban settings conflict with human interests, not just individuals or businesses are affected: city management becomes involved where public property and services are implicated. Solid waste disposal sites (garbage dumps or landfills), sewer plants, water reservoirs, and airport facilities—all necessary to the urban community—involve land uses which characteristically provide wildlife habitat, thus inviting vertebrate pest problems. City planners, architects, engineers, landscape specialists, structural pest control operators, vector control specialists, animal control officers,

Table 11-1. Vertebrates categorized as pests by how they affect people or their interests, with examples.

Category	*Example*
Economic loss	
Losses of food and fiber	Rats and mice in food warehouse
Degradation or destruction of man-made structures	Gnawing of doors, walls, and electrical wiring by rats; bird droppings defacing buildings
Destruction of other man-made materials	Rats and house mice gnawing goods in storage
Destruction of vegetation	Rabbits, pocket gophers, and deer damaging gardens
Health or potential health problems	
Threat of disease	Pigeon droppings and histoplasmosis; bats, and potential rabies
Parasites and host-related biting insects	Pigeons and sparrows and mite infestations
Physical injury	Rat bites to people and pets; coyote attacks on children
Animal phobias	Fear of rodents, bats, or snakes in homes; fear of almost all animals
Nuisance problems	
Noise pollution	Starlings roosting on building ledges; rats or bats in attics and walls
Odor pollution	Skunks beneath houses; odor of bird droppings in air-conditioning system
Unsightly conditions	Bird droppings on sidewalks, boats, or cars; raccoons scattering garbage
Bird attacks	Nesting blackbirds and mocking birds diving at people
Pet harrassment	Rodents stealing pet food or causing dogs to bark; birds diving at dogs and cats
Degradation or destruction of natural resources	
Damage to natural vegetation	Nitrogen-killed trees in starling and blackbird roosts
Pollution of water	Waterfowl and gulls in potable water reservoirs
Pollution of air	Skunks in parks and schoolgrounds
Predation on more desirable wildlife	Coyotes killing waterfowl; foxes taking eggs of ground-nesting birds
Competition with more desirable wildlife	Starlings displacing nesting sites of hole-nesting birds; introduced fox squirrels displacing western gray squirrels

Table 11-2. Vertebrate pest species and examples of some major urban problem situations.

Pest animal	*Public health and well-being*	*Commercial and residential buildings*	*Food storage, processing, and distribution facilities*	*Vegetable and fruit gardens*	*Turf and ornamental plantings*	*Transportations corridors and roadside right-of-ways*	*Overhead utility lines and power poles*	*Airport facilities (aircraft hazards)*	*Other areas of importance*
				Problem situation					
Amphibians									
Toads and frogs	-	-	-	-	X	-	-	-	Swimming pools, ponds
Reptiles									
Snakes	X	X	-	X	X	-	-	-	Children's play areas
Birds									
Blackbirds	X	X	X	X	X	-	-	X	
Crows	X	X	-	X	X	-	-	X	City parks
Gulls	X	X	X	-	X	-	-	X	Garbage dumps
House sparrows	X	X	X	X	X	-	-	X	
Pigeons	X	X	X	-	X	-	X	X	City squares and parks
Starlings	X	X	X	X	X	X	X	X	Garbage dumps
Swallows	X	X	X	-	-	-	-	-	
Waterfowl	X	-	-	-	X	-	-	X	Reservoirs
Woodpeckers	-	X	-	X	X	-	X	-	
Mammals									
Armadillos	X	-	-	X	X	X	-	-	
Bats	X	X	X	-	-	-	-	-	
Chipmunks	X	-	-	X	X	-	-	-	
Cotton rats	-	-	-	X	X	X	-	X	
Coyote	X	-	-	-	X	X	-	X	Around residences

public health officials, and others are often intimately concerned in their professional activities with wildlife, particularly those that become pests. Table 11-2 provides a list of vertebrates and examples of some of their urban problem situations. Because vertebrate pest problems cut across such a diverse array of urban situations, the breadth of such problems often goes unrecognized.

	Problem situation								
Pest animal	*Public health and well-being*	*Commercial and residential buildings*	*Food storage, processing, and distribution facilities*	*Vegetable and fruit gardens*	*Turf and ornamental plantings*	*Transportations corridors and roadside right-of-ways*	*Overhead utility lines and power poles*	*Airport facilities (aircraft hazards)*	*Other areas of importance*
Deer	X	-	-	X	X	X	-	X	
Deer mice	-	X	X	X	X	-	-	-	
Fox	X	-	-	X	X	X	-	X	
Ground squirrels	X	-	-	X	X	X	-	X	Undermining buildings
House mice	X	X	X	-	-	-	-	-	
Moles	-	-	-	X	X	X	-	X	
Muskrats	-	-	-	-	X	-	-	-	Ponds, reservoirs, levees
Norway rats	X	X	X	X	X	X	X	-	Garbage dumps
Opossums	-	-	-	X	X	-	-	-	
Pocket gophers	-	-	-	X	X	X	-	X	Underground cables
Rabbits	-	-	-	X	X	X	-	X	
Raccoons	X	-	-	X	X	X	-	X	Around residences
Roof rats	X	X	X	X	X	X	X	-	
Skunks	X	X	-	X	X	X	-	X	Around residences
Tree squirrels	X	X	X	X	X	-	X	-	
Woodchucks	-	X	-	X	X	X	-	X	
Wood rats	-	X	-	X	X	-	-	-	
Feral pets									
Cats	-	X	X	X	X	-	-	-	
Dogs	X	X	-	X	X	X	-	X	

ORIGINS OF MAJOR PEST SPECIES

Urban vertebrate pests are not limited to those native species that adapt easily and find the city environment quite conducive to their needs; they also include native species, such as pocket gophers, moles, skunks, woodpeckers, and bats, which are largely displaced when a community or city evolves and grows.

Some of our most serious pests are species that have been accidentally or deliberately introduced into this country. The most serious of rodent pests, the Norway rat (*Rattus norvegicus*), the roof rat (*Rattus rattus*), and the house mouse (*Mus musculus*), which collectively are often referred to as commensal (living with people) rodents, are examples of pests accidentally introduced by our forefathers from infested sailing ships or imported unintentionally in cargo.

Other introductions include the European starling (*Sturnus vulgaris*) and the house or English sparrow (*Passer domesticus*); however, these were deliberately introduced. The pigeon (*Columba livia*) is also not native to this country; its introduction was likely due to escapes from domestication. The origin of vertebrate pests is relevant to pest control because it reflects significantly on our attitudes toward the pests and on how we manage them as pests.

Animals that escape from domestication or are deliberately released to roam freely are referred to as feral animals. Hence we refer to feral pigeons or feral dogs and cats that have no home. Feral or uncontrolled dogs represent an important vertebrate pest in city and urban settings; feral cats are not far behind. To many residents, unmanaged or uncontrolled dogs are considered to be their most significant vertebrate pest, tipping garbage cans, tearing open plastic garbage bags, fouling sidewalks, and damaging landscaping and gardens. More important are the incidences of dog bites and concern over contracting rabies. It seems paradoxical that man's best friend can also be a pest of substantial economic and public health importance.

Some of our pest species are native to the United States but not native to the region where they currently reside. For example, two tree squirrels, the fox squirrel (*Sciurus niger*) and the Eastern gray squirrel (*S. carolinensis*), have been introduced into the West. Opossums (*Didelphis virginiana*) now exist in many regions where they did not originally exist. These three species were transported and released for their value as furbearers or game animals and now at times are considered urban pests.

MAJOR PEST SPECIES

Nationwide, the eight leading urban pest species are house mice, Norway and roof rats, feral or uncontrolled dogs and cats, feral pigeons, house sparrows, and starlings. Locally, of course, other species may be much more important than these. Many other species, including tree squirrels, may at times be considered significant pests, although rarely will their public health or economic importance approach those of rats and mice, except possibly in very limited urban areas.

Urban pest management requires a good knowledge of the identification, biology, behavior, and ecology of the pest species. Books are available on the birds, mammals, reptiles, and amphibians of nearly every state or geographic region and are good starting points for learning about native species. Major references on the management of specific vertebrate pest are more limited. One of the newest and most outstanding loose leaf volumes is entitled *Prevention and Control of Wildlife Damage,* edited by R. M. Timm (1983). Several others are of value for more specialized problems; however, some of the older volumes are out of print and may be found only in libraries (Bateman, 1979; Baur and Jackson, 1982; Chapman and Feldhamer, 1982; Chitty and Southern, 1954; Eadie, 1954; Fitzwater, 1979; Mallis, 1982; Morton and Wright, 1968; National Academy of Sciences, 1970; National Research Council, 1980; National Pest Control Association, 1982).

INTEGRATED MANAGEMENT STRATEGIES FOR VERTEBRATE PESTS

Two basic strategies are used in resolving vertebrate pest problems: indirect control, sometimes referred to as preventive control, which includes such techniques as environmental sanitation, habitat manipulation, rodent- or bird-proofing of buildings, fencing, netting, the use of repellents, and providing an alternative food; and direct control or reductional control, which includes such techniques as capture and removal, shooting, trapping, fumigants, and poison baits. The latter strategy is the opposite of preventive control, since it is used when a problem already exists or is imminent; hence this may be referred to as corrective control. The terms "control" and "management" are used interchangeably in this chapter, as they are in most vertebrate pest literature.

Direct and indirect control strategies are commonly combined and used in resolving vertebrate problems, with one strategy aimed at a relatively speedy solution to the problem and the other used to prevent recurrence. The variety of control techniques and options available for many vertebrate pest problems makes integrated pest management (IPM) an effective and logical management concept.

Direct control generally removes the offending individuals or species relatively rapidly, but with some species it may have little long-lasting effect. Indirect or preventive control, on the other hand, is aimed at preventing damage or conflict, and the density of a population may or may not be decreased. If a population change occurs as a result of indirect control, it is slow and generally appears more natural or akin to natural events. For example, a chemosterilant or antifertility agent would usually be considered an indirect control measure. Methods of indirect control such as habitat modification are generally more lasting, but some are also much less predictable in their results than direct control methods.

Many considerations enter into the selection of the most appropriate IPM strategy and control techniques to be used. Sometimes the best strategy involves but a single technique, e.g., the rodent-proofing or bird-proofing of a building or the modification of habitat.

MANAGEMENT METHODOLOGY

The methods and materials used in the integrated management of vertebrate pests are diverse and include such major approaches as environmental sanitation, habitat modifications, rodent- and bird-proofing of buildings, exclusion with fences and various types of protective barrier, a wide range of chemical and physical repellents, trapping, chemosterilants, fumigants, contact toxicants, and poison baits.

An extensive methodology is needed because of the large variety of pest species and the complexity of the pest situations. Vertebrates and their control are covered extensively by legislation, and this becomes most significant in their management. No other pest group is regulated by so many laws, regulations, and ordinances; these primarily relate to direct control methods but also touch upon nearly every aspect of vertebrate pest management. They often dictate how and when certain animals can be managed. Vertebrate pesticides, which include the toxicants as well as a large group of nontoxic chemical repellents, must be registered by the U.S. Environmental Protection Agency (EPA) and state regulatory agencies for the species for which they are used.

There are four fundamental bodies of information needed for properly managing urban vertebrate pests: a thorough knowledge of the pest species; a working knowledge of the techniques and materials used in their control; an understanding of IPM principles and concepts to formulate appropriate control strategies; and an up-to-date awareness of all applicable laws and regulations governing the pest species, control methodology, and pest situations.

MAJOR CONTROL METHODS AND MATERIALS

Some major methods and materials used in integrated vertebrate pest management are given in Tables 11-3 and 11-4 to illustrate the diversity of techniques and materials used. They have been categorized according to use patterns; however, these categories are not rigid, and some techniques could logically be placed under one of several headings. Some of the less-used techniques such as nest destruction, limiting natural food supplies or water,

Table 11-3. Major pesticidal options and approaches for urban vertebrate pest management with examples.

Pesticide methodology	*Example*
Lethal pesticides	
Poison baits	Bait to control rodents and pest birds such as house sparrows and pigeons
Contact toxicants	Toxic perches for pigeons, house sparrows and starlings
Grooming toxicants	Tracking powders for house mouse and rat control
Fumigants	Material for fumigating rodents in their burrows and building fumigation for rodents
Stressing agents	Surfactants for roost spraying against starlings and blackbirds
Chemosterilants	
Baits	Used to inhibit reproduction in pigeons and Norway rats
Chemical frightening agent	Used in baits to repel birds by creating a flock-frightening response
Nonlethal chemical repellents	
Area (olfactory) repellents	To keep dogs, cats, bats, and skunks from traveling into or occupying buildings or dooryards
Repellents to protect man-made materials	Sprayed on garbage bags to keep dogs and cats from tearing them open
Repellents to protect plants	To protect plants from rabbit and deer damage
Seed protectants	To protect garden seed from seed-eating rodents and birds
Sticky or tacky (tactile) repellents	To repel birds from building ledges or rafters

Note: See also Table 11-2, p. 256.

hand-capturing, snares, denning, cannon netting, mist netting, falconry, tranquilizer guns, tranquilizers and immobilizing agents, tracking powders, liquid and gels, oral rabies vaccines, surgical neutering, and catch or trap crops are not discussed but may be useful for resolving specific animal problems.

Both pesticidal and nonpesticidal methodology are discussed only briefly. More detailed in-depth information on vertebrate pest management materials, techniques, and IPM strategies can be found in cited references.

Lethal or Direct Control Measures

Methods that remove the animal or are outright lethal to the pest species include both pesticidal and nonpesticidal approaches. Trapping falls into this

Table 11-4. Major nonpesticidal options and approaches for urban vertebrate pest management with examples.

Nonpesticidal methodology	*Example*
Lethal control	
Shooting	Shooting of pigeons
Trapping	Trapping of pocket gophers, moles, rats, and mice
Capture	Catching and euthanizing of feral dogs
Exclusion	
Pest-proofing of building	Rodent, bird, and bat proofing of buildings
Fences	Fences to exclude deer, skunks, and opossums
Tree guards	Protecting young trees from rabbits and deer
Netting	Protecting tree crops and gardens from birds
Repellents	
Visual frightening devices	Colored flags and whirling devices to scare birds
Mechanical or physical repellents	Wire porcupine spines to keep birds from building ledges
Electrical repellents	Shocking wires to keep birds from buildings
Frightening sounds	Automatic acetylene exploders to frighten birds
Habitat manipulation	
Landscape modification	Pruning of trees to make them less suitable as roost sites
Sanitation	Improving garbage pickup to reduce food source for rats
Supplemental feeding	Providing feed for birds to keep them from feeding on garden
Selecting less-preferred plants	Planting of deer-resistant plants
Biological control	
Introduced predators	Cats may provide some minor benefits through mole or pocket gopher catches in the garden
Native predators	Gopher snakes for pocket gopher control
Introduced pathogens	Not recommended because of hazards to people

category. However, trapping may be nothing more than a capture technique, and the animals may be removed and released elsewhere or put to death.

Poison baits. Poison baits are used to control commensal rodents as well as such species as populations of meadow voles, deer mice, pocket gophers, woodrats, pigeons, starlings, and house sparrows.

Toxicants are usually divided into two categories: acute (single-dose) toxicants such as zinc phosphide, red squill, and strychnine; and chronic (multiple-dose) toxicants. The anticoagulant rodenticides generally fall into the second category. Poison baits are often the most cost-effective approach to commensal and field rodent control.

Fumigants. Fumigants are gaseous toxic pesticides that kill as a result of inhalation. Materials such as calcium cyanide, methyl bromide, chloropicrin, aluminum phosphide, and gas cartridges are used to control burrowing rodents such as marmots (groundhogs or woodchucks), ground squirrels, and Norway rats. Fumigants are sometimes used by specially certified applicators to treat entire buildings to rid them of insects and rodents.

Toxic perches. The control of such birds as pigeons, house sparrows, and starlings can be accomplished with strategically placed toxic perches. Toxicant-bearing wicks in the hollow perches transfer the contact toxicant (endrin or fenthion) through the feet of the perched bird.

Euthanizing agents. Euthanasia is the action of putting to death unwanted animals in a painless way, using one of various euthanizing agents, usually under very controlled conditions. This approach is used by dog pounds primarily to destroy unwanted dogs and cats but is sometimes used for destroying various pest animals once captured. Euthanizing agents are not considered pesticides because of the manner in which they are used.

Shooting. Firearms or air rifles are occasionally a useful control tool where they are legal and not hazardous to others. They are sometimes used to remove a few pigeons or sparrows from a warehouse or to remove skunks, foxes, or coyotes from parks or other recreational areas. The use of firearms is illegal in most urban areas, although permits may be issued under special circumstances.

Trapping. Types of traps and their uses are so extensive that entire books have been written on the subject (Bateman, 1979; McCracken and Van Cleve, 1967). The object of trapping is to eliminate the pest by killing it; or to eliminate the pest from the vicinity by live-trapping and relocating it elsewhere, as is sometimes done with raccoons, tree squirrels, rabbits, skunks, and beavers. Kill traps are commonly used for commensal rats and mice and for such species as pocket gophers, ground squirrels, and moles.

Special Chemical Measures

This category includes three control techniques that do not conveniently fit into any other group.

Chemosterilants or reproductive inhibitors. An aza-steroid (20, 25 diazacholesterol dihydrochloride) is registered for pigeons and is the only reproductive inhibitor (chemosterilant) available for bird control. The combination rodenticide and chemosterilant alpha-chlorohydrin is registered for Norway rat control. At low doses it causes temporary sterility in adult male rats; at slightly higher doses it causes irreversible sterility; and at still higher doses it is lethal and works as an acute rodenticide for both males and females.

Chemical frightening agent or psychochemicals. The chemical frightening agent 4-aminopyridine is applied to baits for the control of such birds as pigeons, sparrows, blackbirds, starlings, crows, and gulls. The few birds that ingest the treated bait display erratic behavior and emit distressing calls that, in turn, frighten the other members of the flock. By dosing a relatively small number of birds, a few of which may die, the material can produce flock-alarm reactions that repel the rest of the birds from the area. The area avoidance may last for a long time.

Stressing agent. Tergitol is a nontoxic surfactant and is registered as PA-14 for controlling blackbirds and starlings at roosting sites. This surfactant, when sprayed on birds, destroys the insulation value of feathers by cutting the natural oils. Its effectiveness relies on rain occurring during or shortly following the PA-14 application, with cold weather (45°F or below) following that same night. Rain penetrates the feathers, and in cold temperatures the sprayed birds die of exposure or stress.

Exclusion

Exclusion includes rodent- and bird-proofing of buildings, fencing, tree guards, and exclusion netting. Collectively, they represent some of the most effective and predictable indirect control measures available (Scott and Borom, 1976).

Rodent and bird proofing. Rodent or bird proofing of buildings is a very cost-effective form of vertebrate control. It involves the deliberate exclusion of rodents, birds, and other animals from buildings by screening, sealing, or plugging off all possible entries to the building.

Fences. Deer- and rabbit-proof fences are used to protect backyards, gardens, and landscaped areas from damage. Deer fences are also useful to keep deer confined to greenbelts or cemeteries and away from freeways or airport runways, where they can be a traffic menace. Some fences can be electrified for an added effect. Fences are also effective in excluding armadillos, skunks, snakes, stray dogs, and certain other pests. Effective fence design must take into consideration the abilities of the pest species and how they will attempt to circumvent such fences (Fitzwater, 1972).

Tree guards. Tree guards are made of gnaw-resistant material and are used to wrap around the trunks of young trees to prevent gnawing damage by rabbits, woodchucks, and meadow mice. Physical guards are also used around various tree limbs being attacked by sapsuckers or other members of the woodpecker family. Wide metal bands are sometimes placed around tree trunks to prevent squirrels and rats from climbing a tree to feed on fruit or nuts.

Netting. Netting made of plastic, fiber, or wire is a widely used control technique to protect flowers, vegetables, and berries from bird depredation. Lightweight netting is used to completely cover backyard tree crops from birds. It can also prevent swallows, woodpeckers, house sparrows, and pigeons from defacing or nesting on buildings and statues. When done properly, the netting is hardly visible.

Repellents

Methods of repelling vertebrate pests or for preventing animal damage are many and varied. Chemical repellents are classified as pesticides, whereas other repellent methods are not. The following represent the major types of repellents.

Seed protectants. Chemical seed protectants, such as thiram and copper oxalate are generally taste or odor repellents used to coat seeds to protect them from seed-eating birds and rodents.

Chemical repellents to protect plants. There are several kinds of repellent used to protect growing plants from bird or mammal damage. Such chemical repellents are applied to seedlings, young trees, and other plants to protect them from deer, rabbits, and hares. Thiram, bone tar oil, ammonia mixtures, and putrescent egg solids are examples of taste or odor repellents. Methiocarb, which causes conditioned aversion, is used to prevent birds from damaging several fruit crops. A number of odor repellents are marketed for application around gardens to repel cats and dogs. Others are used to repel skunks from beneath buildings and bats from attics. The effectiveness of chemical repellents varies greatly and is influenced by a number of factors that are often uncontrollable.

Sticky or tacky repellents. Various sticky-type chemical repellents are applied to building ledges, rafters, and sometimes dormant trees to discourage roosting birds. These sticky repellent materials are objectionable to the pest but harmless and do not entrap birds or mammals.

Visual frightening devices (repellents). A wide variety of visual repellents such as replicas of snakes, hawks and owls, colored flags, balloons, revolving and flashing lights, whirling devices, and scarecrows have been used for bird control. Few are effective for more than a short period of time. Visual repellents are even less effective for mammal species.

Mechanical or physical repellents. Rows of permanent "porcupine" upward-projecting metal spines are fastened on building ledges to physically repel roosting or nesting birds. Other types of physical barrier are also used on building ledges. They are all designed to make it physically impossible for the birds to alight on the surface.

Electrical repellents. Uninsulated wires are strung along building ledges and are charged much as an electric cattle or deer fence. One wire serves as a ground; another wire carries a high voltage charge. However, the amperage is very low, so birds are unharmed but frightened away by the shock.

Frightening sound repellents. Various types of acoustical scaring device have been used to repel pest animals, particularly birds. Automatic acetylene, propane, or butane gas exploders, which make a loud report, are one of the most commonly used devices for bird control. Bioacoustics, or the use of recorded bird distress or alarm calls that are played back over a loudspeaker system, have also been effective against certain bird species (Frings, 1964). Firecrackers, rockets, and other pyrotechnics are sometimes also used to disrupt birds at roosts. Exploding shells (shellcrackers) fired from shotguns or pistols are effective in some situations but are mostly used in protecting agricultural crops. Various other types of sound-producing equipment have also been used with varying results. Generally both birds and mammals habituate to sounds very rapidly, so if good results are achieved, they are usually short-lived.

Habitat Manipulation or Modification

This technique of vertebrate pest management is so varied that space does not permit a full discussion. It can be the removal of weed cover for a space of 6 ft or more from the exterior of a warehouse to eliminate house mouse or meadow vole cover, the pruning of low branches on shrubbery to reduce rabbit cover, the thinning of dense trees or topping of tall trees to make them less suitable as bird roosts, or the removal of trees that may be particularly attractive as bird roosts and the planting of substitute species not used as roosts.

Landscape design and landscape maintenance can greatly influence wildlife either positively or negatively. For example, mowing grassy areas short will make the areas less suitable for such native rodents as meadow voles and cotton rats. At airports, the grassy areas near the runways may be deliberately mowed higher to make the areas less attractive to certain birds as areas for loafing. Habitat manipulation requires a thorough knowledge of the pest species involved and what it requires or prefers in the way of habitat for all aspects of life (feeding, roosting, denning, and nesting) so that appropriate actions can be taken.

One important point to remember is that habitats are not abolished, they are only modified. Thus what may be done to modify a particular habitat to make it less suitable for one pest species may at the same time make it more suitable for another. Habitat modification must be compatible with other landscaping or land-use practices.

Several specific aspects of habitat modification follow.

Sanitation. In the control of commensal rodents, building and area sanitation are always stressed. This reduces the food supply and habitat and makes the use of poison bait more effective because of increased bait consumption, since less other food is available. Sanitation includes good housekeeping as well as good garbage and refuse disposal (Pratt and Johnson, 1975). Properly secured garbage cans with tight lids reduce raccoon, dog, and coyote problems. Pigeons are often attracted to locations where grain is unloaded from railway cars. Sparrows are a nuisance around warehouses, fast food restaurants and malls where spilled food or garbage may be abundant. The daily clean-up of these food sources will do much to reduce bird problems.

Supplemental feeding. To deliberately provide the pest with some highly preferred food in the hope that the pest species will be attracted to another area or leave your growing garden alone is not unlike using a trap or catch crop, except the preferred food is not grown on the site. A gardener may set up bird feeders in the hope that the birds will then leave a newly planted garden alone. Suet has been used to attract woodpeckers away from wood siding they may be damaging on a house. Supplemental feeding, however, often results in attracting more of the pests than would normally be present.

Selecting less preferred plants. While horticulturalists have not developed ornamental plants resistant to vertebrate pests, they have developed lists of plants for gardeners and landscapers that are least preferred as food by deer (Cummings, Kimball, and Longhurst, 1980). Plants that should be avoided in landscaping because they are particularly attractive as habitat for the roof rat are also known.

Biological Control

Biological control in the traditional sense can be viewed as indirect population control, for it involves such tactics as the deliberate introduction of fatal or debilitating pathogens or exotic predators. Although biological control has been quite effective against certain insects and weed pests and is an important aspect of IPM, it has unfortunately met with little success in vertebrate control. Predators of vertebrates are not species-specific in their food selection. Once they are released into an ecosystem, people have little if any control over their future effect on the biota. Effective pathogens likewise are often not pest-specific; thus biological control in the classical meaning plays a negligible role in integrated management of vertebrate pests.

Introduced predators. Numerous examples exist of predators being introduced to control agricultural vertebrate pests. With few exceptions, the original pest problem has not been resolved, and the predators frequently become pests themselves by preying on desirable vertebrates (Howard, 1967).

Domestic cats are sometimes used to help control commensal rodents. If rodents are eliminated or reduced to a low level, using traps or rodenticides first, good hunter cats may be effective in preventing reinfestations or in keeping the rodent populations—especially mice—low. But generally speaking, cats in a home do not guarantee rodent-free premises. Active dogs on long sliding leash runs can sometimes be effective in scaring birds, rabbits, and deer from gardens.

Native predators. It has been suggested that gopher snakes will rid your garden of pocket gophers. However, although many snakes live on small rodents, there is little evidence they keep rodent species under control to the degree that would be acceptable in backyards. Barn owl nest boxes have been installed in some residential areas in the hope of reducing roof rat populations. Though barn owl populations have in some cases increased, predictably there has been no measurable adverse effect on roof rat numbers.

Native vertebrate predator-prey relationships are often misunderstood. We forget that they have evolved together, and in countless situations the number of prey present determines the number of predators—not vice versa. In certain situations, however, native predators may have some adverse effect on certain vertebrate pests, but they are unpredictable and cannot be relied upon to prevent animal damage.

Falconers with their trained falcons or hawks, on their other hand, have been effective in scaring certain pest birds from airport runways to help prevent aircraft-bird strikes.

Introduced pathogens. The use of pathogens for the control of pest vertebrates was first suggested by Pasteur. In the late 1800s *Salmonella* bacteria were used against voles (*Microtus* spp.) in Europe and later for the control of rats and mice (Wodzicki, 1973). Unfortunately, in a number of instances people fell victim to the introduced nonspecific bacterial diseases. Because of the potential hazard to people, and nontarget animals, the use of diseases to control commensal or native rodents is not advocated. The most classic successful example of using an introduced pathogen for a vertebrate pest is the use of myoxmytosis virus for the control of the introduced European rabbit (*Oryctolagus cuniculus*) in Australia, but it, too, was not without its problems.

Based on present knowledge, it is highly unlikely that we will ever introduce predators or pathogens for the control of vertebrate pests in urban situations. This is in part because many vertebrates that concern us as pests are pests in only a few situations, and the same species may be viewed as desirable or neutral much of the time.

KEY URBAN VERTEBRATE PESTS AND MANAGEMENT OPTIONS

Important urban vertebrate pests and principal options used in IPM are presented in Table 11-5 along with select references. The table is not all-

Table 11-5. Major urban vertebrate pests and examples of pest problems, with management strategies and techniques and references.

Pest	*Example of problem*	*Management strategies and techniques*	*References*
Amphibians			
Toads and frogs	Swimming pool contamination, landscape degradation	Habitat modification, barriers, and fences	Fitzwater, 1974; Mallis, 1982; Marsh and Howard, 1978*b*; National Academy of Sciences, 1970
Reptiles			
Snakes	Health hazard to children, fear of snakes in yards and homes	Habitat modification, snake-proof fencing, trapping	Fitzwater, 1974; Mallis, 1982; Timm, 1983
Birds			
Blackbirds	Structure and landscape degradation	Bird barriers, repellents, roost spray, pruning trees, toxic perches, frightening devices	Good and Johnson, 1978; National Academy of Sciences, 1970; Stefferud, 1966; Timm, 1983
Crows	Roost in park and yard trees, landscape degradation	Frightening devices, bioacoustics, pruning trees, psychochemical agent	Morton and Wright, 1968; Timm, 1983
Gulls	Pond pollution, structure degradation	Bird barriers, frightening devices, psychochemical agent	Morton and Wright, 1968; Stefferud, 1966
House sparrows	Enter buildings, environmental degradation	Bird barriers, repellents, exclusion, toxic perches, toxic baits, psychochemical agent, traps	Bauer and Jackson, 1982; Mallis, 1982; Morton and Wright, 1968; Summers-Smith, 1967; Timm, 1983

(continued)

Table 11-5 *(continued)*

Pest	*Example of problem*	*Management strategies and techniques*	*References*
Pigeons	Enter buildings, environmental	Exclusion, bird barriers, repellents, trapping, shooting, toxic perches, toxic baits, chemosterilants	Bauer and Jackson, 1982; Benton and Dickinson, 1966; Mallis, 1982; Morton and Wright, 1968; Stefferud, 1966; Timm, 1983
Starlings	Structure and landscape degradation	Frightening devices, bird barriers, bioacoustics, pruning trees, repellents, roost spray, toxic perches	Bauer and Jackson, 1982; Frings, 1964; Mallis, 1982; Morton and Wright, 1968; Stefferud, 1966; Timm, 1983
Swallows	Degradation of buildings	Bird barriers, repellents	Timm, 1983
Waterfowl	Pond pollution, landscape degradation	Habitat modification, frightening devices, fences, trapping, immobilizing agents	Morton and Wright, 1968; Stefferud, 1966; Timm, 1983
Woodpeckers	Damage to buildings and poles, damage to trees	Bird barriers, repellents, frightening devices, shooting, traps	Mallis, 1982; Stefferud, 1966; Timm, 1983
Mammals			
Armadillos	Environmental degradation, garden pests	Trapping exclusion	Chamberlain, 1980; Mallis, 1982; Timm, 1983

Bats	Enter buildings, environmental degradation	exclusion, repellents, trapping, contact toxicants	Champman and Feldhamer, 1982; Corrigan and Bennett, 1983; Greenhall, 1982; Mallis, 1982; Timm, 1983
Chipmunks	Enter buildings, landscape and garden pests	Exclusion, repellents, trapping	Eadie, 1954; Fitzwater, 1979; Mallis, 1982
Cotton rats	Landscape, turf and garden pests	Habitat modification, toxic baits, trapping	Eadie, 1954; Mallis, 1982; Timm, 1983
Coyotes	Attack people, garden pests, garbage scavengers	Trapping, shooting, denning	Eadie, 1954; Howell, 1982; McCracken and Van Cleve, 1967; Timm, 1983
Deer	Landscape and garden pests	Exclusion, repellents, resistant plants, and tranquilizer guns	Bashore and Bellis, 1982; Caslick and Decker, 1978; Chapman and Feldhamer, 1982; Cummings, Kimball, and Longhurst, 1980; Timm, 1983
Deer mice	Enter buildings, garden pests	Exclusion, trapping, toxic baits	Eadie, 1954; Mallis, 1982; Timm, 1983
Foxes	Garden pest, garbage scavengers	Trapping, shooting, denning	Eadie, 1954; McCracken and Van Cleve, 1967; Timm, 1983
Ground squirrels	Landscape, turf, and garden pests	Fumigants, trapping, toxic baits	Chapman and Feldhamer, 1982; Eadie, 1954; Mallis, 1982; Timm, 1983
House mice	Enter buildings	Sanitation, habitat modification, toxic baits, fumigants, trapping, tracking powder	Brooks, 1973; Chitty and Southern, 1954; Mallis, 1982; Marsh and Howard, 1981; National Academy of Sciences, 1970; Pratt and Brown, 1976; Scott and Borom, 1976; Timm, 1983

(continued)

Table 11-5 *(continued)*

Pest	*Example of problem*	*Management strategies and techniques*	*References*
Moles	Landscape, turf, and garden pests	Trapping	Bateman, 1979; Chapman and Feldhamer, 1982; Eadie, 1954; Marsh and Howard, 1978*a*; Timm, 1983
Muskrats	Pond and landscape degradation	Habitat modification, trapping, fumigants, toxic bait	Chapman and Feldhamer, 1982; Eadie, 1954; Errington, 1961; Fitzwater, 1979; Timm, 1983
Norway rats	Enter buildings, environmental degradation	Sanitation, habitat modification, toxic bait, fumigants, trapping, tracking powder	Brooks, 1973; Chitty and Southern, 1954; Fitzwater, 1979; Howard and Marsh, 1981; Mallis, 1982; National Research Council, 1980; Pratt and Brown, 1976; Pratt and Johnson, 1975; Scott and Borom, 1976; Timm, 1983
Opossums	Landscape degradation, garden pests	Exclusion, trapping	Eadie, 1954; Mallis, 1982; McCracken and Van Cleve, 1967; Timm, 1983
Pocket gophers	Landscape, turf, and garden pests	Trapping, toxic baits	Bateman, 1979; Eadie, 1954; Fitzwater, 1979; Howard, 1978*b*; Timm, 1983
Rabbits	Landscape, turf, and garden pests	Fences, repellents, trapping, habitat modification	Chapman and Feldhamer, 1982; Eadie, 1954; National Academy of Sciences, 1970; Timm, 1983

Raccoons	Garbage scavengers, landscape and garden pests	Exclusion, trapping, shooting, repellents	Eadie, 1954; Mallis, 1982; McCracken and Van Cleve, 1967; Timm, 1983
Roof rats	Enter buildings, landscape and garden pests	Sanitation, habitat modification, toxic baits trapping, tracking powder	Brooks, 1973; Chitty and Southern, 1954; Howard and Marsh, 1981; Mallis, 1982; National Academy of Sciences, 1980; Pratt and Brown, 1976; Scott and Borom, 1976
Skunks	Reside beneath houses, landscape and garden pests	Exclusion, trapping, shooting, repellents	Chapman and Feldhamer, 1982; Eadie, 1954; Mallis, 1982; Timm, 1983
Tree squirrels	Landscape and garden pests, enter buildings	Exclusion, trapping, toxic baits	Eadie, 1954; Mallis, 1982; Timm, 1983
Woodchucks	Landscape, turf, and garden pests	Fumigants, trapping, toxic baits	Eadie, 1954; Fitzwater, 1979; Mallis, 1982; Timm, 1983
Wood rats	Enter buildings, landscape and garden pests	Exclusion, habitat modification, toxic baits, trapping	Eadie, 1954; Mallis, 1982; Timm, 1983
Feral Pets			
Cats	Landscape and garden pest, pest of birds	Exclusion, repellents, trapping	Eadie, 1954; Neville, 1983; Timm, 1983
Dogs	Bite people, garbage scavengers, landscape and garden pests	Exclusion, repellents, capture	Beck, 1973; National Research Council, 1980; Nowell, 1978; Timm, 1983

inclusive; many minor pests are not mentioned, some of the less-used management options are omitted, and the cited references have been limited to a few of the most significant ones.

Case Study: Urban Pigeon Control

As an example of IPM, the following composite case study is offered involving a serious pigeon problem in a city of about 35,000 inhabitants, which is the county seat.

The pigeons had been a nuisance for years, but recently the pigeon population had grown substantially and finally reached an intolerable level. The problem was brought to a head by two cases of histoplasmosis in a 3-month period. News coverage heightened concern over the health problem. The city council decided something had to be done and directed the county health department to study the problem and present a plan of action.

The following are the steps that should be taken in developing a pest management strategy or plan of action to be undertaken by a city:

1. Identify the pest species causing the problem. In this situation it was obvious, but this is not always the case with urban vertebrate pest problems.
2. Assess the problem or problems for which solutions are needed and establish key objectives. In this situation it was both a public health problem from droppings and filth caused by pigeons and a nuisance factor. Some economic losses at the local grain elevator were ongoing but were difficult to document, as was the defacing of structures.
   Such an assessment should include a rough approximation of the number of pigeons in the population, their distribution within the city, and identification of those key places or properties that contribute in a major way to providing food and water, nesting, loafing, and roosting sites.
3. Determine the city's legal and financial constraints that must be considered before commencing pigeon management, i.e., available funds, available personnel, public relations (public sentiment), pertinent city ordinances, cooperation from those directly affected, possible liability if nothing is done, and liability that may occur in carrying pigeon management.
4. Study all the relevant management options available, including costs for materials and manpower required for the control techniques.
5. Study the essential biology and behavior of pigeons relative to possible control options. Determine what nontarget species might be at risk.
6. Determine what wildlife, pesticide, or pest management-related (i.e., local, state, or federal) laws and regulations may influence control options and decide which options can be carried out under existing regulations.

For example, can firearms be discharged in the city limits, or has the city been declared a wildlife or bird preserve, thus affording protection to all birds?

7. Determine if laws or regulations exist that can be used to aid in control efforts; for example, health regulations by which a property owner can be forced to take corrective measures to reduce a health threat.
8. Develop a control strategy that best meets the desired objectives and constraints that must be adhered to. The control strategy may be a sequence of actions or concurrent actions or both. Different but equally effective strategies are possible for many vertebrate pest problems.
9. Develop a follow-up plan to institute when principal objectives have been met, incorporating long-term solutions such as pigeon-proofing of buildings, monitoring of the pigeon population, public education concerning the potential for future pigeon problems, and steps that can be taken to avoid such problems.

It was decided that the main objective would be to reduce the pigeon population to an acceptable level (a level at which the potential public health and nuisance problem would be minimal). A clean-up of existing accumulated droppings would be emphasized to reduce the present health hazards. Where economically feasible, pigeon-proofing of affected buildings and building ledges used by the birds would be pursued.

A control strategy was established whereby pigeon population reduction was to be achieved primarily by trapping, and a contract was let by the city to a pest control firm with a good local reputation. The contracting firm bid on the job and supported its bid with a very detailed and professionally organized trapping plan. A member of the health department was designated project leader to coordinate the pigeon control program and serve as the liaison with the pest control firm.

Individual building or property owners were called upon and encouraged to take action to clean up pigeon filth and pigeon-proof their buildings where possible and economically practical (Baur and Jackson, 1982; National Pest Control Association, 1982). An existing but rarely used health ordinance that could mandate such clean-up would be used if voluntary compliance failed. Through the news media, a county health officer provided information to the public on how the clean-up of droppings could be accomplished with minimal exposure to dustborne diseases.

In the downtown area, several owners of buildings, including publicly owned buildings, used plastic or nylon netting to net off some of the complex architectural configurations that provided both nesting and roosting sites. Others used prong-type physical barriers (Nixalite®) to keep birds off the ledges. A few property owners utilized a more temporary solution involving the use of tacky or sticky bird repellents on ledges (National Pest Control Association, 1982; Timm, 1983).

Trapping by the pest control firm was undertaken on rooftops of six key buildings used frequently by the pigeons for loafing, roosting, and nesting. Several sizes of bob-type traps were used on those rooftops that could accommodate them and where servicing was convenient (National Pest Control Association, 1982).

At the start of the project in an isolated area near a grain elevator, strychnine pigeon bait was used following an appropriate period of prebaiting (Baur and Jackson, 1982; National Pest Control Association, 1982; Timm, 1983). In this instance it was an intelligent management option. Because this feeding area was used by pigeons from throughout the city, the number of pigeons was rapidly reduced, as was the magnitude and expense of trapping efforts needed to control the remaining population.

It took approximately 6 months to reduce the pigeon population to about one tenth its original level. At that time the number of traps was reduced by two thirds and left at the most needed locations. A bimonthly visual monitoring program was instigated to keep tabs on pigeon numbers so that necessary corrective measures could be taken to prevent recurrence.

Case Study: Meadow Mouse Control in a Park

Meadow mice (voles), which belong to the Genus *Microtus,* have a tendency to fluctuate in numbers, with many species reaching exceptionally high populations periodically (Chapman and Feldhamer, 1982). When densities are high, they can do substantial damage to landscaped areas. This case study represents one of those periods when the population had reached intolerable levels, destroying turf, ornamental shrubs, and girdling susceptible trees on a golf course and an adjoining relatively large suburban park. Numerous burrows and runways were found in unmowed grassy areas and were becoming more common even in the mowed turf area. The bark of the trunks of a dozen or more trees had been completely girdled at ground level, and many deciduous and evergreen ornamental shrubs had bark removed as high as 18 in above the ground. Glimpses of darting mice could be observed as the area was surveyed for damage.

Management options in this situation were few as the problem was near crisis level. After carefully assessing the problem, it was decided that baiting would be the quickest and most cost-effective solution. Registered rodenticides were limited, but of those available, a zinc phosphide oat groat bait was selected (Mallis, 1982; Timm, 1983). Zinc phosphide presents very little secondary hazard to pets, mammalian or avian predators. Dyed oat bait presents little hazard to small ground-feeding, seed-eating birds, and large game birds seldom frequented the area. Public access to the park could be temporarily curtailed, and golf course use could be restrictively managed if need be. The bait was broadcast with a hand-cranked cyclone-type seeder in the most densely and uniformly mouse-infested areas at a rate of 15 lb per acre, according to label directions. In more sparsely infested areas, teaspoon amounts of bait were placed directly in burrow openings or runways (Eadie, 1954).

Pre- and post-treatment censusing of the mouse population with snap traps indicated that it was reduced by 87%. Some subsequent follow-up baiting was conducted by hand in a few spots around shrubbery and trees that apparently had been missed. The control program was successful in reducing most of the damage, with no observed or known loss of nontarget wildlife. Two weeks following control, the densely grassed areas were mowed as closely as possible with a tractor-drawn rotary-type mower to reduce cover for the mice and prevent population resurgence. Weed management around trees and shrubbery areas was increased, and where feasible, the grass was mowed away from landscape plantings.

Case Study: An Urban Raccoon Problem

In an effort to reduce the scattering of garbage by raccoons in a small coastal city, the city council passed a law that only metal garbage cans, with lids adequately secured to prevent opening by raccoons, would be legal for household garbage and refuse (Leedy and Adams, 1984; Timm, 1983). The law further made feeding of raccoons illegal. Garbage pickup was increased from once to twice a week. Where raccoons continue to be a problem, large wire-mesh live-catch traps could be borrowed from the city maintenance department and trapped raccoons picked up by the local humane society (Mallis, 1982; McCracken and Van Cleve, 1967; Timm, 1983).

MAJOR PROBLEMS CONFRONTING EFFORTS IN VERTEBRATE PEST MANAGEMENT

There are many problems confronting vertebrate pest control. Some are similar to those of other types of pest control; however, some are unique to vertebrate pest control. One of the most serious problems is the limited research conducted to develop new or improved pest management strategy and new vertebrate pest control materials. Little research is conducted by federal or state agencies, and research conducted at universities is minute compared with the man-years of scientist and funding resources devoted to insect, plant disease, and weed control. The research and development efforts by industry directed toward new vertebrate pesticides and repellents are also very limited.

To the practitioners, the lack of research translates into a lack of effective and economical solutions to certain pest problems, especially those that are regional in scope or occur relatively infrequently. The lack of new vertebrate pesticides is also due to the high costs of development of a new product through registration. These costs are not related to whether or not the end product has a large or small potential market, and thus industry is often reluctant to venture into vertebrate control material research and development (Marsh, 1981).

Other major problems confronting vertebrate pest management are the diverse philosophies and attitudes toward animals and their welfare. The current trend of animal rights and humaneness movements hampers vertebrate management efforts by ruling out some useful techniques for resolving specific pest problems, e.g., opposition to leg-hold traps and toxicants for use against coyotes and other animal pests.

Vertebrate pest management suffers from the fact that vertebrate pest problems, depending on the species and the situation, are the concern of or come under the responsibility of different agencies, and there is often a lack of coordination and communication between involved agencies. For example, rat and mouse problems may be the concern of public health, dog and cat problems may be handled by the dog pound or animal control officer, and deer and raccoon problems may be the responsibility of the fish and game conservation agency. Departments and agencies responsible for these matters vary from city to city and state to state.

The need for urban vertebrate pest management predictably will continue to increase. Solutions to these major problems confronting vertebrate pest management will not come easily nor rapidly.

MAJOR OPPORTUNITIES AVAILABLE FOR PROGRAMS IN VERTEBRATE PEST MANAGEMENT

There is little doubt that habitat management or modification offers the most untapped or underused opportunities for vertebrate pest management. Although there is much to be learned concerning habitat management, we are not in many instances utilizing what we do know. For example, improved sanitation by using rodent-proof dumpsters, more frequent garbage pickup, and refuse clean-up in vacant lots have proven effective in reducing Norway rat problems. Yet often urban rat control efforts concentrate on the use of poisons, and only a token effort is directed at improved sanitation. This may not hinge on a lack of knowledge or an oversight but rather on the fact that different departments usually administer the two functions (rodent control and solid-waste disposal), with little coordination or cooperation between the two.

Architects and biologists should work more closely together so that buildings can be designed to better exclude rodents and birds and so that building exteriors will be less inviting to birds for nest-building and roosting. More attention should be given by engineers to potential animal problems when planning and designing highways, rapid transit systems, underground communication lines, reservoirs, flood levees, and airports to minimize wildlife problems. The opportunities are substantial for architects, engineers, city planners, and vertebrate pest specialists to collaborate at the planning stages.

Landscape design and maintenance to reduce or avoid certain vertebrate pest problems also represents habitat modification, and many opportunities for improvement exist in this area.

FUTURE DIRECTIONS FOR PRACTITIONERS

Based on past history and current information, techniques in urban vertebrate pest management will probably advance only slightly in the next decade. Revolutionary changes in materials and methodology are not anticipated. The changes that do occur will probably be relatively subtle, reflecting predictable advances.

Structural pest control operators (PCOs) are major vertebrate pest management practitioners, and they conduct a great amount of urban commensal rodent control along with insect control. The efforts of an estimated 30,000 PCOs in the United States (National Research Council, 1980) are primarily aimed at pests in and around buildings. Anticipated changes in the design, materials, and construction of buildings to make them more energy efficient and less costly to build may alter future directions. Some PCOs are specialized and undertake the control of tree squirrels, bats, and pest birds. However, fewer PCOs will take on the less frequent vertebrate pest problems such as snakes, skunks, raccoons, and opossums, or garden- and turf-related pests like pocket gophers and moles. In the future, structural PCOs may branch out and do more vertebrate pest management outside their traditional roles.

Commercial pest control operators other than structural PCOs may also branch out more from garden and landscape control of insects, diseases, and weeds and include vertebrate pests such as pocket gophers, moles, roof rats, skunks, and raccoons in urban environments. These PCOs are more akin to those who conduct agricultural pest control, and some crossover in future pest control activities is anticipated.

Urban vertebrate pest managers (i.e., city, county, state, or federal employees) fall into two general groups. (1) Some pest managers are public health-oriented and frequently are employed by health departments or a special rodent control unit. Public health issues involving commensal rodents are their major concern. Less frequently they are confronted with bat, pigeon, house sparrow, and starling health-related problems. (2) Some pest managers are wildlife-oriented, may be employed by park departments, fish and game conservation agencies, or departments of agriculture. This group with wildlife responsibilities is generally called upon for resolving problems with birds, such as gulls, woodpeckers, and coots (mudhens), and animals such as deer, skunks, raccoons, armadillos, and alligators, rather than to address commensal rodent problems. The future direction of the latter of these two groups will probably change the most because of increased problem situations.

Animal control officers or dog catchers (i.e., dog and cat control) play a very important role in urban vertebrate pest management. Usually the control of stray or unwanted dogs and cats is handled by city or county employees. However, in some situations a humane organization may undertake these activities with financial support from the city and public donations.

The control of stray and feral dogs is a subject unto itself, differing substantially from other types of vertebrate pest management. Such activities generally are governed by a set of laws and regulations quite independent from other vertebrate pest problems. In some urban communities, the animal control officer will also assist in handling other animal problems such as skunks, raccoons, foxes, and opossums where the animal can be captured and removed. It is anticipated that future activities will remain essentially the same for this group of practitioners.

Self-help practitioners are those individuals who undertake to solve their own vertebrate pest problems, which might be anything from controlling rats or mice in their residence or business to excluding rabbits or deer from their vegetable or fruit garden or trapping a pocket gopher or mole in their lawn. If the trend continues, these individuals should have a wider array of materials to work with in the future, as materials that were once available only to commercial applicators are becoming available through the consumer market, e.g., glue boards, repellents, rodenticide baits, and bait boxes. Through the use of a growing number of available self-help guides and leaflets, this group of vertebrate pest managers is becoming more knowledgeable and confident in resolving their own problems.

Extension wildlife specialists, although not generally considered vertebrate pest management practitioners, nonetheless play such a useful role that some discussion seems essential. The U.S.D.A.'s Cooperative Extension Service is a part of every state land grant university, and it is this group that is active in the dissemination of "how to" information on a wide variety of subjects. Many states have developed leaflets and bulletins on a variety of vertebrate pest management subjects as well as wildlife subjects in general to assist all practitioners. It is anticipated that the number of specialists in animal damage control or vertebrate pest management will grow as the demand for their assistance is increasing.

FUTURE DIRECTIONS FOR RESEARCH

There is a substantial need for basic and applied-oriented research on both pesticidal and nonpesticidal approaches to integrated vertebrate pest control. Many areas of vertebrate pest management have received relatively little attention in the past (vertebrate pheromones, chemosterilants, bioacoustics, immunosuppresants (Benjamini, 1982), genus- and species-specific toxicants, and species-specific diseases for biological control). The opportunities for new methodology and management strategy approaches are many, but only a few scientists are engaged in vertebrate pest research, and funding for such research is limited.

Future research with the greatest economic and public health impact will be that directed at the control of house mice and rats. Because rodenticides (rodent baits) are the most cost-effective approach to direct control, research emphasis will undoubtedly continue in this direction. There is a need for more selective rodenticides, such as the rodenticide norbormide to safeguard children and nontarget animals. Norbormide is the most selective rodenticide ever marketed, being specific to the genus *Rattus*, but unfortunately, it was not well accepted in baits by rats, and for this and other reasons it lost out in the marketplace.

Research on effective chemosterilants or antifertility agents for pest rodents, birds, and canid predators needs to be revived. Considerable research was conducted in the United States and England starting in the late 1960s and continued for about ten years (Marsh and Howard, 1973). Little research is now under way, yet the possibilities for their effective use in urban pest management are substantial for some species such as tree squirrels, raccoons, skunks, foxes, and coyotes.

Much could be done toward formulating more effective rodent baits and baits designed to safeguard humans and nontarget species through the incorporation of group-specific repellents; through imaginative uses of emetics or antidotal substances (Marsh, 1983); and through special innovative packaging and delivery approaches. The rodenticide manufacturers and formulators have not capitalized on the many advancements made in the food development and processing industries to develop highly effective wet or moist baits with a long storage life or freeze-dried baits. The advancements in adhesives, plastics, and package design also hold considerably more promise than has been utilized to date. For example, it may be anticipated that bait blocks will be prepared in the future using a binder other than paraffin.

Research is needed to develop more effective space or area chemical repellents for a number of pest species, particularly raccoons, bats, dogs, cats, and deer. There are few good taste and odor repellents for protecting subterranean or overhead utility and communication lines from damage by gnawing rodents such as pocket gophers, tree squirrels, and roof rats. Research is needed for the development of deer, rabbit, raccoon, and bird repellents that are safe to humans and can be directly applied to growing fruit and vegetables. Safe and effective repellents are needed to repel rats and mice from packaged foods, especially those being shipped to underdeveloped countries. Research on vertebrate chemical repellents offers many interesting challenges.

Vertebrate pheromones are receiving some attention relative to vertebrate control, but much more research is needed. Possibly vertebrate pheromones may never attain the importance that insect pheromones now play in their control. However, evidence suggests they can play some effective role once we learn more about them and their role in communication and behavior (Christiansen, 1976).

Continued research is needed on anticoagulant resistance in the United States, with more emphasis placed on the house mouse. Some attention also needs to be paid to species of field rodents, such as meadow mice (voles), which are being more extensively controlled with anticoagulants. While second-generation anticoagulants have temporarily resolved the concerns over warfarin-resistance in rats and house mice, recent evidence suggests that this may not prevail for long.

There are also innumerable research needs concerning the nonpesticidal side of vertebrate control. We lack a good, yet reasonably priced multiple-catch trap for rats that will compare favorably with the efficacy of the Ketch-All® and the Tin Cat®, which are used for house mice.

Some substantial inroads have been made in Canada on the engineering of more humane kill-traps (Manthorpe, 1981). More such research, however, is needed on designing of live-catch and kill-traps that are used for all pest species. Additional research on the factors that influence trap-associated behavior in vertebrates is critically needed for increasing efficacy and minimizing stress of the trapped animal.

Bioacoustics, the playback of recorded bird distress or alarm calls to frighten conspecifics, has been researched for some bird species, but this area deserves much more study. To date, vertebrate pest control has capitalized little on the significant advancements in the electronic field. This area may hold great promise, especially in the area of pest detection and monitoring.

Landscape design and maintenance represents an area that has received little research directed toward preventing potential wildlife problems, yet many opportunites exist. Much more attention has been given to how landscaping can encourage wildlife to our urban environments than on means to discourage others that may become pests. Additional research should permit us to design landscapes to provide the beauty and diversity of desirable wildlife and yet minimize man-wildlife conflicts.

REFERENCES

Bashore, T. L., and E. D. Bellis, 1982. Deer on Pennsylvania airfields: Problems and means of control. *Wildl. Soc. Bull.* **10:**386-388.

Bateman, J. A., 1979. *Trapping: A Practical Guide.* Stackpole Books, Harrisburg, Pa.

Baur, F. J., and W. B. Jackson, eds. 1982. *Bird Control in Food Plants.* American Association of Cereal Chemists, St. Paul, Minn, 90p.

Beck, A. M. 1973. *The Ecology of Stray Dogs: A Study of Free-Ranging Urban Animals.* York Press, Baltimore, Md., 98p.

Benjamini, L. 1982. The potential use of corticosteroid hormones in rodent biocontrol. *Phytoparasitica* **10:**215-228.

Benton, A. H., and L. E. Dickinson. 1966. Wires, poles and birds. In *Birds in Our Lives,* A. Stefferud (ed.). U.S. Department of the Interior, Washington, D.C., pp. 390-395.

Brooks, J. E. 1973. A review of commensal rodents and their control. *Critical Reviews Environ. Control.* **3**(4):405-453.

Caslick, J. W., and D. J. Decker. 1978. *Control of Wildlife Damage in Orchards and Vineyards,* Information Bulletin 146. Extension Service, Cornell University, Ithaca, N.Y., 16p.

Chamberlain, P. A. 1980. Armadillos: Problems and control. In *Proceedings of the 9th Vertebrate Pest Conference, Fresno, California,* J. P. Clark (ed.). University of California, Davis, pp. 163-169.

Chapman, J. A., and G. A. Feldhamer, eds. 1982. *Wild Mammals of North America: Biology, Management, Economics.* Johns Hopkins University Press, Baltimore, 1148p.

Chitty, D., and H. N. Southern. 1954. *Control of Rats and Mice,* 3 vols. Oxford University Press, 532p.

Christiansen, E. 1976. Pheromones in small rodents and their potential use in pest control. In *Proceedings of the 7th Vertebrate Pest Conference, Monterey, California,* C. C. Siebe (ed.). University of California, Davis, pp. 185-195.

Corrigan, B., and G. W. Bennett. 1983. Bats: Managing a nuisance. *Pest Control Technol.* **11**(6):68-69, 71-72, 74.

Cummings, M. W., M. H. Kimball, and W. M. Longhurst. 1980. *Deer-resistant Plants for Ornamental Use,* Leaflet 2167. Division of Agricultural Science, Cooperative Extension, University of California, 7p.

Eadie, W. R. 1954. *Animal Control on Field, Farm, and Forest.* Macmillan, New York, 258p.

Errington, P. L. 1961. *Muskrats and Marsh Management.* Stackpole, Harrisburg, 184p.

Fitzwater, W. E. 1972. Barrier fencing in wildlife management. In *Proceedings of the 5th Vertebrate Pest Conference, Fresno, California,* R. E. Marsh (ed.). University of California, Davis, pp. 49-55.

Fitzwater, W. E. 1974. Reptiles and amphibians: A management dilemma. In *Proceedings of the 6th Vertebrate Pest Conference, Anaheim, California,* W. V. Johnson (ed.). University of California, Davis, pp. 178-183.

Fitzwater, W. D. 1979. Vertebrate pests. In *Encyclopedia of Structural Pest Control,* vol. 4. National Pest Control Association, Dunn Loring, Va., pp. 34-37.

Frings, H. 1964. Sound in vertebrate pest control. In *Proceedings of the 2nd Vertebrate Pest Control Conference, Anaheim, California.* University of California, Davis, pp. 50-56.

Good, H. B., and D. M. Johnson. 1978. Nonlethal blackbird roost control. *Pest Control* **46**(9):14-16, 18.

Greenhall, A. M. 1982. *House Rat Management,* Resource Publication 143. U.S. Department of the Interior, Fish and Wildlife Service, 33p.

Howard, W. E. 1967. Vertebrate pests: Biocontrol and chemosterilants. In *Pest Control: Biological, Physical, and Selected Chemical Methods,* W. E. Klingore and R. L. Doutt (eds.). Academic Press, New York, pp. 343-386.

Howard, W. E., and R. E. Marsh. 1981. *The Rat: Its Biology and Control,* Leaflet 2896. Division of Agricultural Science, Cooperative Extension, University of California, 31p.

Howell, R. G. 1982. The urban coyote problem in Los Angeles County. In *Proceedings of the 10th Vertebrate Pest Conference, Monterey, California,* R. E. Marsh (ed.). University of California, Davis, pp. 21-23.

Leedy, D. L. 1979. *An Annotated Bibliography on Planning and Management for Urban-Suburban Wildlife,* FWS/OBS-79/25. U.S. Department of the Interior, Fish and Wildlife Service, 256p.

Leedy, D. L., and L. W. Adams. 1984. *A Guide to Urban Wildlife Management.* National Institute for Urban Wildlife, Columbia, Md., 42p.

Leedy, D. L., R. M. Maestro, and T. M. Franklin. 1978. *Planning for Wildlife in Cities and Suburbs,* FWS/OBS-77/66. U.S. Department of the Interior, Fish and Wildlife Service, 64p.

McCracken, H., and H. Van Cleve. 1967. *Trapping: The Craft and Science of Catching Fur-Bearing Animals.* A. S. Barnes, Cranbury, N.J., 196p.

Mallis, A., ed. 1982. *Handbook of Pest Control,* 6th ed. Franzak and Foster, Cleveland, Ohio, 1104p.

Manthorpe, D. 1981. Research program for the development of humane trapping systems. In *Proceedings of the Worldwide Furbearer Conference, Frostburg, Maryland, 1980,* J. A. Chapman and D. Pursley (eds.). R. R. Donnelly and Sons, Fall Church, Va., pp. 1579-1587.

Marsh, R. E. 1981. Future of pesticides in vertebrate pest control. In *Proceedings of the 5th Great Plains Wildlife Damage Control Workshop,* Lincoln, Nebraska, R. M. Timm and R. J. Johnson (eds.). University of Nebraska, Lincoln, pp. 54-59.

Marsh, R. E. 1983. Rodenticide selection and bait composition to minimize potential primary hazard to nontarget species when baiting field rodents. In *Proceedings of the First Eastern Wildlife Damage Control Conference, Ithaca, N.Y.*, D. J. Decker (ed.). Cornell University, Ithaca, N.Y., pp. 155-159.

Marsh, R. E., and W. E. Howard. 1973. Prospects of chemosterilant and genetic control of rodents. *Bull. World Health Organization* **48:**309-316.

Marsh, R. E., and W. E. Howard. 1978*a*. Vertebrate control manual: Moles. *Pest Control* **46**(4):24-27.

Marsh, R. E., and W. E. Howard. 1978*b*. Vertebrate control manual: Pocket gophers, toads and frogs. *Pest Control* **46**(3):30-34.

Marsh, R. E., and W. E. Howard. 1981. *The House Mouse: Its Biology and Control,* Leaflet 2945. Division of Agricultural Sciences, Cooperative Extension, University of California, 31p.

Morton, R. K., and E. N. Wright, eds. 1968. *The Problems of Birds and Pests.* Academic Press, London, 254p.

National Academy of Sciences. 1970. *Vertebrate Pests: Problems and Control.* Washington, D.C., 154p.

National Pest Control Association. 1982. *Bird Management Manual.* NPCA, Dunn Loring, Va., 118p.

National Research Council. 1980. *Urban Pest Management.* Report prepared by the Committee on Urban Pest Management, Environmental Studies Board, Commission on Natural Resources. National Academy Press, Washington, D.C., 272p.

Neville, P. 1983. Humane control of an urban cat colony. *Int. Pest Control* **25**(5): 144-145, 152.

Nowell, I. 1978. *The Dog Crisis.* St. Martin's Press, New York, 270p.

Noyes, J. H., and D. R. Progulske, eds. 1974. *Wildlife in an Urbanizing Environment,* Planning and Resource Development Series No. 28. Cooperative Extension, University of Massachusetts, Amherst, 128p.

Pratt, H. D., and R. Z. Brown. 1976. *Biological Factors in Domestic Rodent Control,* CDC No. 76-8144. U.S. Department of Health, Education and Welfare, Public Health Service, Center for Disease Control, Atlanta, Ga., 30p.

Pratt, H. D., and W. H. Johnson. 1975. *Sanitation in the Control of Insects and Rodents of Public Health Importance,* CDC No. 76-8-138. U.S. Department of Health, Education and Welfare, Public Health Service, Center for Disease Control, Atlanta Ga., 42p.

Scott, H. G., and M. R. Borom. 1976. *Rodent-borne Disease Control Through Rodent Stoppage.* U.S. Department of Health, Education and Welfare, Public Health Service, Center for Disease Control, Atlanta, Ga., 34p.

Stefferud, A., ed. 1966. *Birds in Our Lives.* U.S. Department of the Interior, Fish and Wildlife Service, Washington, D.C., 562p.

Summers-Smith, D. 1967. *The House Sparrow.* Collins, London, 270p.

Timm, R. M., ed. 1983. *Prevention and Control of Wildlife Damage.* Cooperative Extension Service, University of Nebraska, Lincoln, 660p.

Wodzicki, K. 1973. Prospects for biological control of rodent populations. *Bull. World Health Organization* **48:**461-467.

12

Managing Pests of Food

Robert Davis

U.S. Department of Agriculture
Agricultural Research Service
Savannah, Georgia

Pest management in and around food stores, processing and preparation facilities, and in the home is based on the need to maintain both quality and quantity food as demanded by the processors and consumers and to conform to the rules and regulations promulgated by government action agencies (Anonymous, n.d.; Baur, 1984). Pest management approaches are usually programs encompassing several of the available pest management tools. However, the basis of all good pest management approaches begins with a program of sanitation (Osmun, 1984). Good sanitation is not just a well-recognized good manufacturing or handling practice but a requirement. The Food, Drug and Cosmetic Act, Section 402(a)(5), states that a food is adulterated, "if it has been prepared, packed, or held under unsanitary conditions whereby it may become contaminated with filth" (Food and Drug Administration, 1976, 1981). In fact, it can be seriously questioned whether or not any pest management program can be effective without good sanitation as a prerequisite.

The management of pests of food in urban situations offers many unique challenges to the responsible manager, sanitarian, and pest management specialist. Many of these challenges will affect the quality of the end product. Other challenges will impact upon possible health and safety considerations because of the close proximity that exists between people and their food from harvest to consumption. Of course, economics will always be an important consideration because of the increased inherent value of a harvested commodity as it moves through its processing and subsequent marketing and distribution channels. Therefore, some important considerations for every pest manager are as follows:

1. What is the pest species involved?
2. Is there time at this point in the food processing, marketing or distribution channel to effect a pest management technique?
3. What will be the value lost if a pest management technique is not made or is ineffective?
4. Is the available management technique(s) efficacious?
5. What pest management options are legal and safe?
6. What effect will the available pest management technique have on the commodity to be treated and its packaging, warehousing, and transportation structures and the environment?

All recent advances and those potential pest management techniques on the horizon will still have to meet the rigid regulations of today's regulators and possibly future regulations imposing more limiting restrictions on the use and application of pest management techniques. The remainder of this chapter is devoted to a discussion on some recent developments in areas offering potentially new pest management techniques for use around foods.

PROBLEM RECOGNITION

Methodology for establishing and identifying that a pest problem exists is still in its infancy. The establishment of federal and state defect action levels (DALs) and other standards as may be set forth in the Good Manufacturing Practice statements for various food storage and processing industries are still largely labor intensive (Food and Drug Administration, 1977, 1981). When associated with food storage, processing, or preparation, the presence of any evidence of life, past or present, that is foreign to the commodity is evidence of a potential problem.

The presence of pests, the results of their contamination, or their damage to our food is unacceptable and embarrassing to most people and often make feeds unacceptable by our domestic animals. This presence of insects in our food products is exceeded in importance only by those insects that attack our person and cause discomfort or transmit diseases.

The number and variety of pest species found attacking food in warehouses and processing plants and other urban areas are quite extensive (Mallis, 1982; Redlinger and Davis, 1982; World Food Program, 1970). Such pests are to be considered contaminants, as are the arthropod parasites and predators of these pests. The following is a list of principal insects that will illustrate some of the variety of insects infesting just stored grain and grain products. (The common name has been approved by the Entomological Society of America Committee on Common Names of Insects, 1978, College Park, Maryland.)

Latin name	Common name
Plodia interpunctella (Hubner)	Indianmeal moth
Ephestia cautella (Walker)	almond moth
Tribolium confusam Jacquelin du Val	confused flour beetle
Tribolium castaneum (Herbst)	red flour beetle
Oryzaephilus surinamensis (L.)	sawtoothed grain beetle
Oryzaephilus mercator (Fauvel)	merchant grain beetle
Cryptolestes pusillus (Schoenherr)	flat grain beetle
Tenebroides mauritanicus (L.)	cadelle
Tenebrio molitor (L.)	yellow mealworm
Liposcelis sp.	psocids
Sitotroga cerealella (Olivier)	Angoumois grain moth
Rhyzopertha dominica (F.)	lesser grain borer
Sitophilus granarius (L.)	granary weevil
Sitophilus oryzae (L.)	rice weevil
Sitophilus zeamais Motschulsky	maize weevil

The importance of pest identification cannot be overemphasized. It is essential to establishing an effective pest management procedure. The use of most of the newer management techniques such as pheromones, biological controls, modified atmospheres, and specific chemicals will require a knowledge of the pest to be controlled that will only be available by having a correct identification. Also the label and labeling for most pest control substances usually require that the application methodology and treatment site be stated for the management of each specific pest or group of pests (Okumura, 1984).

Securing correct identification is not difficult in most instances. The U.S.D.A. Cooperative Extension Service located throughout the respective states and the states' land grant universities can usually provide or direct one to an authoritative identification. For the reader who wishes to try his or her hand at identifying insects, there are a number of good references (Borror and DeLong, 1964; Jacques, 1947; Pratt, Littig, and Marshall, 1960; Truman, Bennett, and Butts, 1976).

Problems of insecticide resistance are ever present and often a major contributing cause for pest management failures. Presently, the existence of pesticide resistance is beginning to cause concern and may soon complicate the export of both raw and processed foods (Yoshida, 1980). Where current quarantines presently have been based exclusively on species that are considered to be nonexistent in the receiving nation, future quarantine may well be based on whether or not there is a pesticide-resistant strain of that species that may be imported.

The importance of good pest management takes on added significance when one considers that pesticide resistance is attributed almost entirely to poor or incomplete application of the pesticides in question. Several organizations are monitoring levels of pesticide resistance, including international organizations, such as the Food and Agricultural Organization (FAO) and World Health Organization (WHO) of the United Nations (Champ and Dyte, 1976). These organizations are currently most interested in the insect pests of food grains and in rodent reservoirs and insect vectors of human diseases.

CHEMICAL PEST MANAGEMENT

The use of chemical pesticides remains the mainstay in the arsenal of those attempting to manage pests in food (Davis, 1981). However, the use of these pesticides in the United States requires that they be applied in strict accord with their U.S. Environmental Protection Agency (EPA) approved manufacturer's or formulator's label and labeling materials. Also, some pesticides may be designated as restricted-use pesticides and thereby carry the additional requirement that they be applied by or under the supervision of an individual certified to use these pesticides. This certification is vested in the states by EPA. Those interested in this certification program should direct their inquiries to their state Department of Agriculture or comparable agency. All other questions regarding the use of any pesticide that are not answered on the pesticide label or labeling should be directed to the extension entomologists at the state's land grant university.

The use of pesticides to manage pests of foods can be divided into either a preventive or a corrective approach. Prevention involves the use of pesticides to limit the spread of existing infestation and limit the possibility of acquiring an infestation from outside sources. Correction involves the use of pesticides to kill or eliminate an existing infestation but usually does not provide protective value against possible reinfestation.

Preventive Approaches

The preventive approaches will usually involve one or more of the following: spraying a residual-type pesticide onto the walls and floor of the empty storage facility before it is filled; spraying or dusting a residual-type pesticide on the grain as it is being placed into storage; dispensing a pesticidal vapor or aerosol into the headspace over the grain during storage or into a warehouse or food processing facility; spraying or dusting a residual-type pesticide onto a surface during storage or undertaking a spot or crack and crevice treatment of the food storage, transportation, or processing facility.

Corrective Approaches

The corrective approach is required when assurance is needed that a commodity is free of live insect infestations. Here the use of fumigants offers the only practical approach. Fumigants, when properly handled and applied, are a safe, efficient, and effective means of eliminating infestations of live insects. However, in the hands of an inexperienced person, they can be very hazardous and often ineffective. Efficacy will depend upon achieving fumigant gas distribution throughout the food mass or structure and then retaining the required concentration of the gas for the proper time interval. The serious reader is referred to the excellent manual by Monro (1969) for a complete discussion on most aspects of fumigation.

Next to retaining the fumigant gas for the required time interval, which is largely a matter of sealing the structure with gasproof tarpaulins, plastic films, and tapes, the selection of an appropriate fumigant is of prime consideration. This selection will depend on a number of factors, some of which will only be gained through experience with particular situations. However, some factors that can and should be considered are: legal uses of the fumigant; type of commodity and its quantity; moisture content and temperature of commodity; kind or type of facility; spot or bulk type of treatment; time available to effect fumigation; allowable residue tolerances; proper fumigant distribution; and ability to aerate properly and safely.

The legality of using a particular fumigant is entirely within the providence of federal, state, and local regulatory agencies. In most instances, the EPA's approved label and labeling materials will provide information on all federal requirements and restrictions. However, other governmental agencies may have additional restrictions. The use of an approved fumigant may also be vested in individuals or firms involved in the ownership of the commodity or storage facility. It is always a good practice to inform all parties involved, even those with a remote interest, that the commodity has been or is to be fumigated.

Insecticide resistance to the organophosphates is now worldwide for many pests, and responsible management should continually assess whether or not their problem involves resistant species or not. In the selection of a replacement pesticide, it is also important to have data on the potential for cross-resistance. As yet, resistance has not developed to levels of concern for the carbamates or the pyrethroid insecticides.

The recently reported resistance to methyl bromide and phosphine is most disturbing (Champ and Dyte, 1976; Mills, 1983; Tyler, Taylor, and Reed, 1983). The use of fumigants should always be considered as the last resort. Again, it is quite evident that inappropriate use, i.e., allowing for insufficient exposure to the toxic gas, has allowed the survivors to breed and develop resistant strains.

We also have the added problem that several of our pesticides are not as safe for the applicator or the consumer as we once thought. In the future, we will most assuredly not have many of our present-day pesticide chemicals. Those that are retained will probably be much restricted over today's allowable uses. Possible methods for continued use of toxic pesticides that promise less exposure or lower doses of these toxicants would be the greater use of insecticidal space treatments and the application of fumigants in-transit (Davis, 1982; Gillenwater et al., 1971). Very small amounts of quite toxic chemicals have been used very successfully in warehouses and some selected processing facilities during periods of little or no operation such as at night or on holidays. In some instances, particularly where long-term storage is anticipated, preventive programs using periodic ULV or ULD (ultra low volume or dose) applications of nonresiduals as space treatments also have been quite successful (Gillenwater et al., 1971). The use of in-transit fumigation particularly in ships, containers, and railcars has provided an opportunity for pest management that is usually away from populated areas with a minimum exposure to the applicator (Davis, 1982; Jay, Davis, and Zehner, 1983; Ronai and Jay, 1982).

NONCHEMICAL PEST MANAGEMENT

It is the many new areas of nonconventional chemicals and the nonchemical methods for pest management that have recently received more attention and offer some exciting potentials. These methods, not all new as some are being reexamined again after years of dependence on conventional pesticides, do not offer the opportunity to entirely escape from the use of the more traditional use of chemical pesticides. However, they do promise to be effective adjuncts for integration and provide means to minimize the use of chemical pesticides and possibly eliminate their use all together in some selected situations. It is the development of integrated pest management (IPM) plans that is most urgently needed to encourage and stimulate continued research on these novel pest management approaches.

Irradiation

There has been a great deal of information accumulated on the use of irradiation to control pests of foods, particularly in grain and grain products

(Cornwell, 1966; Josephson and Peterson, 1982). Irradiation or ionizing radiation may be used in two ways (direct treatment and genetic control) to manage pest populations. Both types of treatment of infested commodities to control pests provide a residue-free process. Genetic control involves the release of irradiated individuals that have been rendered sterile by irradiation to compete with the normal individuals. Subsequent pest generations will experience population declines as a result of sterile progeny or infertile eggs. Presently, 19 nations allow the use of irradiation on 26 foods or food groups (Table 12-1).

The effects of irradiation have been evaluated for two sources of ionizing energy, i.e., gamma radiations produced by nuclear disintegration of radio isotopes (cobalt-60 and cesium-137) and high speed or accelerated electrons emitted from a heated cathode. Gamma radiations have much greater penetration than do accelerated electrons and have received the most research attention, particularly on bulk raw commodities. However, the use of accelerators is now beginning to receive more attention (Watter, 1979), and it has been reported that a grain elevator in the Soviet Union has been equipped with an electron accelerator (Zakladnoi et al., 1982).

The use of ionizing radiations has both promise and some limitations. The method is very effective and can be very inexpensive over the long run. However, the initial cost is very high, and the cost of transporting sufficient commodity to the irradiator for maximum utilization may be prohibitive. Nevertheless, even with these economic considerations, we can probably expect to see irradiation used in the not too distant future for disinfestation of high-value commodities such as fresh fruits, nuts, dried fruits, and some meats. An economic problem encountered in many countries producing and exporting agricultural surpluses is that they are the larger countries with several to many ports, which would each require an irradiation facility. Rather than build an irradiation facility at every export location, it has been suggested that it may be more feasible to install irradiation facilities at ports for treatment of incoming commodities only (Bailey, 1979).

Controlled Temperatures

Both low and high temperatures have been used in selected situations and offer promise for future consideration. Insects are cold-blooded (poikilothermic), so their body temperatures closely follow that of their environment. When temperatures are lowered, insect activity decreases until all activity stops. Further decreases in temperature will result in death (Ashina, 1966). Similarly, as temperature increases, activity will increase to a point where some vital process is inhibited and activity will cease. Continued exposure or further increase in temperature will result in death.

Table 12-1. Present status of national acceptance of food irradiation.

Foods	*Argentina*	*Belgium*	*Canada*	*Chile*	*Denmark*	*France*	*Hungary*	*Israel*	*Italy*	*Japan*	*Netherlands*	*Norway*	*Philippines*	*South Africa*	*Spain*	*Thailand*	*Uruguay*	*United States*	*USSR*
Blood proteins											X								
Chicken				X				X			X			X					
Cocoa beans				X															
Dates				X															
Dried fruits																			X
Dry food concentrates																			X
Egg powder											X								
Fish and fish products				X							X								
Froglegs											X								
Frozen fish											X								
Frozen shrimp											X								

Garlic		X				X			X					X					
Grain																			X
Mango				X										X					
Mushrooms											X								
Onions		X	X	X		X	X	X	X		X			X	X	X			X
Papaya				X										X					
Paprika		X																	
Potatoes	X	X	X	X	X	X		X	X	X	X		X	X	X		X	X	X
Pulses				X															
Rice and ground rice products				X							X								
Rye bread											X								
Shallots		X				X													
Spices		X		X		X	X				X	X						X	
Strawberries		X		X										X					
Wheat, flour, whole wheat flour			X	X														X	

Source: From van Kooij, 1983

Table 12-2. Estimated minimal and optimal temperatures for population increase for selected stored-grain insects.

Species	Temperature (°C)	
	Minimum	Optimal
Almond moth	17	28-32
Angoumois grain moth *Sitotroga cerealella* (Olivier)	16	26-30
Broadhorned flour beetle *Gnatocerus cornutus* (F.)	16	24-30
Confused flour beetle	21	30-33
Cowpea weevil *Callosobruchus maculatus* (F.)	22	30-35
Flat grain beetle *Cryptolestes pusillus* (Schronherr)	22	28-33
Grain mite *Acarus siro* (L.)	7	21-27
Granary weevil	15	26-30
Indianmeal moth	18	28-32
Khapra beetle *Trogoderma granarium* Everts	24	33-37
Lesser grain borer	23	32-35
Mediterranean flour moth *Anagasta kuehniella* (Zeller)	10	24-27
Merchant grain beetle	20	31-34
Red flour beetle	22	32-35
Rice weevil	17	27-31
Rusty grain beetle *Cryptolestes ferrugineus* (Stephens)	23	32-35
Sawtoothed grain beetle	21	31-34

Source: From Howe, 1965; Mills, 1979

The use of low temperature to manage a pest population does not necessarily require lethal temperatures. In most cases, insect pests species of foods are largely of tropical or subtropical origin, and therefore minimum temperatures for growth and other activity cease at 15-22°C (Table 12-2). Maximum lethal temperatures are usually near the upper limit (within 3-8°C) of the optimum temperature range and in most instances do not exceed 45°C. A longstanding procedure for insect control in flour mills has been the "superheating" of the mill for a minimum of 30 hours to 54°C (130°F) (Dean, 1913). There are economic considerations when one must shut down, close up, and heat an entire processing plant for a sufficient period to obtain 30 hours of 54°C temperature. Another limitation of this controlled temperature procedure is the time required for penetration of the heat or cold into the commodity. This can be a major obstacle when treating large amounts of

Table 12-3. Chilling times for selected commodities. All commodities were exposed in a 0.76m³ (27 ft³) freezer filled to capacity.

Commodity	*Freezer setting (°C)*	*Time to 0°C (hr)*	*Time to equilibrium (hr)*
Cornflakes (twenty-eight 3.2 lb [1.45 kg] cases)	−10	7	30
	−15	6	30
	−20	5	35
Flour (seven 100 lb [45.45 kg] bags)	−10	55	160
	−15	29	130
	−20	25	145
Elbow macaroni (fifteen 24 lb [10.91 kg] cases) and blackeyed peas (fifteen 24 lb [10.91 kg] cases)	−10	29	130
	−15	18	95
	−20	19	100

Source: From Mullen and Arbogast, 1979

commodity (Table 12-3). Thermal conductivity of bulks of grain have been discussed by several authors (Disney, 1954; Hall, 1979; Navarro and Calderon, 1982; Oxley, 1944). However, it is a pesticide residue-free method available in the arsenal of the pest management specialist and could be utilized if the economic consideration of time can be accommodated.

Protective Packaging

Food-processing and distributing industries package infestible foods that are shipped and stored under a wide variety of conditions and environments. Protecting foods from subsequent infestation is a prime concern of the package industry and the packager. Neither will be able to maintain any control over handling and storage environments throughout marketing and distribution channels after a commodity leaves their facility. The use of packaging, which through good construction offers few avenues of access to the packed commodity, should be a prime consideration of management along with sanitation and other preventive pest management procedures. The use of appropriate strength materials will also aid in preventing damage to the commodity by penetration of the packaged materials by pests. The addition of synergized pyrethrins to the outer paper layer of multiwalled paper bags containing at least 50 lb of product, or between film laminates for dried fruit packages of 2 lb, has been approved by the EPA (Anonymous, 1966, 1974). Indications are that other less toxic and more effective chemicals are available for treating packages (Highland, 1973, 1984). None has been registered as yet for this purpose, but some may be approved for use in the future either here in the United States or elsewhere in the world. Insect-resistant packaging offers one of the very few promising methods of maintaining a pest-free product after the product leaves a manufacturer's control.

Biological Control

The increased use of biological control through the utilization of parasites, predators, and pathogens offers promise for the future. Great advances were made in our understanding of the use of parasites and predators during the early half of this century. Then the era of organic pesticides put a virtual halt on this research, and we waited until the mid-1960s and the 1970s for the continuation of research on this promising management technique. Droutt (1972) stated that "the ingenious and resourceful use of parasites and predators to suppress pest populations should be given primary consideration in building safe, economical and enduring pest control programs consistent with the maintenance of environmental quality." Of course, Droutt's discussion was primarily directed at the control of field infestations and not pests infesting stored commodities. Nevertheless, there are many possible applications for the use of parasites and predators in protecting post-harvest agricultural commodities, particularly in raw commodity storage. Considerable research has been conducted to demonstrate that parasites and predators are effective, and the technique really only needs limited implementation to demonstrate the economics (Press, Cline, and Flaherty, 1982; Press, Flaherty, and Arbogast, 1975). There is also the problem of predator and parasite availability, but such will best be solved in the marketplace and not by research.

It is with pathogens that the greatest potential seems to avail. The research on the use of protozoa has resulted in a rather "mixed bag" of results. The orders Gregarinida, Coccidia, and Microsporidia are commonly found among insects attacking foods. However, many gregarines (Eugregarina) do little damage to their hosts and apparently exist largely as commensals. The Neogregarina gregarines are quite pathogenic. The genus *Mattesia* is known to infect the flat grain beetle, the rusty grain beetle, and members of the genus *Trogoderma.* Shapas, Burkholder, and Bousch (1977) have shown that this protozoan can be used quite successfully in conjunction with a pheromone lure and a bait station.

The bacteria, *Bacillus thuringiensis,* and various granulosis viruses offer promise for the control of stored-product moths, particularly for the Indianmeal moth, almond moth, and the Angoumois grain moth (McGaughey, 1975*a,* 1975*b,* 1976).

Considerable research still needs to be done in this area when one considers the myriad of viruses, rickettsia, bacteria, fungi, and protozoa that exist in varying degrees of association with insect pests. However, more practical use is needed of the available pathogens before we can expect too much more commercialization of new pest management agents from this group.

Modified Atmospheres

Using modified or controlled atmospheres is an adaptation of the ancient practice of hermetic storage (Shejbal, 1980). Hermetic storage involves sealing raw agricultural commodities, such as grain, beans, or oilseeds, generally in underground pits, and allowing the respiration of the commodity plus that of any pests present to deplete the oxygen to a level that will asphyxiate the pests. The reduced amount of oxygen in the atmosphere of a hermetically sealed storage also protects the commodity from fungal attack; and thus, the condition tends to maintain higher quality over extended storage periods. Hermetic storage in pre-industrial times was probably the only means of keeping large quantities of grain free from insect attack for significant lengths of time in areas with mild winters.

The use of modified atmospheres offers a more practical method of pest management in stored foods. It will control insect pests much quicker than hermetic storage and, like hermetic storage, does not leave chemical residues. It usually does not require extensive modification of existing storage facilities, although some additional sealing may be needed. The method simply involves changing the existing atmosphere in the storage structure to one lethal to insects by purging with carbon dioxide (CO_2), nitrogen (N_2) or with the combustion gas products from a modified ("inert") atmosphere generator. All three of these materials have been granted exemptions from the requirement of tolerances by the EPA for raw (40 CFR 180, subpart D, sec. 180.1049, 180.1050, 180.1051) and processed (21 CFR, part 193, sec. 193.323, 193.45, 193.65) agricultural products. However, research in the U.S.D.A.'s Agricultural Research Service on these materials has shown that CO_2 is preferred because of generally faster insect mortality, lower costs, less stringent sealing requirements for structures treated, and less influence on performance caused by slight fluctuations in concentration (Jay, 1980). For these reasons, most research in the United States has been directed toward the use of CO_2 for stored-product insect control.

Except for laboratory studies, little interest was shown in this technique of using modified atmospheres or in hermetic storage until 1970. This lack of interest was undoubtedly due to the success of conventional fumigants and grain protectants. It was then that Department of Agriculture researchers from Savannah, Georgia, attained and maintained a 35% CO_2 concentration for 2, 4, and 7 day periods in an upright concrete silo containing 68,000 bushels of in-shell peanuts at Columbus, Georgia (Jay, Redlinger, and Laudani,

1970). Later, in 1973 researchers from Savannah successfully controlled natural infestations of the rice weevil complex (*Sitophilus* spp.) and the Angoumois grain moth in 28,000 bushels of corn in an upright concrete silo at the terminal elevators in North Charleston, South Carolina (Jay and Pearman, 1973). In this test, a CO_2 concentration of about 60% was successfully attained and maintained for a 96 hour period. The success of this test showed more than a 99.9% reduction of all species of insects was obtained with over a 99% reduction in damaged kernels.

Interest in the use of N_2 for controlling stored-product insects began in Australia in the early 1970s and continued through 1976. Results of these studies and Australian recommendations for the use of N_2 to protect grain in storage are best described by Banks and Annis (1977). Also, considerable research has been conducted in Italy on the use of N_2 to protect grain during long-term storage. This Italian research led to the development of a total marketable system including metal silos, conveyors, and equipment for applying N_2 by Laboratori Ricerche di Base, Monterotondo a Assoreni in Rome, a member of the Association of ENI Companies for Scientific Research. Considerable amount of information on N_2 and its application is available in the proceedings of an international symposium held in Italy (Shejbal, 1980).

Modified atmospheres offer a residue-free method of controlling insects in stored grain, oilseeds, and other raw and some processed commodities. It also shows potential for use in in-transit situations such as hopper cars, ocean-freight-over-the-road type containers, river and ocean-going barges and ships in the export trade. All these vehicles and their potential for holding CO_2 for insect control have not been studied. However, we should expect to see increasing use of this pest management technique in the near future.

Pheromones

The use of various semiochemicals (signal chemicals) has stirred excitement among everyone from the less informed among the general public to the most knowledgeable scientists. Use of this potential management tool allowed some early successes in reducing numbers of agricultural pests in cropping situations. Here the use of pheromones showed the ability to increase trap collections and to reduce the occurrence of mating by confusing the males as to the presence of females.

The methodologies for effective pheromone utilization have eluded many of those working in the post-harvest areas of pest management. A review of the literature shows that it was not lack of interest because many scientists were busy isolating and synthesizing these semiochemicals (Burkholder, 1979). Apparently because such good success was being achieved with available chemical pesticides, including fumigants, and the desire for those needing and using pest management procedures to have 100% control, the practical application has not been readily seen.

Research has shown now that, while pheromones may not necessarily lead to a control or management technique in themselves, they can be used as effective tools for monitoring infestations and could therefore be considered for integration into overall management programs (Burkholder, 1979; Shorey, 1977). A recent innovative use of pheromones that may well expand their use is as lures for inoculation of pathogens into pest populations (Shapas, Burkholder, and Boush, 1977).

It should be remembered that acceptance of such survey tools as pheromone traps may be slow, especially by those in the processing and distribution of foods where the presence of regulators and customers are felt with regularity. To many, the presence of a trap with insects will be an indictment that a facility has a pest problem. Also, an empty trap will imply to many that management has just had or suspects a pest problem. This philosophy extends also to other attractant or repellent devices. Full acceptance of such devices will probably have to wait for changes in the laws and regulations governing the food industry and for their acceptance by the consuming public.

FUTURE CONSIDERATIONS IN RESEARCH

No one has a really good crystal ball for predicting the kinds of tools that will be available for future use in pest management. In the food industry, there do appear to be some promising areas where research breakthroughs would further our technology considerably.

The newest innovation that stands ready to solve some of our most pressing problems is the computer. Its use is only now beginning to make its appearance in the area of managing the business aspects of pest management firms. Some uses have been made in various food industries, but these are largely limited to the analysis of data for quality control and the evaluation of scientific data in the laboratories in food-processing industries. The sanitarian and other urban pest managers need to begin to emulate the modest degree of success that the agricultural pest management consultants are beginning to sustain. Programs for the analyses of sampling results of pest surveys and perhaps day degrees for pest development would be a modest and practical beginning. Refinements in sampling procedures will also be required before maximum results can be achieved, particularly to move this methodology from destructive to nondestructive approaches. Pest thresholds will need to be refined and defined as we learn more about how various pests are affected by the environment and how they in turn affect a commodity. Much of this, of course, will of necessity need to wait for the required software.

It is quite obvious to many that we will require the use of conventional chemical pesticides for some years into the future. This continued use will depend upon using these materials in the right place and the right time with a minimum of hazard to all. Research into application technology that will allow for minimization of pesticide residues is urgently needed.

The development of predictive models of pest populations in storing, marketing, and distributing are needed. Such models will provide greater assurance that our pest management treatments are applied with proper timing and minimize residues.

Simple, reliable, fast, and nondestructive insect detection tools and techniques are needed to turn our pest management efforts from a preventive to a corrective approach. Such devices should be not only available for detecting pests inside foods such as grains, nutmeats, dried fruit, and so on, and their packaging but also for detecting the presence of the more exposed pest populations.

Of course, there is need for continued and expanded research in the search for alternatives to conventional chemical pesticides.

CONCLUSION

Only a limited discussion on pest management techniques has been presented in the space available. Table 12-4 presents a quick summary of some of these and other management techniques. Still, many other potential pest control techniques have of necessity been omitted from this chapter. The future is bright with a great variety of possible management techniques. Their use will provide an array of new approaches and each with its own good and bad points and needing the selection and manipulation of the professional pest manager. It is questionable whether we will ever see uniform approaches to pest management, even for the same insect pest in similar types of situations. A lesson might be learned from our associates in production agriculture, where pest management consultants are designing programs for individual farms and their unique conditions and requirements. In this regard, two case studies are presented at the end of this chapter to illustrate some of the variety of conditions experienced in managing pest of foods.

Case Study 1

A manufacturer of a bottled condiment sauce has been notified several times in recent weeks by his quality control personnel that insect fragments have

Table 12-4. Management techniques for use on pests of foods.

Method	*Possible application site*	*Selected references*
Conventional chemicals		
Space treatments	In warehouses containing only packaged foods; also, in food preparation facilities, but all food surfaces must be washed before use	Gillenwater et al., 1971; U.S. Department of Agriculture, 1982
Residual sprays	Anywhere crack and crevice or spot treatments are needed; also, as a grain protectant	Cotton, 1963; U.S. Department of Agriculture, 1982
Fumigants	In the treatment of structures and raw and processed foods and feeds	Monro, 1969; U.S. Department of Agriculture, 1982
Ionizing radiation		
Gamma	In the treatment of wheat and wheat products, spices and papaya; little, if any, current commercial use	Cornwell, 1966; Davis, 1972; Josephson and Peterson, 1982
Accelerated electrons	Same as for gamma radiation	Watter, 1979
Nonionizing radiation		
Infrared	Insect control is usually achieved as grain is dried.	Dermott and Evans, 1976*a*, 1976*b*; Kirkpatrick, Brower, and Tilton, 1972
Microwaves	Same as for infrared	Kirkpatrick, Brower, and Tilton, 1972
Radiofrequency	Potential not adequately evaluated	Nelson, 1973

(continued)

Table 12-4 (*continued*)

Method	*Possible application site*	*Selected references*
Controlled temperatures		
Heat	Heating commodity to temperatures of 50°C-80°C for short periods of time (less than 15 min); treating mills and other food-processing structures with temperatures of 54°C for 30 or more hr	Dermott and Evans, 1976*a*, 1976*b*; Pepper and Strand, 1935
Cold	Maintain commodities below the temperature threshold (15°C to 20°C); freezing to kill all stages of the insects. (0°C to −20°C)	Howe, 1965; Mills, 1979; Mullen and Arbogast, 1979
Protective packaging	Providing well-sealed packages to prevent easy entry by insects; also, the use of insecticide-treated packages to minimize damage by boring insect pests	Highland, 1984
Biological control		
Insect and mite parasites and predators	Subject not yet well investigated but offers potential for use with raw products that will require further cleaning before processing	Bruce, 1983; Droutt, 1972; Jay, Davis, and Brown, 1968; Press, Cline, and Flaherty, 1982; Press, Flaherty, and Arbogast, 1975; Williams and Floyd, 1971
Protozoa	Same as for insect parasites and predators (some research indicates that protozoan diseases of insects can be spread by use of insect baits and lures)	Schwalbe, Boush, and Burkholder, 1973; Shapas, Burkholder, and Boush, 1977

Bacteria and virus	*Bacillus thuringiensis* is approved for use as a spray on the surface of grain as a control for moth pests	Heimpel, 1972; McGaughey, 1975*a*, 1975*b*, 1976
Controlled atmosphere		
Hermetic storage	Not being utilized in the United States to any extent and not applicable to most urban situations	Shejbal, 1980
Carbon dioxide	Excellent control if the commodity container can be adequately sealed to retain the gas; treatments are temperature dependent and usually will require longer times than conventional fumigants	Davis and Jay, 1983; Jay, 1980
Nitrogen	Same as for carbon dioxide but with time requirements greatly extended	Banks and Annis, 1977
Combustion gases	Same as for carbon dioxide	Storey, 1973
Pheromones	The use of pheromones to date has largely been limited to their use as lures in traps; they offer great promise as survey tools in integrated pest management programs	Burkholder, 1979
Aeration	The use of aeration (ambient temperatures) to dry and cool commodities has been used effectively in both temperate and subtropical climates	Navarro and Calderon, 1982
Food attractants (baits)	Some use as a survey tool for some quarantine pests	Shapas, Burkholder, and Boush, 1977

been found in the sauce. Following an unsuccessful investigation by the firm's sanitation, a consultant was employed.

An examination of the recovered infested commodity revealed several bottles containing parts of psocids and one bottle with a portion of a wing of a caddisfly. An inspection of the processing plant did not reveal any weak points in the firm's sanitation or pest management program. An inspection of the structure used to store the new bottles and packing materials provided the most probable source of infestation. This structure was an unheated metal warehouse with wooden floor and framework. The new bottles were received in their cardboard shipping containers on wooden pallets. The psocids were found breeding in this warehouse apparently on the fungi growing on the wood in the warehouse and on some of the pallets and also possibly on some older cardboard cartons. The fact that the warehouse was unheated allowed for conditions of high humidity to develop, which enhanced the breeding of the psocids particularly in the spring and fall. Examination of the cases of bottles showed that they were shipped upright and separate from their tops. Psocids apparently simply dropped into the open bottles and were unable to escape. During times of high infestation, the caustic washes that the bottles were submitted to did not completely remove the psocids from all the bottles. The consultant recommended that the humidity be reduced by either temperature control or fans or both.

The presence of the caddisfly required further investigation. This investigation failed to reveal an acceptable rationale, so a second investigation of the immediate vicinity was made in the evening hours. Caddisflies fly in the evening and are often attracted to lights. The warehouse was not in operation in the evening, but there was outside lighting by each outside door. The consultant felt that the presence of the caddisfly found in the condiment was a spurious event to the case at hand but recommended that the lights be removed from the building and put on poles and directed at the building. This action was designed to prevent insects that are attracted to lights from being attracted to the building's doorways and thereby possibly gaining entrance to the building.

Case Study 2

A supplier of health foods such as nuts and dried fruit, and so on, has received complaints that the foods he is supplying to a chain of health food stores is infested. An investigation of these complaints reveals that these foods are often infested in health food stores with moth larvae (Indianmeal moth and almond moth) and cigarette beetles. However, he does not believe that he is responsible for these infestations. He asks his corporate sanitarian to investigate these complaints.

The sanitarian visits several health food stores and their corporate headquarters. He learns that the stores are all located in shopping centers where the lights are on from 9 AM to 9 PM for shopping six days a week with Sunday operations from 12 noon to 6 PM. At all other times, there are security lights on throughout shopping centers. In most instances, the shopping centers are one large building, and shops are separated by walls and all have a common head space area. Dried produce such as mushrooms, dried fruit, nuts, and various seeds are displayed in large open bins of 2 × 2 × 2 ft in size. Most store managers know of their problem and have been using dichlorvos strips at night or in out-of-the-way locations in an effort to solve the problem. Little, if any, success has been achieved, and several have stated they wish to try fumigation.

The corporate sanitarian recommended that the store managers discontinue the use of the dichlorvos strips as they would not generate the needed concentration with the stores open all day and with a common headspace for the vapors to escape. He also recommended against fumigation as it would be dangerous and impossible to retain the fumigant because of the common headspace. He did recommend periodic residual spraying of cracks and crevices in the stores and on display racks. Other recommendations were nonchemical and included buying smaller lots of produce or storing produce in a local cold storage facility until needed, develop the practice of first-in, first-out for infestible produce, secure bins with see-through lids and develop the practice of visually inspecting the infestible produce daily.

REFERENCES

Anonymous. No date. *A Guide to Good Manufacturing Practices for the Food Industry.* Lauhoff Grain, Danville, Ill.

Anonymous. 1966. Piperonyl butoxide and pyrethrins in the outer ply of multiwall paper bags. *Federal Register* **31**(171, Sept. 2), pt. 121, pp. 11608-11609.

Anonymous. 1974. Piperonyl butoxide and pyrethrins. *Federal Register* **39**:38224-38225.

Ashina, E. 1966. Freezing and frost resistance in insects. In *Cryobiology,* H. T. Meryman (ed.). Academic Press, London, pp. 451-486.

Bailey, S. W. 1979. The irradiation of grain: An Australian viewpoint. In *Australian Contributions to the Symposium on the Protection of Grain against Insect Damage during Storage, Moscow, 1978.* Division of Entomology, Commonwealth Scientific and Industrial Research Organization, Australia, pp. 136-138.

Banks, H. J., and P. C. Annis. 1977. *Suggested Procedures for Controlled Atmosphere Storage of Dry Grain,* Technical Paper No. 13. Division of Entomology, Commonwealth Scientific and Industrial Research Organization, Australia, 23p.

Baur, F., ed. 1984. *Insect Management for Food Storage and Processing.* American Association of Cereal Chemists, St. Paul, Minn.

Borror, D. J., and D. M. DeLong. 1964. *Introduction to the Study of Insects,* 4th ed. Holt, Rinehart and Winston, New York, 820p.

Bruce, W. A. 1983. Current status and potential for use of mites as biological control agents of stored-product pests. In *Control of Pests by Mites,* M. A. Hoy, G. F. Cunningham, and L. V. Knutson (eds.), Special Publication 3304. Division of Agriculture and Natural Resources, University of California, Berkeley, pp. 76-78.

Burkholder, W. E. 1979. Application of pheromones and behavior-modifying techniques in detection and control of stored product pests. In *Proceedings of the 2nd International Working Conference, Stored-Product Entomology,* R. Davis (ed.). Ibadan, Nigeria, pp. 56-65.

Champ, B. R., and C. E. Dyte. 1976. *Report of the FAO Global Survey of Pesticide Susceptibility of Stored Grain Pests.* U.N. Food and Agricultural Organization, Rome, Italy.

Cornwell, P. B. 1966. *Entomology of Radiation Disinfestation of Grain.* Pergamon Press, Elmsford, N. Y., 236p.

Cotton, R. T. 1963. *Pests of Stored Grain and Grain Products.* Burgess Publishing, Minneapolis, Minn., 318p.

Davis, R. 1972. Some effects of relative humidity and gamma radiation on population development in *Acarus siro* L. (Acarina: Acaridae). *J. Georgia Entomol. Soc.* 7(1):57-63.

Davis, R. 1981. Stored-grain insects and their control. In *Handbook of Transportation and Marketing in Agriculture,* vol. 2, *Field Crops,* E. E. Finney, Jr. (ed.). CRC Press, Boca Raton, Fla., pp. 111-123.

Davis, R. 1982. In-transit shipboard fumigation of grain. *Pest Control* **50**(12):22, 24-27.

Davis, R., and E. G. Jay. 1983. An overview of modified atmospheres for the control of stored-product insects. *AOM Tech. Bull.,* March, pp. 4026-4029.

Dean, G. A. 1913. Mill and stored-grain insects. *Kansas Agric. Exp. Sta. Bull.* **189:**146-147.

Dermott, T., and D. E. Evans. 1976*a.* High temperature disinfestation: A new approach. *Bulk Wheat* **10:**61-62.

Dermott, T., and D. E. Evans. 1976*b.* Thermal disinfestation of wheat in a fluidized bed. In *Proceedings of the 1st International Conference of Food Engineering,* Section 6, *Storage of Food Grains,* Boston, p. 43.

Disney, R. W. 1954. The specific heat of some cereal grains. *Cereal Chem.* **31:**229-239.

Droutt, R. L. 1972. Biological control: Parasites and predators. In *Pest Control Strategies for the Future.* National Research Council, National Academy of Science, Washington, D.C., pp. 288-297.

Food and Drug Administration. 1976. Federal Food, Drug and Cosmetic Act. Public Health Service, U.S. Department of Health, Education and Welfare, Washington, D.C.

Food and Drug Administration. 1977. *Training Manual for Analytical Entomology in the Food Industry,* J. R. Gorham (ed.), FDA Technical Bulletin No. 2. Washington, D.C.

Food and Drug Administration. 1981. *Principals of Food Analysis for Filth, De-*

composition, and Foreign Matter, J. R. Gorham (ed.), FDA Technical Bulletin No. 1 (rev). Washington, D.C.

Gillenwater, H. B., P. K. Harein, E. W. Loy, Jr., J. F. Thompson, H. Laudani, and G. Eason. 1971. Dichlorvos applied as a vapor in a warehouse containing packaged foods. *J. Stored-Prod. Res.* **7:**45-56.

Hall, C. W. 1979. *Dictionary of Drying.* Marcel Dekker, New York, 350p.

Heimpel, A. M. 1972. Insect control by microbial agents. In *Pest Control Strategies for the Future.* National Research Council, National Academy of Science, Washington, D.C., pp. 298-316.

Highland, H. A. 1973. Synthetic pyrethroids as package treatments to prevent insect penetration. *J. Econ. Entomol.* **66:**540-541.

Highland, H. A. 1984. Insect infestation of packages. In *Pest Problems in the Food Industry Are Manageable,* F. Baur (ed.). American Association of Cereal Chemists, St. Paul, Minn., pp. 309-317.

Howe, R. W. 1965. A summary of estimates of optimal and minimal conditions for population increase of some stored-product insects. *J. Stored-Prod. Res.* **1:**177-184.

Jacques, H. E. 1947. *How to Know Insects,* 2nd ed. W. C. Brown, Dubuque, Iowa, 206p.

Jay, E. G. 1980. *Methods of Applying Carbon Dioxide for Insect Control in Stored Grain,* Advances in Agricultural Technology, Southern Series S-13. Science and Education Administration, U.S. Department of Agriculture, Washington, D.C., 7p.

Jay, E. G., and G. C. Pearman, Jr. 1973. Carbon dioxide for the control of an insect infestation in stored corn (maize). *J. Stored-Prod. Res.* **9:**25-29.

Jay, E. G., R. Davis, and S. Brown. 1968. Studies on the predaceous habits of *Xylocoris flavipes* (Reuter) (Hemiptera: Anthocoridae). *J. Georgia Entomol. Soc.* **3**(3):126-130.

Jay, E. G., L. M. Redlinger, and H. Laudani. 1970. The application and distribution of carbon dioxide in a peanut (groundnut) silo for insect control. *J. Stored-Prod. Res.* **6:**247-254.

Jay, E. G., R. Davis, and J. M. Zehner. 1983. *Development of Safe and Efficacious Fumigation Procedures for In-Transit "Truck-Ship Type" Containers: A Feasibility Study,* Advances in Agricultural Technology, Southern Series AAT-S-28/April. Agricultural Research Service, U.S. Department of Agriculture, Washington, D.C., 13p.

Josephson, E. S., and M. S. Peterson, eds. 1982. *Preservation of Food by Ionizing Radiation,* vols. 1-3. CRC Press, Boca Raton, Fla.

Kirkpatrick, R. L., J. H. Brower, and E. W. Tilton. 1972. A comparison of microwave and infrared radiations to control rice weevils (Coleoptera: Curculionidae) in wheat. *J. Kansas Entomol. Soc.* **45:**434-438.

McGaughey, W. A. 1975*a.* Compatibility of *Bacillus thuringiensis* and granulosis virus treatments of stored grain with four grain fumigants. *J. Invertebr. Pathol.* **26:**247-250.

McGaughey, W. A. 1975*b*. A granulosis virus for Indianmeal moth control in stored wheat and corn. *J. Econ. Entomol.* **68:**346-348.

McGaughey, W. A. 1976. *Bacillus thuringiensis* for controlling three species of moths in stored grain. *Can. Entomol.* **108:**105-112.

Mallis, A., ed. 1982. *Handbook of Pest Control,* 6th ed. Franzak and Foster, Cleveland, Ohio, 1,104p.

Mills, K. A. 1983. Resistance to the fumigant hydrogen phosphide in some stored-product species associated with repeated inadequate treatments. *Mitt. Ges. Allg. Angew. Entomol.* **4**(1-3):98-101.

Mills, R. B. 1979. Potential and limitation on the use of low temperatures to prevent insect damage in stored grain. In *Proceedings of the 2nd International Working Conference on Stored-Product Entomology,* R. Davis (ed.). Ibadan, Nigeria, pp. 244-259.

Monro, H. A. U. 1969. *Manual of Fumigation for Insect Control,* FAO Agricultural Studies No. 79. Rome, Italy, 382p.

Mullen, M. A., and R. T. Arbogast. 1979. Time temperature mortality relationship for various stored-product insect eggs and chilling time for selected commodities. *J. Econ. Entomol.* **72:**476-478.

Navarro, S., and M. Calderon. 1982. *Aeration in Subtropical Climates,* FAO Agricultural Service Bulletin 52. Rome, Italy.

Nelson, S. O. 1973. Insect studies with microwave and other radio-frequency energy. *Bull. Entomol. Soc. Am.* **19:**157-163.

Okumura, G. T. 1984. Insect pest identification. In *Insect Management for Food Storage and Processing,* F. Baur (ed.). American Association of Cereal Chemists, St. Paul, Minn., pp. 44-50.

Osmun, J. V. 1984. Insect pest management and control. In *Insect Management for Food Storage and Processing,* F. Baur (ed.). American Association of Cereal Chemists, St. Paul, Minn., pp. 17-24.

Oxley, T. A. 1944. The properties of grain in bulk, III, The thermal conductivity of wheat, maize, and oats. *J. Soc. Chem. Ind.* **63:**53-55.

Pepper, J. H., and A. L. Strand. 1935. Superheating as a control for cereal-mill insects. *Montana State Coll. Agric. Exp. Sta. Bull.* **197:**13-19.

Pratt, H. D., K. S. Littig, and C. W. Marshall. 1960. *Introduction to Arthropods of Public Importance.* Center for Disease Control, U.S. Department of Health, Education and Welfare, Atlanta, Georgia, 35p.

Press, J. W., B. R. Flaherty, and R. T. Arbogast. 1975. Control of the red flour beetle, *Tribolium castaneum,* in a warehouse by a predaceous bug, *Xylocoris flavipes. J. Georgia Entomol. Soc.* **10:**76-78.

Press, J. W., L. D. Cline, and B. R. Flaherty. 1982. A comparison of two parasitoids, *Bracon hebetor* (Hymenoptera: Braconidae) and *Venturia canescens* (Hymenoptera: Anthocoridae) in suppressing residual populations of the almond moth, *Ephestia cautella* (Lepidoptera: Pyralidae). *J. Kansas Entomol. Soc.* **55:**725-728.

Redlinger, L. M., and R. Davis. 1982. Insect control in post-harvest peanuts. In *Peanut Science and Technology,* H. E. Pattee and C. T. Young (eds.). American Peanut Research and Education Society, Yoakum, Texas, pp. 520-570.

Ronai, K. S., and E. G. Jay. 1982. Experimental studies on using carbon dioxide to replace conventional fumigants in bulk flour shipments. *Assoc. Operative Millers Bull.,* August, pp. 3954-3958.

Schwalbe, C. P., G. M. Boush, and W. E. Burkholder. 1973. Factors influencing the pathogenicity and development of *Mattesia trogoderma* infesting *Trogoderma glabrum* larvae. *J. Invertebr. Pathol.* **21:**176-182.

Shapas, T. J., W. E. Burkholder, and G. M. Boush. 1977. Population suppression of *Trogoderma glabrum* by using pheromone luring for protozoan pathogen dissemination. *J. Econ. Entomol.* **70:**469-474.

Shejbal, J., ed. 1980. *Proceedings of the International Symposium on Controlled Atmosphere Storage of Grains.* Elsevier, Amsterdam, 608p.

Shorey, H. H. 1977. Manipulation of insect pests of agricultural crops. In *Chemical Control of Insect Behavior: Theory and Application,* H. H. Shorey and J. J. McKelvey, Jr. (eds.). Wiley, New York, pp. 353-367.

Storey, C. L. 1973. Exothermic inert-atmosphere generators for control of insects in stored wheat. *J. Econ. Entomol.* **66:**511-514.

Truman, L. C., G. W. Bennett, and W. L. Butts. 1976. *Scientific Guide to Pest Control Operations,* 3rd ed. Purdue University/Harvest, Cleveland, Ohio, 276p.

Tyler, P. S., R. W. Taylor, and D. P. Reed. 1983. Insect resistance to phosphine fumigation in food warehouses in Bangladesh. *Int. Pest Control* **25**(1):10-13, 21.

U.S. Department of Agriculture. 1982. *Guidelines for the Control of Insect and Mite Pests of Food, Fiber, Feeds, Ornamentals, Livestock and Households,* Agricultural Handbook 584. Agricultural Research Service, Washington, D.C.

van Kooij, J. G. 1983. Present status of international and national standardization and regulation of food irradiation. *Food Irradiation Newsletter* **7:**3-10.

Watter, F. L. 1979. Potential of accelerated electrons for insect control in stored grain. In *Proceedings of the 2nd International Working Conference, Stored-Product Entomology,* R. Davis (ed.). Ibadan, Nigeria, pp. 278-286.

Williams, R. N., and E. H. Floyd. 1971. Effects of two parasites, *Anisopteromalus calandrae* and *Choetospila elegans,* upon populations of the maize weevil under laboratory and natural conditions. *J. Econ. Entomol.* **64:**1407-1408.

World Food Program. 1970. *Food Storage Manual,* parts 1-3. Tropical Stored Products Centre, Ministry of Overseas Development, Slough, England.

Yoshida, T. 1980. Plant quarantine and intraspecific variation in insect pests. *Plant Protection* **34**(1):35-41.

Zakladnoi, G. A., A. I. Men'shenin, E. S. Pertsovskii, R. A. Salimov, V. G. Cherepkov, and V. S. Krshreminskii. 1982. Industrial application of radiation deinsectification of grain. *Atomyava Energiya* **52**(1):57-59.

13

Pest Management of Wood-Destroying Organisms

Harry B. Moore

North Carolina State University, Raleigh

It has been suggested that the proper terminology for wood protection activities should be "integrated protection against wood-infesting structural pests" rather than integrated pest management (Williams, 1983*b*). This is because there is no acceptable level of infestation (economic threshold), particularly in a house, which is the average person's largest investment of a lifetime. Total elimination and exclusion of pests is sought rather than managing populations that may still exist in the wood of the structure. The same is generally true for insect pests in wood products that are for sale or in use. Some level of biodeterioration caused by bacteria or fungi is much more likely to be tolerated as natural by users.

Another aspect of protection from wood-destroying organisms that differs particularly from management of agricultural and horticultural pests is that there are many more people involved in the chain of events that leads up to the use of the commodity to be protected. From the time the tree is cut until the products made from the harvested wood are in use, there are numbers of individuals who must actively contribute to the protection of the wood. When exposed to the natural environment, most wood is a perishable commodity. Beginning with the loggers, continuing with the sawyers, lumber yards, and manufacturers (in the case of wood structures, the designers and builders), there must be a different, though coordinated, effort by each if damage to the wood is to be prevented. The total process includes the maintenance of the wood after it is in use. The maintenance aspect is the major thrust of this chapter, but it is impossible to deal with it in isolation from the complete chain of events. The deterioration that begins at tree harvest or later may not be evident until after the wood is in use, so protection of the wood prior to use may be the best and most economical management strategy. Cooperation among all those involved in the production and use of wood is the key to successful pest management of wood-destroying organisms. This cooperation becomes a matter of motivating people as well as a biological problem. I will cover this issue to a limited extent as well as the more traditional aspects of pest management.

Wood has been a favored material for many purposes since man first used tools and constructed shelters. It is easier to work with than stone or metal, and it has a beauty and warmth that give it aesthetic value. However, from its first use, wood has required special consideration as to its durability. Our forefathers learned through experience which woods were durable and which were not. There often was an abundance of wood, and the limiting factor was the labor required to harvest and process the wood when replacement was required. As society became more organized and the age of machines evolved, wood became easier to obtain but more difficult to replace in many of the complex situations where it was in use. Wood with natural resistance was not always available or appropriate to use. It then becomes necessary to find ways to protect wood from biodeterioration.

When wood is used and maintained properly, it will last a long time. Some wooden structures have given a century or more of good service. Unfortunately, many wooden buildings only a few years old have suffered severe damage by decay fungi and insects. The key factor involved in the biodeterioration of wood in buildings is its moisture content (DeGroot and Esenther, 1982). The major causes of wood moisture content that is favorable to wood-destroying organisms in buildings are water leaks, design features, and construction practices that cause a buildup of moisture or humid air under or within buildings (Amburgey, 1983). If adequately seasoned wood is used in the construction of properly designed, built, and maintained buildings, the moisture content of the wood will remain low enough to prevent some and slow the development of other wood-destroying organisms. These good practices alone will not prevent all damage, but they are achievable with current technology and would not significantly increase construction costs. It will continue to be necessary in the near future to also use chemicals to protect existing buildings and most of those that will be constructed.

With present practices, prevention, remedial control, and repair of damage to buildings from subterranean termites alone cost U.S. citizens about $750 million annually (Mauldin, 1982). Decay fungi prevention, control and damage repair cost that much or more. Wood-destroying beetles probably add another $50 million annually. Thus, we have reason to be concerned over the economic problem. The major dependence on chemicals to manage wood-destroying organisms has also led to concern about the environment and the health of individuals exposed to those chemicals. A more integrated approach will lead to better, more economical results and less dependence on chemicals.

REVIEW OF PEST MANAGEMENT PROCEDURES THROUGH THE PRESENT

Pesticide Use Patterns

Prior to World War II, before the development of many new synthetic chemicals, protection of wood from decay fungi and subterranean termites, the major pests, consisted primarily of creosote impregnation of wood destined for use in ground contact or in damp locations. In the late 1930s, pentachlorophenol became available as a preservative against decay fungi. It is applied in various oil carriers under pressure or by brush, dip, or spray, the final use determining the type of oil used and the method of application. From the 1940s until the present, a wide range of chemicals were introduced for use as fungicides to prevent wood decay. Some of those are copper naphthenate, zinc naphthenate, borax-boric acid, tributyltin oxide, chromated-copper-arsenate, ammoniacal-copper-arsenate, acid-copper-chromate, chromated-zinc-chloride, and copper-8-quinolinolate. The wood preservatives used most widely today in buildings are chromated-copper-arsenate and ammoniacal-copper-arsenate. Common names include salt preservative, Boliden CCA, Osmose K33, and Wolman CCA (Levi, 1978; Graham, 1973). The chemicals are applied under pressure in a water solution and precipitate in the wood so that they cannot be leached out when exposed to water. The wood preservatives are effective in preventing damage by insects as well as decay fungi (Coulson and Lund, 1973). Water-repellent wood preservatives are effectively used on millwork such as windows, exterior doors, trim, and so on, to prevent deterioration by fungi (Verrall and Amburgey, 1979). Alternatives to the use of broad-spectrum preservatives have been sought. Some of the measures now being tested are impregnation of wood products with resins, bulking with polyethylene glycol, heat treatments, and chemical modifications of wood (De Groot and Esenther, 1982). Wood thus treated has increased resistance to brown-rotting fungi.

The use of chemicals, applied in the soil adjacent to foundations of buildings to form a barrier that subterranean termites find toxic or repellent, began in this country in the 1930s. At that time soil treatment was largely experimental. Solutions of pitch, tar, creosote, and other materials dissolved in used motor oil were commonly used, as were water solutions of arsenic salts (Smith, 1982). Also used in soil barriers were oil solutions of paradichlorobenzene, orthodichlorobenzene, trichlorobenzene, and chlorinated naphthalenes. Often, the applicators did not understand the principles of termite biology and the barrier concept of structural protection (Snetsinger, 1983). The technique has been used and improved since that time by the pest control industry, and it is reported in detail in the National Pest Control Association's *Approved Reference Procedures for Subterranean Termite Control* (Rambo, 1980). This technique involves saturating the soil with a termiticide adjacent to all piers, chimneys, pipes, and foundation walls from the top of the soil to the top of the footing. The termiticide is applied by trenching and rodding exposed soil, but holes must be drilled in concrete that covers soil to be treated. Voids in foundation walls are also drilled and injected with termiticide to flood the tops of the footings and prevent termite entry through cracks or joints there. In the late 1930s, pentachlorophenol, under development as a wood preservative, was used in oil solution to treat soil to prevent subterranean termite access to buildings. It is still registered for that purpose but is little used, since it is not as long-lasting or as inexpensive as most other termiticides available.

In 1939 the U.S. Department of Agriculture Forest Service began research on chemicals for termite control at what is now known as the Forestry Sciences Laboratory at Gulfport, Mississippi. The tests conducted between 1939 and 1943 did not find any chemicals that would control termites for more than three years. In 1944, at the request of the U.S. Army Corps of Engineers, the laboratory began tests to find a treatment that could be used to protect caches of ammunition stored on the soil in wooden boxes. This series included the then new insecticide, DDT, and led to it being recommended to protect buildings from termites (Smith, 1982). This was the beginning of tests at the laboratory to find new and better chemicals to use for soil treatments (Smith, Beal, and Johnston, 1972). Thus several chlorinated hydrocarbon insecticides emerged as the primary termiticides in the United States from the 1950s through the present day. They are aldrin, chlordane, dieldrin, and heptachlor. All of them have been 100% effective more than 30 years in field tests (Beal and Howard, 1982). In 1981 chlorpyrifos, an organophosphate, was labeled as a soil insecticide for subterranean termite control and is now being used. Since then, isofenphos, another organophosphate, and endosulfan, a chlorinated hydrocarbon, and permethrin, a pyrethroid, have been registered as soil termiticides. A number of other compounds have passed the laboratory requirement that they last five years in field tests and are now eligible to be submitted for registration.

In 1974 Esenther and Beal (1978) and Beard (1974) reported success with the use of an attractant-mirex bait in suppressing or controlling subterranean termites. In this bait-block method of termite control, small wood blocks, partially decayed by a fungus (*Gloeophyllum trabeum* [Pers. ex Fr.]) that makes them preferred by termites, are impregnated with a slow-killing toxicant and placed in the soil around a termite infestation. When the termites feed on the poisoned wood bait, they carry the toxicant back to the colony, and through food sharing, the colony is gradually killed. Mirex has since been removed from the market by the U.S. Environmental Protection Agency (EPA), and search for a replacement continues. So far, laboratory tests indicate that methoprene, a growth regulator (Haverty and Howard, 1979); chlortetracycline, an antibiotic (Mauldin and Rich, 1980); fluorinated lipids (Prestwich et al., 1983); and avermectin, a macrocyclic lactone (R. H. Beal, pers. comm.) show promise as replacements for mirex in the bait system and are undergoing field testing.

Direct application of insecticide into subterranean termite galleries in wood will speed the demise of the colony when there is some residual moisture in the wood that allows them to survive without access to soil moisture. In years past, the chlorinated hydrocarbon termiticides as water emulsions were used for wood injection. They are no longer labeled for this use. Presently, bendiocarb, boric acid, chlorpyrifos, and propoxur are labeled for this purpose.

A bodied, greaselike, oil emulsion of pentachlorophenol, with heptachlor added, is available for topical application to wood to allow deep penetration of the toxicants to control termites and wood decay. The status of this use of pentachlorophenol is presently in a state of uncertainty. The EPA is reviewing the classification of all uses of pentachlorophenol, and the judicial and administrative process will ultimately determine the availability of this particular formulation.

Some wood species have heartwood that is naturally resistant to termites and decay. Solvent extracts from resistant heartwoods have been tested by treating susceptible wood with them and exposing the treated wood to subterranean termites. Results indicate a potential use of antitermitic wood extractive components in termite control (Carter and Beal, 1982).

Drywood termites in limited infestations are sometimes controlled by drilling small holes into the interior of infested wood and injecting insecticidal dusts. The dusts follow the galleries for some distance but are also picked up on the bodies of termites and carried to untreated areas. Some dusts are contact poisons, and some are stomach poisons. Beginning in the 1930s, such inorganic dusts as Paris green, sodium fluosilicate, sodium fluoride, and calcium arsenate were used. This drill-and-treat method was difficult, dangerous, and often not completely effective. Recently, boric acid powder, a much safer material, has been registered for this purpose.

This drill-and-treat method of drywood termite control has also involved liquid formulations of insecticides. Orthodichlorobenzene and paradichlorobenzene were used in the 1930s. From the late 1940s until the mid 1970s oil solutions of pentachlorophenol, chlordane, and lindane were used. Recently, chlorpyrifos in methylene chloride solution, applied from a self-pressurized cylinder, has been labeled for this purpose in the United States.

Until 1983 the fumigant ethylene dibromide, in oil solution with a residual insecticide, was applied into drilled wood or brushed on the surface to control drywood termites. Ethylene dibromide is no longer registered in the United States for these applications. The bodied oil emulsion of pentachlorophenol with heptachlor, as mentioned for subterranean termite control, has also been used successfully to control drywood termites when it is applied to the wood surface with a paint brush or caulking gun (Ebeling, 1975).

Silica aerogels have been used to effectively prevent and control drywood termites (Ebeling, 1975). When a light coating is blown onto wood surfaces, it clings to the drywood termites as they explore for nest sites. By absorbing the protective lipid layer on the body parts that contact the dusted surfaces, fatal desiccation results. In dry climates the material should remain effective for many years.

None of the materials applied to wood by injection or topically are as effective as fumigants applied in the buildings under gas-proof tarpaulins. Over the years, hydrogen cyanide, acrylonitrile mixtures, methyl bromide, and sulfuryl fluoride have been used. Presently, methyl bromide and sulfuryl fluoride are used; sulfuryl fluoride is most common.

Wood-boring beetles, prior to World War II, were treated by the use of petroleum oil dips and sprays and after the war with residual insecticides incorporated. Lindane, dieldrin, and chlordane became the most common active ingredients used in the United States. Dieldrin and chlordane are no longer registered for this purpose, and lindane may only be applied as a water emulsion in nonliving areas of buildings or in oil solution to wooden articles such as furniture. Chlorpyrifos is registered for use against wood-boring beetles, and it may be applied as a water emulsion to interior surfaces in occupied areas of buildings as well as in unoccupied areas. Within the past few years a number of pyrethroid compounds have been tested as replacements for chlorinated hydrocarbons (Baker and Berry, 1980). Permethrin in particular has been tested in England and found effective against a number of important species. Permethrin has a persistent odor, a significant limiting factor when used inside structures. Laboratory experiments in Germany (Doppelreiter, 1983) have recently shown that the growth inhibitor, diflubenzuron, may prove useful in the control of wood-boring beetles. It has several years of persistance when impregnated into wood and is relatively nontoxic to warm-blooded animals. Neither permethrin nor diflubenzuron is labeled for this use in the United States. During processing of green hardwood lumber, boron dip-diffusion may now be used to control or prevent lyctid powderpost beetles (Williams, 1984). Oil solution of pentachlorophenol is still registered for beetle control, but it has limited effectiveness as a toxicant. Pentachlorophenol is apparently repellent to some of the beetle species and remains active in that capacity. In very heavy infestations of beetles, fumigation of buildings under gas proof tarpaulins with methyl bromide or sulfuryl fluoride is very effective. There is, however, no residual effect, and it is costly.

Carpenter ants over the years have been treated with all of the insecticide sprays, dusts, and baits used for any house-invading ant species. There are a number of formulations currently available and effective, though not for as long as the previously standard material, chlordane dust or spray. The long residual life of chlordane made it the material of choice until its registration for ant control was canceled by the EPA. Currently available residual insecticides include organophosphates and carbamates, as well as boric acid powder.

Available Nonpesticidal Procedures

Protection of buildings from all wood-destroying organisms requires that regular inspection of the likely points of attack be made. For most parts of the United States this means at least once a year. In some geographic regions, where there is a high incidence of attack by one or more organisms, inspections should be made twice annually. A proper inspection cannot be done without adequate knowledge of the identification, biology, and habits of the pest organisms, as well as a thorough understanding of building construction. There is not space available here to detail the information. Readers are urged to consult one or more references for additional information in this area (Ebeling, 1975; Levi, 1983; Mallis, 1982; Moore, 1979; Verrall and Amburgey, 1979). Visual inspection and sounding of wood suspected of infestation are the standard procedures. Very recently the use of termite-detecting dogs has been introduced and is slowly being accepted by the pest control industry (Moreland, 1982). Properly trained dogs extend the capability of finding some infestations that would not be detectable by an experienced technician.

The first, and still viable, procedure to prevent decay fungi and insects from destroying wood was and is to use naturally durable wood varieties. Very little of the wood most resistant to wood-destroying organisms is available in commercial quantities, so that method is of limited practical value.

Wood not naturally resistant to decay and not chemically treated to make it resistant will decay unless its moisture content is kept below 20%. There are many ways in which buildings can be designed, built, and maintained to provide this needed dryness (Amburgey, 1983; Verrall and Amburgey, 1979). Some of the features involved include proper drainage around the building, adequate wood-to-soil clearance to prevent wetting of the wood by capillarity, cross-ventilation in crawl spaces, soil covers to reduce humidity in crawl spaces, wide roof overhang, proper flashing, sealing and caulking joints exposed to wetting, and preventing plumbing leaks.

When water-conducting fungi (*Poria* spp.) are the problem, even more strenuous management efforts must be made. In addition to the usual environmental changes in the building vicinity, it is necessary in some cases to physically excavate soil to provide 8 in of vertical clearance between untreated wood and soil. The insertion of sheet metal shields between the foundation and the wood may also be necessary (Verrall and Amburgey, 1979).

Recent research on the biological control of decay fungi has shown that certain species of bacteria directly attack decay fungi with enzymes, others produce substances toxic to the fungi (Preston et al., 1982). There are also wood-inhabiting fungi that are antagonistic to some of the wood-decaying species. Even their dead remains in previously infested wood provide residual protection from decay (Bruce and King, 1983). These biological control methods are still experimental.

As early as 1934 construction recommendations for the prevention of damage by subterranean termites (Kofoid, 1934) stressed a number of important practices that are still being followed. These include removal of all stumps and large roots on the site to be occupied by the building, removal of all wood forms used in pouring concrete in the foundation, removal from direct ground contact of wood and cellulosic debris generated during construction, construction of foundation walls and piers of poured concrete or masonry units laid in Portland cement mortar (to extend 6 in above finish grade), and ventilation of crawl spaces to provide cross-ventilation. These methods also help prevent decay.

Good building practices alone will not prevent subterranean termite attack, but they are an essential part of the integrated management of these important pests. The first strategy listed in the National Pest Control Association's recommendations (Rambo, 1980) is mechanical alterations. For many years not enough attention has been given to correcting structural faults, and greatest reliance has been placed on chemical control measures. The result is control failures that would not have occurred if the accessibility of the wood and the environmental conditions had not been favorable to the termites. Beginning in the late 1920s or early 1930s metal shields placed on the top of foundation walls were used to deter subterranean termite entry. These were recommended for many years by the Forest Service and the Federal Housing Administration as an alternative to soil treatment. Shields are so seldom installed and maintained correctly that they have limited value, while giving the property owner a false sense of security. The purpose of the shields is to prevent hidden entry by the termites through cracks, voids, or joints in the masonry foundation walls or piers. The shields force the termites to build their mud shelter tubes on the surface where they can be seen before the termites eventually breach the shields. Chemical treatment can then be employed to stop their activity.

On an experimental basis, certain plastics have been impregnated into wood to render it resistant to subterranean termite attack. Depending on the hardness of the wood-plastic composite, impregnation has been shown to be effective (DeGroot and Esenther, 1982). It is not yet a process that is practical to use commercially.

Within the past year, a nematode species parasitic to subterranean termites has been marketed under the trade name Spear® (Weidner, 1983). The nematode-containing material is to be applied in water suspension onto or into soil infested by termites. Based on laboratory tests only, nematodes are said to seek out the termites and enter their bodies. Once they have penetrated the host, a toxic bacterium emerges from the nematode, multiplies rapidly, and kills the insect within 24-48 hours. The Forestry Sciences Laboratory at Gulfport, Mississippi, has run laboratory tests and installed field tests with this new biological control organism (R. H. Beal, pers. comm.). No recommendation will be forthcoming until full field testing is completed and results are analyzed.

Nonchemical control of drywood termites has consisted primarily of physically excluding the reproductives as they attempt to establish new colonies by entering wood crevices or holes, where 90% of infestations begin (Ebeling, 1975). This has included keeping exposed wood surfaces painted, joints caulked, and access openings to attics and crawl spaces screened. None of these methods is very successful (Mallis, 1982).

Recently a device for drywood termite control which generates high frequency electricity (100 kHz) and high voltage (90,000 volts) but low current (90 watts) has been marketed (Ebeling, 1983). Known commercially as the Extermax®, the system operates when the generator wand is moved over the surface of termite-infested wood. The current penetrates the wood where there are termite galleries and kills the termites outright or by starvation when the even more susceptible gut fauna of the termites are killed. The symbiotic protozoans and bacteria are essential to the termites in the digestion of cellulose.

Wood-boring beetles have specific moisture requirements in order to develop in wood. The manipulation of wood moisture content has for many years been a preventive and control measure for these beetles. Kiln-drying wood below 20% moisture content quickly renders it too dry for reinfestion by borers, which initiate their attack in unseasoned wood. If sufficiently high temperatures are employed in the kiln schedule, any beetles present are killed. Research has recently shown that anobiid powderpost beetles (Williams, 1983*a*) and old house borers (Vongkaluang, Moore, and Farrier, 1982) might be controlled in buildings by careful attention to humidity control, which translates into moisture content of exposed wood. Proper clearance, drainage, and ventilation in buildings are an integral part of woodborer prevention and control, as they are for termites and decay fungi.

Small wooden articles, and even furniture, infested by wood-boring beetles can be placed in a deep freeze for several days to kill the larvae (Moore, 1983).

Many of the beetles attacking seasoned wood need the stimulus afforded by the wood surface and texture in order to oviposit. By coating wood surfaces with finishes of various kinds, the wood is rendered unattractive to the females. This practice is more often applied to furniture and other manufactured wooden articles than to structural wood.

In 1974 (Hightower, Burdette, and Burns, 1974) the use of microwave energy was shown to be effective in controlling wood products insects. This is not the first attempt at the use of radiation to control these insects, but it was shown to be economically feasible, a factor missing in previous reports.

Carpenter ants generally prefer wood that has been softened by decay, and some species require a very high humidity in their nest. Reducing wood moisture content by stopping rain leaks, providing clearance between wood and soil, and other mechanical procedures to accomplish this purpose are an important part of carpenter ant management (Mallis, 1982). Related to that is the manipulation of the environment surrounding the buildings to be protected. Placing stored firewood above ground to keep it dry and removing dead tree limbs and stumps near the building will reduce potential sources of future infestation. Closing ant entry points in foundation walls and siding and trimming shrubs and trees that contact the building are also a part of this procedure.

PEST MANAGEMENT TODAY

It is not this book's purpose to exhaustively review all areas of pest management for wood-destroying organisms. Table 13-1 will allow readers an entry point into the pest management literature on all the major wood-destroying pest groups, while subsequent discussion will focus on particular pest groups.

Putting Pest Management Principles into Practice

In the preceding discussion of chemical and nonchemical pest management procedures, it can be discerned that several different strategies are com-

Table 13-1. A summary of the most important wood-destroying pests and management strategies, with references.

Key pests	*Pest management strategies*	*References*
Brown rot and white rot fungi	Cultural, mechanical, chemical, and legal	Ebeling, 1975; Levi, 1973, 1983; Mallis, 1982; Mampe, 1974; Truman, Bennett, and Butts, 1976; Verrall and Amburgey, 1979
Water-conducting decay fungi	Cultural, mechanical, chemical, and legal	Ebeling, 1975; Levi, 1973, 1983; Mallis, 1982; Truman, Bennett, and Butts, 1976; Verrall and Amburgey, 1979
Subterranean termites	Cultural, mechanical, chemical, biological, and legal	Beal, Mauldin, and Jones, 1983; Ebeling, 1975; Levi, 1983; Mallis, 1982; Moore, 1979; Moore and Haverty, 1979; Rambo, 1980; Weidner, 1983
Drywood termites	Cultural, mechanical, physical, chemical, and legal	Ebeling, 1975, 1983; Levi, 1983; Mallis, 1982; Moore, 1979; Moore and Haverty, 1979; Truman, Bennett, and Butts, 1976
Powderpost beetles	Cultural, mechanical, physical, chemical, and legal	Ebeling, 1975; Levi, 1983; Mallis, 1982; Moore, 1979; Moore and Haverty, 1979; Truman, Bennett, and Butts, 1976
Old house borer	Cultural, mechanical, physical, chemical, and legal	Ebeling, 1975; Levi, 1983; Mallis, 1982; Moore, 1979; Moore and Haverty, 1979; Truman, Bennett, and Butts, 1976
Carpenter ants	Cultural, mechanical, chemical, and legal	Ebeling, 1975; Hansen, 1984; Levi, 1983; Mallis, 1982; Moore, 1979; Moore and Haverty, 1979; Truman, Bennett, and Butts, 1976

monly recommended for managing each type of wood-destroying organism. This is not to say that recommended combinations are generally practiced. In some cases, the property owner has a management role to play that is neglected (Williams and Smith, 1983). Competition in the structural pest control industry has often held prices down to the point that long-term, nonchemical procedures have been omitted from management practices. This has been possible because many of the pesticides, as used in the past, have performed well enough to at least partially compensate for the deletion of other strategies in the management program. A much more critical public concern over pesticides inside buildings has led to new regulatory activity (Moreland, 1983*a*, 1983*b*; National Academy of Sciences, 1982; U.S. Environmental Protection Agency, 1983) that will drastically change the way pesticides are used.

Many pest management strategies for wood-destroying organisms are routinely incorporated into seemingly unrelated operations. Often the practitioners have no specific knowledge that biodeterioration of wood is at least a part of the reason for certain practices having evolved. This lack of knowledge leads to omission of sometimes critical attention to specifications. Just one example of this is the construction of buildings on foundations raised well above the ground level. Foundations are designed primarily to support buildings. If, however, soil moisture did not create conditions conducive to biodeterioration of wood, building codes would not necessarily require that ventilation, clearance, and drainage be incorporated into design and construction of foundations.

The control of decay fungi is based squarely on the concept that dry wood will not decay, and if wood cannot be kept dry, it must be adequately treated with a fungicide. In buildings the greatest part of this strategy could be accomplished through proper design, construction, and maintenance, including specifying properly dried wood for construction and keeping it dry before, during, and after construction. Remedial control of decay fungi in buildings is carried out primarily by correcting construction faults that allow moisture to gain access to untreated wood (Verrall and Amburgey, 1979). Where it is not economically or aesthetically feasible to alter design to keep wood dry, the wood that will be exposed to moisture should be replaced with pressure-treated wood. The strategies for the protection of wood from decay fungi are nothing more than good building and maintenance practices. This has, however, not prevented the public from presently having to spend a large portion of the nearly $1 billion a year wood decay bill for remedial treatment and wood replacement.

Management of carpenter ants in buildings involves a thorough understanding of the biology, habits, and ecology of the ants. For many years, a long-lasting residual insecticide (chlordane) was available and extensively applied in and around buildings. This use allowed less attention to specific areas of ant activity and more reliance on attrition to accomplish ultimate control and to prevent reinfestation. With present restrictions on the number and types of pesticides available for carpenter ant management, more careful attention must be given to specific placement of insecticides and greater concern for altering the environment to make it more hostile to the ants. Locating the nest and concentrating control efforts in that area has always been a key recommendation for carpenter ant control. Many species require a high humidity in the nest and tend to be associated with damp wood. An electronic moisture meter may help in pinpointing potential nest sites. Removing the source of water will reduce the moisture required to maintain favorable nest conditions. Carpenter ants tend to begin their nests in wood softened by decay, but they do not always require that. Sometimes a colony or portion of a colony moves *en masse* from an outdoor nest site to an available, protected cavity indoors. This leads to the requirement that all favorable nesting sites in the vicinity of the building to be protected be located and eliminated if feasible. It has recently been reported that carpenter ants make visible trails to their nests outdoors. These trails are vital clues when seeking out nests (Hansen, 1984). Likewise, points of entry for carpenter ants must be determined and closed or removed. This includes sealing openings in foundation walls, siding, trim. Above-ground access must also be located and eliminated where possible, including branches of trees and shrubs that contact the structure. Power and telephone lines also provide access to the higher portions of buildings. These lines cannot be removed, but their points of contact with the building must be investigated and treated as needed. Instructing the property owner concerning the importance of maintaining the roof, gutters, and so on, to keep structural wood dry is an important aspect of a complete management system.

Anobiid powderpost beetles require more than 12% wood moisture content for the young larvae to survive, even when the wood is very nutritious (Williams, 1983*a*). As wood ages, it becomes less nutritious to the beetles, and a moisture content of about 13% is required. These beetles also benefit nutritionally from wood that is partially decayed. From this understanding of their ecological requirements, the environment can be manipulated by eliminating moisture that causes wood to decay. Wood moisture content can be lowered by placing a vapor barrier over the soil under crawl space houses. This might be an alternative to the application of residual insecticides to infested wood, particularly when pesticide contamination is a concern. As a supplement to chemical control, it also benefits the structure by making it less susceptible to wood decay.

If the NPCA's *Approved Reference Procedures for Subterranean Termite Control* (Rambo, 1980) are followed, the resulting management system is a prime example of integrated pest management that has been recommended for decades and practiced by many. Chemically treated soil barriers are at present the primary method of control used by most applicators. However, to provide the most ecologically sound protection from subterranean termites, all management strategies that are economically feasible must be used. These include site sanitation, physical barriers to termite entry, substantial clearance between wood and soil, and good ventilation and drainage underneath and adjacent to buildings. The lower the humidity and the greater the distance between soil and wood, the more difficult it is for subterranean termites to build and maintain mud shelter tubes they must use when they leave the protection of the soil. Some of the newer methods under development, which should be considered when and if they become available, include parasitic nematodes and toxic bait-blocks. These latter methods will be especially useful where there is danger of fouling an on-site water source or where liquid chemical applications might move into ventilation systems and lead to contamination and vapor movement into the interior of the building. With careful use of chemical termiticides in certain specific parts of buildings and nonchemical alternative procedures in contamination-prone areas, it should be possible to provide good, long-term protection.

FUTURE DIRECTIONS IN PEST MANAGEMENT PROGRAMS FOR WOOD-DESTROYING ORGANISMS

Major Problems

1. *A lack of interaction and cooperation* exists among the individuals and organizations interested in wood used in buildings. Much of the technology needed to provide protection of wood from biodeterioration is available. There is, however, no presently functioning integrated system for reaching each of the interested parties with the necessary technical information to provide the capability to protect the wood while it is in their control. The recently established Wood Protection Council, under the sponsorship of the National Institute of Building Sciences, should help overcome this problem. One important goal of this body is to provide a vehicle for bridging the gaps in communication between the various individuals and organizations interested in wood protection.

2. *Negative and often inaccurate media coverage* concerning termiticides has created an adversarial atmosphere between the frightened public, regulatory agencies, and pest control companies. An already litigious population has found new reasons to go to court. The pest control industry has reacted

by refusing to accept the potential liability involved in treating certain types of problem and the property owners have little recourse. Some state regulatory agencies have established more restrictions on the use of termiticides, restrictions that are having negative effects on the pest control industry. Quite possibly, the public will feel these effects in the long run.

3. *Buildings are designed or built* in ways that often preclude adequate treatment with environmentally safe and economically feasible methods. The slow rate of air exchange in modern, weather-tight structures has led to higher levels of undesirable substances in the air, among them being pesticides applied correctly as to site and amount, and problems are compounded even when construction features are not unusual. Lack of understanding of wood biodeterioration and its prevention among architects and engineers is one part of the problem. Another part is inadequate enforcement of building codes that already include regulations that if followed would go far toward reducing risk of wood deterioration.

4. *Limited research funds and personnel* that are needed to provide definitive information, not only on management strategies but on biology, ecology, and even taxonomy of important wood-destroying organisms. This basic information is becoming more important as our management systems become less dependent on pesticides and more environmentally sensitive.

Major Opportunities Available

L. H. Williams has correctly stated (1983*c*) that pest management programs for wood-destroying organisms in buildings, from a holistic viewpoint, must include all the wood protection efforts of all individuals and organizations who are involved in the production, processing, and use of wood. Progress must be made in educating these users and handlers of wood as to their individual roles in the protection process. Sufficient economic incentives for cooperation must be identified and communicated. Until then, we must deal with existing types of problem and new ones that will inevitably become evident. This means there are almost unlimited opportunities for developing long-term, multi-party programs and also for less comprehensive programs that deal with the problems at various steps or stages in the production and use of wood. Some examples of currently needed programs will illustrate the point.

Log kit homes have become extremely popular all over the United States in the past decade. Many purchasers of such homes have the mistaken notion that log homes are essentially maintenance free. (This idea has been suggested by some sales personnel.) Some manufacturers have been insufficiently aware of the physical and chemical nature of wood. They do not understand how it reacts to loss or gain of moisture and to wood-destroying organisms, before they purchase the logs, during processing and storage, and after sale. Poor building design, inadequate seasoning, and insufficient treatment of logs have resulted in very serious biodeterioration of logs in houses, particularly those that were owner erected. A pest management program designed for the log kit home manufacturers is long overdue and will require the public admission that log homes are, in fact, perishable if not properly protected. The manufacturers have encountered enough problems in the recent past that they should now have the incentive to cooperate fully. The program should include specifications for quality of logs and protective measures to be applied before delivery of the rough logs; proper design requirements; strategies for adequate drying, storage, and safe chemical treatment of logs; and a well-written and illustrated set of instructions for construction and maintenance that includes and emphasizes the procedures to be followed in order to safely prevent and control biodeterioration.

Multi-owner housing such as condominiums and townhouses is becoming more common. Walls are shared, and usually the owners share maintenance costs. When one unit in a building is infested with subterranean termites, pest control firms are faced with the task of trying to adequately treat the infested unit and of convincing owners of adjacent, uninfested units that they must share in the cost. The common walls must be treated. The infestation may be originating from adjacent sites, and untreated units may be quickly infested after the treated unit is isolated by chemical soil treatment. Presently, partial treatment of multi-unit buildings is not generally favored by the pest control industry. The alternative is to insist that an entire building be treated when one unit is infested. If the building was treated during construction, complete retreatment may be unnecessary. A pest management program that addresses the needs of the owner of the infested unit and is fair to the other owners and to the pest controller is much needed.

A pest management program for earth-sheltered houses would be a significant contribution. Energy conservation benefits involved in earth-sheltered houses have made them increasingly attractive to consumers. The obvious problem with water leaking into such structures has been addressed (Lane, 1981). Although moisture management will tend to reduce insect and decay problems, the more extensive strategies needed to protect from these organisms have not been developed and publicized. As with waterproofing, it is imperative that some of these procedures be incorporated during construction. Remedial treatment is vastly more involved than in above-ground construction, particularly when the soil covering the structure and the waterproofing system must be disturbed during treatment. Close cooperation between the architect, the builder, and the pest management practitioner is needed.

Future Directions for Practitioners

There will be a continuing and increasing need for comprehensive training and retraining of pesticide applicators. Closer supervision of personnel in the field by those responsible will be necessary. This supervision will be done by the users of pesticides either voluntarily or because more restrictive regulations will be passed. It is essential that training of practitioners include positive identification of the organisms and their biology, ecology, and habits so that applicators can understand the principles behind all types of management strategies well enough to carry them out in a consistently effective manner.

More concern over the safe use of pesticides should reduce accidents and misuse. Practitioners must also become more aware of the correct methods for monitoring pesticides in and around buildings and the decontamination procedures to be used to correct problems when they occur.

The commercial pest control industry must change its self-image from the primary role of pesticide applicators to advisers to their clients. In this new role, they could and would recommend inspection and nonchemical management strategies as readily as pesticide application.

Property owners must become more involved in the protection of buildings from wood-destroying organisms. The practitioners will need to more carefully outline the role of the client in the pest management system and formally incorporate this in the contract for service. The Cooperative Extension Service has an important role in this area as well. Public awareness of wood-destroying organisms and their management must be a goal for all agencies that deal in technology transfer.

The formal, academic training and the continuing education of architects, construction engineers, builders, and wood products manufacturers must be expanded to include specific instruction in wood-destroying organisms and their management.

Future Directions for Researchers

Pest management of wood-destroying organisms involves the motivation of people as well as technology and scientific data. Research is needed in several disciplines if real progress is to be made.

Entomology, plant pathology, and wood science are the basic sources of information for the technology of wood-destroying organism pest management, so these disciplines will play a vital role in future improvements. There should be a continuation of current research on more effective and less hazardous pesticides and on biological control agents for wood-destroying organisms. The current practical, applied research in field applications of laboratory-developed strategies, chemical and nonchemical, should be continued. Another aspect of field research that should be pursued is more specific determination of why and when pest problems occur. Basic research on wood-destroying organisms in the fields of ecology (including population dynamics), biology, morphology, physiology, and taxonomy should be continued and expanded to the extent of available funding.

Engineering research should develop new equipment and techniques. These are needed for detecting wood deterioration in situ, more precise application of pesticides, and testing for pesticide application quality and safety.

Chemists need to develop techniques, chemicals, and chemical formulations to reduce potential exposure of applicators and building occupants to pesticides. Toxicologists should pursue research designed to define health risks, if any, to occupants of structures chemically treated for termites or wood decay fungi.

Cooperative research between psychologists and pest management specialists is needed to determine attitudes and perceptions of consumers and handlers of wood as related to the importance of wood-destroying organisms and how they may be managed. This is important information in designing educational programs and publicizing them.

There is little data on the economic impact of wood-destroying organisms. Economists should be encouraged to work with entomologists, plant pathologists, and wood scientists to develop this important information that researchers need when they are trying to provide justification for funding.

New and improved methods of technology transfer need to be developed. Education specialists should be enlisted to gather and package wood protection information into programs that will encourage application of management practices. These programs should be designed so that evaluation to measure progress is possible. Technology transfer then becomes a vital aspect of improving the protection of wood. It will require gathering the needed information from all available sources, presenting it in a usable form to those who need it, and determining the impact of what has been presented so that improvements may be made.

REFERENCES

Amburgey, T. L. 1983. Integrated protection against structural wood-infesting pests, part 5. Avoiding structural pest problems by design and construction practices. *Pest Manage.* **2**(5):32-35.

Baker, J. M., and R. W. Berry. 1980. Synthetic pyrethroid insecticides as replacements for chlorinated hydrocarbons for the control of wood-boring insects. *Holz als Roh- und Werkstoff* **38:**121-127.

Beal, R. H., and R. W. Howard. 1982. Subterranean termite control: Results of long term tests. *Int. Biodeter. Bull.* **18**(1):13-18.

Beal, R. H., J. K. Maudlin, and S. C. Jones. 1983. *Subterranean Termites: Their Prevention and Control in Buildings,* Home and Garden Bulletin 64. U.S. Department of Agriculture, Forest Service, Washington, D.C.

Beard, R. L. 1974. *Termite Biology and Bait-block Method of Control,* Bulletin 84. Connecticut Agricultural Experiment, New Haven, Conn.

Bruce, A., and B. King. 1983. Biological control of wood decay by *Lentinus lepideus* (Fr.) produced by *Scytalidium* and *Trichoderma* residues. *Mater. u. Organ.* **18**(3):171-181.

Carter, F. L., and R. H. Beal. 1982. Termite responses to susceptible pine wood treated with antitermitic wood extracts. *Int. J. Wood. Preserv.* **2**(4):185-191.

Coulson, R. N., and A. E. Lund. 1973. The degradation of wood by insects. In *Wood Degradation and Its Prevention by Preservative Treatments,* vol. 1, *Degradation and Protection of Wood,* D. D. Nicholas (ed.). Syracuse University Press, Syracuse, N.Y., pp. 277-305.

De Groot, R. C., and G. R. Esenther. 1982. Microbial and entomological stresses on structural use of wood. In *Structural Use of Wood in Adverse Environments,* R. W. Meyer and R. M. Kellogg (eds.). Van Nostrand Reinhold, New York, pp. 219-245.

Doppelreiter, H. 1983. Zur Langzweitwirkung von Diflubenzuron (Dimilin) gegen de Hausbockkafer (*Hylotrupes bajulus* (L.)) in Kiefernsplintholz. *Mater. u. Organ.* **18**(4):263-268.

Ebeling, W. 1975. *Urban Entomology.* Division of Agricultural Sciences, University of California, Los Angeles.

Ebeling, W. [1983]. *The Extermax System for Control of the Western Drywood Termite,* Incisitermes minor. Etex, Las Vegas.

Esenther, G. R., and R. H. Beal. 1978. Insecticidal baits on field plot perimeters suppress *Reticulitermes. J. Econ. Entomol.* **71**(4):604-607.

Graham, R. D. 1973. History of wood preservation. In *Wood Deterioration and Its Prevention by Preservative Treatments,* vol. 1, *Degradation and Protection of Wood,* D. D. Nichols (ed.). Syracuse University Press, Syracuse, N.Y., pp. 1-30.

Hansen, L. D. 1984. A PCO's guide to carpenter ant control. *Pest Control Technol.* **12**(4):56-58.

Haverty, M. I., and R. W. Howard. 1979. Effects of insect growth regulators on subterranean termites: Induction of differentiation, defaunation, and starvation. *Ann. Entomol. Soc. Am.* **72**(4):503-508.

Hightower, N. C., E. C. Burdette, and C. P. Burns. 1974. *Investigation of the Use of Microwave Energy for Weed Seed and Wood Products Insect Control,* Final Technical Report Project E-230-90-1. Radar Division, Systems and Techniques Department, Engineering Experiment Station, Georgia Institute of Technology, Atlanta.

Kofoid, C. A., ed. 1934. *Termites and Termite Control,* 2nd ed. University of California Press, Berkeley.

Lane, C. A. 1981. Waterproofing earth-sheltered houses: There's a lot more to it than tar on the walls. *Fine Homebuilding* **1**(2):35-37.

Levi, M. P. 1973. Control methods. In *Wood Deterioration and Its Prevention by*

Preservation Treatments, vol. 1, *Degradation and Protection of Wood,* D. D. Nicholas (ed.). Syracuse University Press, Syracuse, N.Y., pp. 181-216.

Levi, M. P. 1978. The standard building code and wood decay in homes. *South. Bldg.*, April-May, pp. 22-28.

Levi, M. P. 1983. *Guide to the Inspection of Existing Homes for Wood-Inhabiting Fungi and Insects,* AG 321. North Carolina Extension Service, North Carolina State University, Raleigh.

Mallis, A., ed. 1982. *Handbook of Pest Control,* 6th ed. Franzak and Foster, Cleveland, Ohio, 1,104p.

Mampe, C. D., ed. 1974. *Wood Decay in Structures and Its Control.* National Pest Control Association, Vienna, Va.

Maudlin, J. K. 1982. The economic importance of termites in North America. In *The Biology of Social Insects,* M. D. Breed, C. D. Michener, and H. E. Evans (eds.), Proceedings of the 9th International Union for the Study of Social Insects. Westview Press, Boulder, Colo., pp. 138-141.

Maudlin, J. K. 1983. The Forest Service research program on protection of wood in use. *Pest Manage.* **2**(1):14, 16, 19.

Maudlin, J. K., and N. M. Rich. 1980. Effect of chlortetracycline and other antibiotics on protozoan numbers in the eastern subterranean termite. *J. Econ. Entomol.* **73**(1):123-128.

Moore, H. B. [1979]. *Wood-Inhabiting Insects in Houses: Their Identification, Biology, Prevention, and Control.* U.S. Department of Agriculture, Forest Service, Department of Housing and Urban Development (Interagency Agreement IAA-25-75).

Moore, H. B. 1983. Controlling insects in furniture provides problems. *Pest Control* **51**(9):44, 46.

Moore, H. B., and M. I. Haverty. 1979. Insects injurious to unfinished and finished wood products. In *Forest Insect Survey and Control,* J. A. Rudinsky (ed.). Oregon State University Book Stores, Corvallis, pp. 263-352.

Moreland, D. 1982. A dog's life. *Pest Control. Technol.* **10**(4):44-46, 50.

Moreland, D. 1983*a*. Massachusetts passes final termiticide regs: PCO's relieved. *Pest Control Technol.* **11**(8):68-69.

Moreland, D. 1983*b*. Congress considers pesticide legislation. *Pest Control Technol.* **11**(11):65-69.

National Academy of Sciences. 1982. *An Assessment of the Health Risks of Seven Pesticides Used for Termite Control.* Committee on Toxicology. Board on Toxicology and Environmental Health Hazards, Commission on Life Sciences, Washington, D.C.

Preston, A. F., F. H. Erbisch, K. R. Kramm, and A. E. Lund. 1982. Developments in the use of biological control for wood preservation. *Proc. Am. Wood-Preserv. Assoc.* **78:**53-61.

Prestwich, G. D., J. K. Maudlin, J. B. Engstrom, J. F. Carvalho, and D. Y. Cupo. 1983. Comparative toxicity of fluorinated lipids and their evaluation as bait-block toxicants for the control of *Reticulitermes* spp. (Isoptera: Rhintotermitidae). *J. Econ. Entomol.* **76**(4):690-695.

Rambo, G. W., ed. 1980. *Approved Reference Procedures for Subterranean Termite Control.* National Pest Control Association, Vienna, Va.

Smith, V. K. 1982. Biology of subterranean termites and their control, 2, Termite control methods. *Pest Manage.* **1**(10):6-9.

Smith, V. K., R. H. Beal, and H. R. Johnston. 1972. Twenty-seven years of termite control tests. *Pest Control* **40**(6):28, 42, 44.

Snetsinger, R. 1983. *The Ratcatcher's Child: The History of the Pest Control Industry.* Franzak and Foster, Cleveland, Ohio.

Truman, L. C., G. W. Bennett, and W. L. Butts. 1976. *Scientific Guide to Pest Control Operations,* 3rd ed. Purdue University/Harvest, Cleveland, Ohio, 276p.

U.S. Environmental Protection Agency. 1983. *Analysis of the Risks and Benefits of Seven Chemicals Used for Subterranean Termite Control,* Report No. EPA-540/9-83-005. Office of Pesticides and Toxic Substances, 70p.

Verrall, A. F., and T. L. Amburgey. [1979]. *Prevention and Control of Decay in Homes.* U.S. Department of Agriculture, Forest Service, Department of Housing and Urban Development (Interagency Agreement IAA-25-75), 148p.

Vongkaluang, C., H. B. Moore, and M. H. Farrier. 1982. Mortality and activity of first-instar larvae of the old house borer, *Hylotrupes bajulus* (L.) (Coleoptera: Cerambycidae), at low wood moisture. *Mater. u. Organ.* **17**(3):233-240.

Weidner, T. 1983. Spear termiticide: A new era in biological warfare. *Pest Control Technol.* **11**(1):72, 76, 79, 80.

Williams, L. H. 1983*a.* Wood moisture levels affect *Xyletinus peltatus* infestations. *Environ. Entomol.* **12**(1):135-140.

Williams, L. H. 1983*b.* Integrated protection against structural wood-infesting pests, part 1. *Pest Manage.* **2**(1):12-14.

Williams, L. H. 1983*c.* Integrated protection against structural wood-infesting pests: Summary and discussion needs for improving integrated wood protection efforts. *Pest Manage.* **2**(9):25-27.

Williams, L. H. 1984. Dip-diffusion treatment of unseasoned hardwood lumber with boron compounds for prevention of beetles in lumber and other products. In *Pathways to Increased Cost Effectiveness and Management and Utilization of Eastern Hardwoods,* Proceedings of the 12th Annual Hardwood Symposium of the Hardwood Research Council, May 8-11, Cashiers, N.C. Hardwood Research Council, Cashiers, N.C., pp. 154-163.

Williams, L. H., and V. K. Smith. 1983. Integrated protection against structural wood-infesting pests, part 7, Homeowner responsibilities for continued wood protection. *Pest Manage.* **2**(7):24-26.

14

Managing Household Pests

Michael K. Rust
University of California, Riverside

Of the millions of insects and arthropods that inhabit the Earth, only a few have successfully adapted to and exploited human environments, accompanying food storage, and forms of transportation over the last several thousand years. Some species, such as the Argentine ant, confused flour beetle, German cockroach, granary weevil, and Pharaoh ant, lost their ability to fly and require human assistance for broad dispersal. Many have adapted to surviving in our artificially manipulated environments, feeding on human processed foods, and utilizing synthetic fibers and fabrics. An increasing number of species have evolved physiological mechanisms to resist pollutants and toxic chemicals, including pesticides. It is this remarkable plasticity that makes these arthropods so unique and such worthy adversaries.

Surveys of homeowners, entomologists, and professional pest control operators indicate that ants, cockroaches, fleas, termites, stored product pests, and fabric pests are the major categories of indoor pests (Table 14-1) in the United States (Rachesky, 1978; National Research Council, 1980; Fowler, 1983). Responses to such surveys are partly influenced by the public's experiences and attitudes (Byrne et al., 1984; Frankie and Levenson, 1978; Levenson and Frankie, 1981; Wood et al., 1981), whereas some arthropod pests have significant economic, medical, or veterinary importance in certain regions of the United States that also influence surveys. Some examples of pests that take on regional economic importance are carpenter ants in the northeast and northwest, dampwood termites in the northwest, drywood

Table 14-1. Important arthropod pests in the United States and a selected bibliography of various pest management strategies other than conventional pesticide applications available for pest control.

Pest groups/pests	Geographical regions[a]	Within structure	Available strategies	References
Ants				Ebeling, 1975; Eckert and Mallis, 1937; Mallis, 1960; Marlatt, 1930; Nuhn and Wright, 1979; Sheltar and Walter, 1982; Thompson, 1983; Truman, Bennett, and Butts, 1976
Argentine	SE, Ca	Nest outdoors	Baits	Beatson, 1968; Eckert and Mallis, 1937; Edwards and Clarke, 1978; Hagmann, 1982; Mallis, 1960; Marlatt, 1907; Newton, 1980; Wright and Stout, 1978
Crazy ant	SE, NE	Nest outdoors; indoors in northeast		
Fire ants	SE, SW	Nest outdoors	Nest treatment	Eckert and Mallis, 1937; Marlatt, 1907; Sheltar and Walter, 1982; Thompson, 1983; Truman, Bennett, and Butts, 1976; Wagner, 1983
Little black	NE, ME	Nest outdoors; indoors in masonry or woodwork		
Odorous house	Cos	Nest outdoors; indoors in wall voids by hot water pipes and heaters, planter boxes	Exclusion	Eckert and Mallis, 1937; Thompson, 1983
			IGRs	Edwards and Clarke, 1978; Newton, 1980; Wilson and Booth, 1981
Pavement	NE, MW, Ca	Nest outdoors; indoors at lower masonry walls		
Pharoah	SE, Cos	Nest outdoors; indoors, wall voids, subfloor, attic, etc.		
Thief	NE, MW, Cos	Nest outdoors; indoors in wall voids, crawl space, masonry		

Cockroaches				Bennett, 1977; Cochran, 1982*b;* Cornwell, 1968, 1976; Ebeling, 1975; Frishman, 1982; Mallis, 1960; Truman, Bennett, and Butts, 1976
American	SE, SW, Cos	Outdoors, sewers, vegetation; indoors, basements, food storage and preparation areas	Baits	Barson, 1982; Barson and Lole, 1981; Bennett, Runstrom, and Bertholf, 1984; D'Agnese, 1984; Farmer and Robinson, 1984; Frishman, 1982; Gupta et al., 1973; Mallis, 1960; Quattrochi, 1968; Reierson and Rust, 1984; Reierson et al., 1983; Rust and Reierson, 1981*b;* Truman, Bennett, and Butts, 1976
Brownbanded	Cos	Indoors, warm areas, widely dispersed, animal rearing facilities		
German	Cos	Indoors in kitchen, bathroom, food preparation and storage areas	Built-in	Cornwell, 1968; Ebeling, 1971, 1975, 1978; Ebeling and Wagner, 1959; Ebeling, Wagner, and Reierson, 1967, 1977; Gupta et al., 1973; Moore, 1972, 1977; Piper and Frankie, 1978; Piper et al., 1975; Slater et al., 1979; Tarshis, 1961, 1964*b*
Oriental	NE, SW, Cos	Indoors in northeast, damp and cool basements and cellars; outdoors in southeast, southwest in vegetation, wood piles, sewers, garages	Exclusion and habitat modification	Farmer and Robinson, 1984; Moore, 1973; Piper and Frankie, 1978; Piper et al., 1975
			Sanitation	Gupta et al., 1973; Owens, 1980; Piper et al., 1975
			Trapping	Barak, Shinkle, and Burkholder, 1977; Bell et al., 1984; Frishman, 1982; Piper and Frankie, 1978; Piper et al., 1975
Smokybrown	SE, SW	Outdoors, vegetation, wood piles, crawl spaces, garages	Biological control	Coler, van Driesche, and Elkinton, 1984; Fleet and Frankie, 1975; Frishman, 1982; Piper and Frankie, 1978; Slater, Hurlbert, and Lewis, 1980
			IGRs	Bennett, Runstrom, and Bertholf, 1984; Runstrom and Bennett, 1984

[a]Cos = cosmopolitan; Ca = California; ME = Mideast; MW = Midwest; NE = Northeast; SE = Southeast; SW = Southwest

(continued)

Table 14-1. *(continued)*

Pest groups/pests	*Geographical regions*[a]	*Within structure*	*Available strategies*	*References*
Fabric pests				Ebeling, 1975; Linsley, 1946; Mallis, 1941, 1960, 1982*a*; Mallis and Carr, 1982; Truman, Bennett, and Butts, 1976
Carpet beetles	Cos	Adults outdoors around flowers, larvae in insect, bird, and rodent nests; stored foods, furs, woolen items	Exclusion, habitat modification, sanitation	Ebeling, 1975; Linsley, 1946; Mallis, 1982*a;* Mallis and Carr, 1982
			Cleaning and protective storage	Ebeling, 1975; Linsley, 1946; Mallis, 1982*b*; Mallis and Carr, 1982
Cothes moth	Cos	Adults prefer dark areas; woolen items, furs, feathers	Trapping	Ebeling, 1975; Ebeling and Reierson, 1974; Mallis, 1941; Mallis and Carr, 1982
Silverfish and firebrats	Cos	Linen, paper, starches, books		
Spiders				Ebeling, 1975; Mallis, 1960; Truman, Bennett, and Butts, 1976
Black widow	SE, SW	Outdoors, vegetation, wood piles, clutter, garages, storage areas	Habitat modification	Ebeling, 1975; Truman, Bennett, and Butts, 1976
			Sanitation	Ebeling, 1975; Mallis, 1960; Truman, Bennett, and Butts, 1976
Brown recluse	MW, SE	Indoors, rags, clutter, storage	Insect control	Ebeling, 1975

Fleas				Bledsoe, Fadak, and Bledsoe, 1982; Ebeling, 1975; Mallis, 1960; Truman, Bennett, and Butts, 1976
Cat and dog	Cos	Indoors, animal bedding, carpets, rodent or oppossum nests	Vacuuming, cleaning	Bledsoe, Fadak, and Bledsoe, 1982; Ebeling, 1975; Osbrink et al., 1983; Truman, Bennett, and Butts, 1976
Human and oriental rat	Cos	Outdoors, in crawl spaces around livestock, associated with rats and rodents	Treatment of pet	Bledsoe, Fadak, and Bledsoe, 1982; Tarshis, 1961
			IGRs	Bledsoe, Fadak, and Bledsoe, 1982; Chamberlain, 1979, 1983
Mites and ticks				Ebeling, 1975; Green, 1982; Mallis, 1960; Truman, Bennett, and Butts, 1976
Tropical rat mite house dust mite brown dog tick	Cos	Indoors, found in association with pets, rodents, birds; attics, crawl spaces; house dust and debris, mattresses	Repellents	Truman, Bennett, and Butts, 1976
			Sanitation, cleaning	Ebeling, 1960; Truman, Bennett, and Butts, 1976
			Built-in	Ebeling, 1960, 1971; Tarshis, 1961, 1964*a;* Tarshis and Dunn, 1959

[a]Cos = cosmopolitan; Ca = California; ME = Mideast; MW = Midwest; NE = Northeast; SE = Southeast; SW = Southwest

termites in the southwest, and smokybrown cockroaches throughout the Gulf states. Approximately, 71% of Arizona residents surveyed reported seeing crickets, and 73% perceived them as pests (Byrne et al., 1984). Northern California residents listed ants, fleas, flies, cockroaches, and moths, whereas Dallas residents listed cockroaches, termites, earwigs, fleas, and ants as their major indoor pests (National Research Council, 1980). New Jersey extension entomologists ranked carpenter ants at the head of their list, followed by subterranean termites, wasps, ants, and cockroaches (Fowler, 1983). Responses must be cautiously interpreted because homeowners are rarely able to properly identify problem arthropods to the genus or species level.

The objectives of this chapter are to discuss some of the broad concepts of indoor household pest management and their incorporation into practical programs. Only after the arthropod has been identified and necessary inspections are completed can implementation of a pest management program proceed. Too often the homeowner or professional pest control operator immediately resorts to chemical aerosols or sprays without considering supplemental or alternative nonchemical procedures. It would be impossible in this chapter to discuss all of the pests, conventional chemical controls, and alternative nonchemical methods that might be successfully used against them. Table 14-1 lists some of the important household pests, alternative pest management strategies, and insecticides that have been successful under field conditions (for stored product pests and wood-destroying organisms see chaps. 12 and 13). Certain less conventional measures, including the use of sorptive dusts, boric acid, baits, insect growth regulators, and pyrethroids that have proven extremely successful against a wide spectrum of pests, will be discussed in some detail in the case studies and as areas for future research.

PESTICIDE USE PATTERNS

Some of the earliest records of indoor pest control mention the use of baits that incorporated phosphorus (Cheng and Campbell, 1940; Cowan, 1865), but not until the turn of the century did inorganic insecticides, namely boric acid and the arsenical and fluorine salts, including sodium fluoride, arsenic pentoxide, paris green, and so on, become popular (Cornwell, 1976; Mallis, 1960). Beginning in the early 1900s and lasting through the 1940s pyrethrum powder was widely used (Cornwell, 1976; Mallis, 1960). The advent of synthetic organic pesticides in the late 1940s, especially DDT and chlordane, signaled a new era in urban pest control. The use of organochlorine compounds, such as DDT, chlordane, lindane, dechlorane, and chlordecone, in baits, sprays, and dusts provided control strategies and control never before available. Even before many of their registrations were revoked, the use of organophosphate and carbamate compounds were gaining wide acceptance. Physiological insecticide resistance, increased public concern of insecticides, and a need for very localized treatments in sensitive areas such as computer facilities, animal rearing rooms, medical laboratories, and health care facilities have stimulated a renewed interest in alternative pest control technology. Consequently, there has been an increased emphasis (particularly by some progressive pest control operators) on integrated pest management utilizing insect growth regulators, biological control agents, new classes of insecticides, and inorganic chemicals. For example, baits incorporating boric acid have provided excellent control against cockroaches and ants (Bare, 1945; Barson, 1982; Wright and Stout, 1978). Results with the photostable and persistent pyrethroids such as cyfluthrin, cypermethrin, and permethrin against cockroaches look extremely promising, and additional compounds will be forthcoming. Recent advancements in pesticide chemistry and an acute need for effective fire ant control have yielded active ingredients belonging to new classes of insecticides (delayed-action toxicants), the amidinohydrazones, avermectins, and formamidines.

The kinds of insecticide, their use patterns, and factors associated with their use in urban settings in Nebraska (Kamble, Gold, and Parkhurst, 1982; Kamble, Gold and Vitzhum, 1978) and Indiana (Bennett, Runstrom, and Wieland, 1983) have been investigated in extensive surveys. In addition, several other extensive studies (Byrne et al., 1984; Frankie and Levenson, 1978; Levenson and Frankie, 1981) have examined some of the psychological, educational, and sociological factors influencing the public's use and perception of pesticides. The results of such studies have tremendous importance for agencies charged with educating the public and the pest control industry. For example, failure by state agencies in California to provide the public accurate information nearly spelled disaster for the Mediterranean fruit fly eradication program several years ago in the San Francisco area.

AVAILABLE NONPESTICIDAL PROCEDURES

Noninsecticidal pest management procedures exist and have been used for years to control or mitigate numerous urban pests (Table 14-1), especially those pests of stored products and wood (see chaps. 12 and 13) and so-called occasional invaders (Ebeling, 1975; Mallis, 1960). Until the advent of many long-lasting residual insecticides in the early 1950s, these nonchemical approaches were the only means of effective pest elimination for many species. Habitat modification around the perimeter of structures such as the removal of certain plants, ground cover, and debris and clutter prevents breeding and build-up of nuisance pests such as spiders, millipedes, earwigs, pillbugs, crickets, and outdoor species of cockroaches (Ebeling, 1975; Mallis, 1960). Fly and mosquito control often involves community cooperation of water, sanitation, street and landscape maintenance departments, plus residents and industry (Ebeling, 1975, 1978; Mallis, 1960; Truman, Bennett, and Butts, 1976). Structural modifications, such as the removal of lights near entrances or screening windows and doors, limit access to flying insects. Sanitation and storage of fabrics, foods, and papers in pest-proof containers help eliminate food sources of insect and rodent pests. In general, these recommendations are designed to eliminate or deny the pest access to food, water, and shelter.

The development of biological control agents for indoor pest management has progressed slowly. The lack of known biological control agents against the major indoor pests may be partly explained because so many of these pests were introduced into the Americas. For example, all the important cockroach pests such as the German cockroach, *Blattella germanica;* the American cockroach, *Periplaneta americana;* the oriental cockroach, *Blatta orientalis;* and the brownbanded cockroach, *Supella longipalpa* were introduced (Rehn, 1945). The American cockroach probably first entered the New World as early as the 1500s, whereas the brownbanded cockroach was a more recent arrival, first appearing in Florida in 1903. Most of the species of flea belonging to the genus *Ctenocephalides,* including the cat and dog flea, are endemic to the Ethiopian region (Hopkins and Rothschild, 1953). Some of the important ant pests introduced into the United States include the Argentine ant, *Iridomyrmex humilis;* crazy ant, *Paratrechina longicornis;* red and black imported fire ants, *Solenopsis invicta* and *S. richteri;* pavement ant, *Tetramorium caespitum;* and Pharaoh ant, *Monomorium pharaonis* (Nuhn and Wright, 1979; Smith, 1965). Pantry pests such as the confused and red flour beetles, *Tribolium confusum* and *T. castaneum;* khapra beetle, *Trogoderma granarium;* granary weevil, *Sitophilus granarius;* and the lesser grain borer, *Rhyzopertha dominica* were probably introduced into the United States in commodities (Ebeling, 1975). Foreign exploration to areas where the major urban pests are endemic needs to be conducted.

There have been some limited successes against the cockroach species that deposit the oothecae. Multiple releases of a parasitic wasp, *Comperia merceti,* that attacks the egg capsules of the brownbanded cockroach, *S. longipalpa,* have shown some promise for control in large buildings (Slater, Hurlbert, and Lewis, 1980). The effectiveness of *C. merceti* may be dependent upon high cockroach densities (Coler, van Driesche, and Elkinton, 1984), and continued release of additional parasites may be necessary at low cockroach densities (Coler, van Driesche, and Elkinton, 1984; Slater, Hurlbert, and Lewis, 1980). Somewhat more ecological and behavioral information is known about the wasp, *Tetrastichus hagenowii,* which parasitizes oothecae of the *Periplaneta* species (Fleet and Frankie, 1975; Piper, Frankie, and Loehr, 1978). Parasitization rates, determined in field sites in Texas and Louisiana, ranged from 0-68% (Piper, Frankie, and Loehr, 1978). Studies suggest that conventional perimeter sprays used for cockroach control decreased the parasite efficacy (Piper and Frankie, 1978). Important correlations between parasitism rates and control and the integration of the parasites with existing control strategies are still lacking.

IMPLEMENTATION OF PEST MANAGEMENT

Identification

The most important initial factor in implementing any pest management program is pest identification. Proper species identification allows one to determine if a problem does exist, the nature of the problem, and the necessary corrective measures. Unfortunately, pest identification is usually made to the categorical level such as ant, beetle, cockroach, or termite. For example, species identification of ants is critical because there may be as many as 50 species commonly reported as pests throughout the United States (Smith, 1965). Strong and Okumura (Strong and Okumura, 1958) listed 135 different pests found in surveys of foods and seeds stored in California. Even though only 4 or 5 cockroach species are frequently encountered by the public, there are abut 26 species that have been reported as potential pests (Cochran, 1982*b;* Cornwell, 1968; Fishman, 1982; Truman, Bennett, and Butts, 1976; Wright, 1965). Arizona, California, Florida, and Texas have at least 10 different termite species that will attack wood (Weesner, 1965).

During the last 10 years in Southern California, some of the most common misidentifications I have encountered have dealt with ants, cockroaches, and beetles found in stored products. These misidentifications often resulted in application of unnecessary insecticides. For example, immature stages of oriental, smokybrown, and American cockroaches were misidentified and not recognized as primarily outdoor pests, resulting in unnecessary applications of insecticide indoors. The outdoor harborages were neglected resulting in poor control and repeated complaints by residents. The odorous house ant was mistakenly identified as the Argentine ant, and potential indoor harborages were never inspected or treated. In the desert southwest, the subterranean termite *Heterotermes aureus*, often misidentified as *Reticulitermes tibialis,* attacks hardwoods and aerial portions of the structure and seems to be tolerant of dry conditions. Infestations may be extremely difficult to control with conventional soil treatments, especially in adobe structures. Failure to properly identify dermestids frequently resulted in applications of malathion, which does not provide good control of the warehouse beetle, *Trogoderma variabile.* Numerous other examples could be cited, illustrating the point that proper pest identification is the first critical step in pest management.

Monitoring and Sampling Techniques

Pest problems generally go unnoticed until populations increase to large numbers and there is evidence of feeding damage or contamination, foraging, or migration of various life stages, or the emergence of highly visible and active adults. For most indoor pest species, reliable survey techniques other than visual inspections do not exist. A knowledge of each pest's habits is the critical element of any inspection. If these habits are not well known to the pest managers, accurate identification to species will be critically important before information on pest habits can be obtained from reference literature. For example, adult clothes moths are not attracted to lights and gravid females are poor flyers so that they are restricted to the site of infestation. The brown dog tick prefers to climb vertical surfaces and is frequently found behind pictures, along window sills, or in the cracks of the roof of the dog house. Odorous house ants and carpenter ants may nest indoors in wall voids, insulation material, hollow doors, and potted plants. Experience with and knowledge of pest behavior will greatly aid pest managers in developing adequate inspection, sampling, and monitoring procedures for many household pests. Some simple monitoring techniques exist for cockroaches and fleas. Most survey techniques for stored product pests have been developed for use in warehouses and factories and are discussed elsewhere (chap. 12).

Probably more work has been done on surveying cockroaches than with any other indoor urban pest. Trapping and visual inspections with the aid of a flushing agent or a flashlight have been used routinely to sample populations for insecticidal efficacy tests. Comparative studies evaluating each technique in apartments infested with German (Baker and Southam, 1977; Owens, 1980; Owens and Bennett, 1983; Reierson and Rust, 1977) and oriental (Baker and Southam, 1977) cockroaches showed that some methods of trapping were not disruptive and provided a more precise relative estimate of the population size than visual or flushing techniques. Each method, however, has certain limitations based on the construction of the structure, cockroach species, economics of the program, and objectives of the survey. For example, visual counts may be extremely time consuming and difficult when there is excessive clutter and debris. Extensive trapping with commercial traps may be too expensive. Flushing with pyrethrins may be disruptive, scattering cockroaches into uninfested areas.

Even though some authors have reported trapping to be an effective means of control (Barak, Shinkle, and Burkholder, 1977; Piper and Frankie, 1978; Piper et al., 1975), most studies show that trapping alone does not provide control but is reliable for monitoring (Barak, Shinkle, and Burkholder, 1977; Ebeling and Reierson, 1974; Owens, 1980; Reierson and Rust, 1977). Comparative studies (Moore and Granovsky, 1983; Owens, 1980) clearly revealed some of the limitations of trapping. For example, commercial sticky traps caught more cockroaches than jar traps but were generally biased toward sampling adults (Owens, 1980; Owens and Bennett, 1983). Several studies show that purported attractants (mentioned with bait attractants in the following section) used in some traps are of limited value, especially against German cockroaches (Ballard and Gold, 1982; Ebeling and Reierson, 1974; Moore and Granovsky, 1983; Reierson and Rust, 1977; Wileyto and Boush, 1983). The availability of food and water had significant effects on trap catch (Reierson and Rust, 1977; Ross, 1981). Trap placement is also a critical factor because moving traps from wall intersections will subsequently reduce trap counts (Ebeling, Wagner, and Reierson, 1966).

The use of flushing agents such as pyrethrins or resmethrin at low concentrations does not generally provide control. However, concentrated pyrethrins provided limited control of German cockroaches (Reierson and Rust, 1977). One negative aspect of inspecting with pyrethrin sprays is the movement of cockroaches away from the treated areas (Ebeling, Wagner, and Reierson, 1966; Owens, 1980; Reierson and Rust, 1977). In addition, flushing agents mixed with residual sprays provided no additional benefit and may increase the repellency of the residual treatment, reducing its efficacy (Ebeling and Reierson, 1973; Ebeling, Wagner, and Reierson, 1966).

Even though considerable emphasis has been focused on surveys of fleas associated with plague, little attention has been devoted to surveys for cat and dog fleas in and around structures. Studies in residences infested with fleas (Bennett and Lund, 1977; Bennett and Robertson, 1978; Bennett and Runstrom, 1980) have generally relied on the "white-sock" inspection technique in which the number of fleas attracted to white knee-socks worn by the inspector is counted. Flea infestations have also been measured by brushing the pet with a fine-toothed comb and counting fleas (Skovmand and Christensen, 1980). Both techniques have severe limitations, since cat fleas are not attracted to white objects (Osbrink and Rust, 1985) and grooming pets is extremely time consuming. The use of a vacuum cleaner with a cloth trap inserted into the hose to collect the adult fleas is the most effective and efficient technique for surveying adult fleas from carpets and furniture (Osbrink et al., 1983; Skovmand and Christensen, 1980). Vacuuming consistently provides higher adult counts than do visual counts, especially with low flea population densities.

Selection of Pest Management Program

The selection of a pest management program and its implementation requires the integration of information concerning the pest, the problem, the non-chemical and chemical strategies available, the site or areas to be covered by the procedure, and its cost effectiveness. Each of these factors is interdependent and may drastically influence the selection of a pest management program. Unfortunately, there are no simple answers or approaches to selecting the appropriate solution, and each pest management procedure must be tailored to the specific problem. The biology or behavior of the arthropod pest may present some unusual problem. For example, the tendency of the Pharaoh ant to bud and fragment colonies prohibits the use of repellent inorganic dusts or sprays. Physiological insecticide resistance in houseflies or German cockroaches may limit the spectrum of effective chemical treatments. Sometimes only one specific stage of an insect's life cycle is responsible for damage or injury. For example, the larval stages of the carpet beetles attack and damage various fabrics and items made from animal fur and feathers. Exclusion of the adult carpet beetle is extremely difficult, and continued surveillance and protection against the larvae is necessary. Another consideration for pest management procedures may include strict requirements in animal rearing facilities or computer rooms that prohibit the use of conventional organophosphate and carbamate insecticides or some non-chemical procedure. Consideration must be given to problems that might arise when small children or exotic pets are present. Sometimes food-handling establishments require that all control procedures be conducted in the late evening or early morning. It is this array of factors to be considered in designing programs that makes this phase of pest management so challenging.

Surveillance and Follow-up of Pest Management Program

Applications of insecticide or the implementation of some control measure are not the end point but just the beginning of a good pest management program. Frequent inspections and monitoring with traps are just as essential to good pest management as is the selection of the proper treatment. It has been my experience that failures in drywood and subterranean termite treatments, with inorganic dusts in built-in pest control programs, and general pest control are often a result of inadequate post-treatment surveillance. It is a critical step that must be included in any good program.

CASE STUDIES

The objectives of the case studies are to outline some general problem areas and provide some specific procedures to supplement conventional sprays and aerosols most frequently used in control programs. These studies are not meant to be all inclusive but examples of alternative strategies that have been shown to benefit existing programs.

Case Study 1

Arthropod pests have access to concealed and secluded areas such as attics, crawl spaces, wall and cabinet voids, appliances, and drop ceilings which are not likely to be accessible to conventional sprays and aerosols.

One of the basic tenets of urban pest management is that all pests require food, water, and shelter. If one or more of these essential requirements can be eliminated, control or even eradication is possible. The pest's access to harborage can be eliminated and pests excluded with the use of sealants, patching materials, and hardware cloth. Harborage can be removed by structural improvements, repairs, and removal of clutter and debris. However, sealants (Farmer and Robinson, 1984) and cleaning (Owens, 1980) alone will not provide improved German cockroach control without other supplemental measures. Besides removal, another effective means of reducing suitable harborage is the use of sorptive dusts such as Dri-Die (95% amorphous silica gel and 2% ammonium fluosilicate) and Drione (38.12% amorphous silica gel, 1.88% ammonium fluosilicate, 1% pyrethrins, 10% piperonyl butoxide).

Sorptive dusts are defined as finely divided powders capable of absorbing waxes and other lipids. When insects and some other arthropods contact these powders, the wax layers coating the cuticle are removed or disrupted, resulting in rapid dehydration (Ebeling, 1971, 1975). The mode of action of sorptive dusts and the physiological mechanisms of the insects responsible for the water loss have been thoroughly reviewed (Ebeling, 1971; Ebeling et al., 1975). Some effective dessicating dusts include the silica aerogels and certain montmorillonite and attapulgite clays. When five insecticidal sprays and dusts were aged for 17 months and tested against drywood termites, the sprayed surfaces failed to produce mortality, and only the clay diluents (about 95% by weight) in the dust formulations provided kill (Ebeling and Wagner, 1959). Some of the advantages of fluorinated silica aerogels (Dri-Die and Drione) are their low absorptivity of water, a positive electrostatic charge (causing them to cling to surfaces), the persistent insecticidal activity, no hazard of silicosis to vertebrates, no or extremely low vertebrate toxicity for those containing pyrethrins, and the lack of physiological resistance mechanisms in target organisms. One disadvantage of Dri-Die and other light amorphous dusts is that they can be difficult to apply and are somewhat messy because of their lightness or low bulk density. In living spaces heavier dusts such as Drione may be more suitable.

Another positive feature of sorptive dusts is their broad spectrum of activity against insects and other arthropods (Ebeling, 1971, 1975; Tarshis, 1961, 1964*b*). Silica aerogels have been shown to be effective against some 30 species of arthropods that attack man, including ectoparasites such as lice, fleas, and mites (Tarshis, 1961). Table 14-2 lists the pests and some of the areas in structures that have been effectively treated.

Sorptive dusts are highly repellent, and insects (especially German cockroaches and ants) will avoid treated areas (Ebeling, 1971; Ebeling, Reierson, and Wagner, 1967, 1968*b;* Ebeling, Wagner, and Reierson, 1966). These materials are ideal preventive treatments for use at the time of construction, often referred to as "built-in pest control" (Ebeling, 1978; Ebeling and Wagner, 1964; Ebeling, Wagner, and Reierson, 1965; Moore, 1973, 1977), before pest populations become established. The elimination of inaccessible spaces, cabinet and wall voids, and other areas likely to serve as harborages can be a positive supplement to other pest management procedures. Cockroaches, silverfish, mites, and other pests are more accessible to conventional sprays and baits when forced from preferred harborages. The use of inorganic dusts has been shown to improve the performance of sprays and baits against German cockroaches (Gupta et al., 1973; Moore, 1973; Slater et al., 1979).

Case Study 2

Cockroaches and silverfish have established heavy infestations in existing multi-unit housing complexes, restaurants, hospitals, and other structures.

Once heavy infestations of cockroaches or silverfish are established indoors, a nonrepellent boric acid dust may prove more effective than sorptive dusts in eliminating potential harborages (Ebeling, Reierson, and Wagner, 1968*b;* Ebeling, Wagner, and Reierson, 1977). One of the unique features of a technical boric acid deposit is its lack or repellency (Ebeling, Reierson, and Wagner, 1967; Ebeling, Wagner, and Reierson, 1977). Consequently, formulated boric acid applied in cabinet or wall voids, attic spaces, and beneath appliances has the potential to kill indefinitely cockroaches that enter the treated areas. Numerous field studies have shown the positive efficacy of such treatments when properly applied, especially against German cockroaches (Ebeling, Reierson, and Wagner, 1968*a,* 1968*b;* Gupta and Das, 1976; Moore, 1972; Wright and Hillman, 1973). Boric acid should be applied in thin films with a powder duster and not in piles or clumped deposits. The number of complaints and retreatments in student housing decreased by 66% when harborage areas were thoroughly treated with boric acid and aqueous sprays were applied in highly visible areas that were inappropriate for dusts (Slater et al., 1979). Numerous studies indicate that combination treatments of boric acid and pyrethroids, each where they would be most effective, provide excellent initial kill and long residual control (e.g., Ebeling, Reierson, and Wagner, 1968*a,* 1968*b;* Gupta and Das, 1976).

Table 14-2. Areas and materials in structures and animals treated with desiccants such as Dri-Die and Drione to provide control of indoor arthropod pests.

Treatment site	*Pest species*	*References*
Attics, false ceilings	Drywood termite, tropical rat and northern fowl mite, cockroaches, firebrats, spiders, and silverfish	Ebeling, 1960, 1971; Ebeling and Wagner, 1959, 1964; Ebeling, Wagner, and Reierson, 1965, 1977; Moore, 1972, 1973, 1977; Tarshis, 1961, 1964*a*, 1964*b*
Sub-floor areas, beneath porches, garages	Cockroaches, fleas, crickets, mites, firebrats, spiders, and silverfish	Ebeling, 1971; Ebeling and Wagner, 1959; Ebeling, Wagner, and Reierson, 1977; Moore, 1973, 1977; Tarshis, 1961
Under appliances, in cabinet and wall voids	Cockroaches, firebrats, pantry pests, silverfish	Ebeling, 1971; Moore, 1973, 1977; Tarshis, 1961
Fabrics, skins, carpets, furniture	Fleas, dermestids, brown dog tick	Ebeling, 1971; Tarshis, 1961
Pets, pet's bedding and housing	Fleas, ticks, lice, mites, brown dog tick	Ebeling, 1960, 1971; Tarshis, 1961; Tarshis and Dunn, 1959

There is little information available concerning the physiological mode of action of borax and boric acid. These inorganic compounds have been used since the early 1900s against ants and cockroaches. Boric acid has been shown to be an orally introduced toxicant (Bare, 1945; Ebeling et al., 1975). In addition, boric acid penetrates the cuticle and is not an absorptive or abrasive dust (Ebeling, et al., 1975). No known instances of physiological resistance have been reported. Added small amounts of adjuvants such as magnesium stearate, tricalcium phosphate, and Cab-O-Sil to prevent the formation of agglomerates and improve flow ability can also increase the speed of acute toxicity (Ebeling, Reierson, and Wagner, 1967). Even small amounts of some adjuvants or flushing agents, however, increased the repellency of the dust deposit, reducing the residual efficacy against German cockroaches (Ebeling and Reierson, 1973; Ebeling, Reierson, and Wagner, 1967; Moore, 1972).

Another inorganic dust, sodium fluoride, was extensively used in the 1930s and 1940s. It also acts as a contact and oral poison (Griffiths and Tauber, 1943) but is considerably more toxic than boric acid to mammals. Even though sodium fluoride provided faster knockdown of German cockroaches (50% kill in 51 min) than did boric acid (50% kill in 1,140 min), sodium fluoride was more repellent in choice tests (Ebeling, Reierson, and Wagner, 1967, 1968*b*). Field studies indicated that sodium fluoride was not as effective as boric acid (Ebeling, Reierson, and Wagner, 1968*a*).

Case Study 3

Cockroaches, silverfish, firebrats, and certain ants infect sensitive areas of the structure or facilities where insecticide odors, deposits, or contamination are not acceptable. Animal-rearing facilities, hospitals and patient care facilities, computer buildings, zoos, and food preparation areas are just some of the possible areas.

Insecticidal bait treatments offer several theoretical and practical advantages over spray formulations. These include generally longer residual activity, safe application in sensitive areas, and easy incorporation into existing control programs, which may include sprays or other nonchemical procedures (Cornwell, 1976; Quattrochi, 1968). Additionally, compounds with poor contact activity (minimal cuticular penetration) and high dosage requirements may be used. Another attractive feature of some of the amidinohydrazone and avermectin baits is their effectiveness against insect populations resistant to sprays containing organophosphates and carbamates. Baits have failed to gain popularity because they contained ineffective toxicants, required a longer time to kill than sprays, had to compete with existing food sources, had to be consumed by the entire pest population, and required additional time to be concealed by the pest control technician (Quattrochi, 1968).

Table 14-3 lists the active ingredients incorporated into baits used in laboratory or field tests against some of the important indoor insect pests. The majority of the research literature on household baits deals with their use for German cockroach control. The most notable successes in field tests against the German cockroach include baits with amidinohydrazone (Bennett, Runstrom, and Bertholf, 1984; D'Agnese, 1984; Reierson and Rust, 1984; Reierson et al., 1983), boric acid (Barson and Lole, 1981; Tsuji and Ono, 1970), chlordecone (Tyler, 1964) and chlorpyrifos (Bennett, Runstrom, and Bertholf, 1984). Even though limited tests have been conducted against other species such as the American, brownbanded, and oriental cockroach, similar results have been obtained. Baits containing amidinohydrazone (R. E. Wagner, pers. comm.), borax (Marlatt, 1907), boric acid (Wright and Stout, 1978), chlordecone (Beatson, 1968; Burden and Smittle, 1975), or the insect growth regulator (IGR) methoprene (Edwards and Clarke, 1978; Wilson and Booth, 1981) have provided satisfactory control of Pharaoh ants. Baits containing borax (Hagmann, 1982) have been successful against pavement ant, little black ant, and thief ant. Liquid baits containing sodium arsenite (Newell, 1909), thallium sulfate (Marlatt, 1930), and boric acid have been successfully used for years against Argentine ants. Amidinohydrazone provided excellent control against a wide spectrum of ants (Wagner, 1983), including the red and black imported fire ants (Williams and Lofgren, 1983; Williams et al., 1980). Sodium fluoride mixed into wheat flour baits produced satisfactory control of firebrats (Wakeland and Waters, 1931) and silverfish (Mallis, 1941).

Various foods have been reported as attractive to cockroaches (Ahmed, 1976; Bare, 1945; Miesch, 1964; Miesch and Howell, 1967; Reierson and Rust, 1984; Rust and Reierson, 1981*b;* Tsuji, 1965; Wileyto and Boush, 1983), earwigs (Fulton, 1923), firebrats (Seiferle et al., 1938; Wakeland and Waters, 1931), and silverfish (Mallis, 1941). Laboratory studies have shown that these insects are not attracted to these materials over more than a few centimeters (Adams, 1933; Ebeling and Reierson, 1974; Miesch and Howell, 1967; Wileyto and Boush, 1983). Semi-solid or liquid baits were more attractive to cockroaches than the identical components in a solid bait, especially with German cockroaches (Bare, 1945; Hawkes, 1974; Lofgren and Burden, 1958; Miesch, 1964; Miesch and Howell, 1967). These results are consistent with my observations that slightly deprived *B. germanica* are attracted to water over a distance of a meter or more.

Fatty acids and their esters, cyclohexyl alkonates, and *n*-alkyl cyclohexaneacetates (Iida, Tominga, and Sugawara, 1981; Sugawara, Kurihara,

Table 14-3. Active ingredients incorporated into baits tested in the laboratory and field situations against various household insect pests.

Species	*Active ingredient*	*References*[a]
Cockroaches		
B. germanica	Acephate	Rust and Reierson, 1981*b*; [Rust and Reierson, 1981*b*]
	Alsystin	Weaver, Begley, and Kondo, 1984
	Amidinohydrazone	Bennett, Runstrom, and Bertholf, 1984; [D'Agnese, 1984]; Reierson and Rust, 1984; [Reierson and Rust, 1984]; Reierson et al., 1983
	Avermectin	Ballard and Gold, 1983
	Boric acid	Bare, 1945; Barson, 1982; [Barson and Lole, 1981]; Reierson and Rust, 1984; Rust and Reierson, 1981*b*; Tabaru, Ono, and Tsuji, 1974; Tsuji and Ono, 1969; [Tsuji and Ono, 1970]
	Chlordecone	[Bills, 1965]; [Burden and Smittle, 1975]; Hawkes, 1974; [Pence, 1961]; Tyler, 1964
	Chlorpyrifos	[Bennett, Runstrom, and Bertholf, 1984]; Reierson and Rust, 1984; [Reierson and Rust, 1984]; [Reierson et al., 1983]; Rust and Reierson, 1981*b*; [Rust and Reierson, 1981*b*]
	Dichlorvos	Meisch, 1964; Meisch and Howell, 1967; [Miesch and Howell, 1967]
	Fenitrothion	Tsuji and Ono, 1969; [Tsuji and Ono, 1970]
	Iodonfenphos	Barson, 1982
	Lindane	Piquett, 1948
	Phosphorus	Cheng and Campbell, 1940; [Cowan, 1865]
	Propoxur	Reierson and Rust, 1984; [Reierson and Rust, 1984]; Rust and Reierson, 1981*b*; [Rust and Reierson, 1981*b*]; Tabaru, Ono, and Tsuji, 1974
	Sodium arsenite	Cheng and Campbell, 1940
	Sodium fluoride	Cheng and Campbell, 1940
	Trichlorfon	[Lofgren and Burden, 1958]; Tabaru, Ono, and Tsuji, 1974; Tsuji and Ono, 1969; [Tsuji and Ono, 1970]

Periplaneta spp.	Carbaryl	Ahmed, 1976
	Chlordecone	[Burden and Smittle, 1975]; Hawkes, 1974; [Neely and Mattingly, 1966]; Wright et al., 1973
	Diazinon	Ahmed, 1976
	Malathion	[Lofgren and Burden, 1958]
	Phosphorus	Cheng and Campbell, 1940
	Propoxur	Ahmed, 1976; [Wright et al., 1973]
	Sodium arsenite	Cheng and Campbell, 1940
	Trichlorfon	Lofgren and Burden, 1958; [Lofgren and Burden, 1958]
B. orentalis	Boric acid	Barson and Lole, 1981
	Chlordecone	Hawkes, 1974
	Mercuric chloride	Cole, 1932
	Phosphorus	Cheng and Campbell, 1940
Firebrats and silverfish		
Thermobia spp.	Arsenicals	Seiferle et al., 1938; [Wakeland and Waters, 1931]
	Barium compounds	Richardson and Seiferle, 1940; Seiferle et al., 1938; Snipes, Hutchins, and Adams, 1936
	Fluorides	Seiferle et al., 1938; [Wakeland and Waters, 1931]
Lepisma and *Ctenolepisma* spp.	Arsenicals	Mallis, 1941; Seiferle et al., 1938; Snipes, Hutchins, and Adams, 1936
	Fluorides	Mallis, 1944; Seiferle et al., 1938
M. pharaonis	Boric acid	[Marlatt, 1907]; [Newton, 1980]; [Wright and Stout, 1978]
	Chlordecone	[Burden and Smittle, 1975]; [Newton, 1980]
	Methoprene	[Edwards and Clarke, 1978]; [Newton, 1980]; [Wilson and Booth, 1981]
I. humilis	Amidinohydrazone	[Wagner, 1983]
	Arsenicals	[Marlatt, 1930]; [Newell, 1909]
	Boric acid	[Newell, 1909]
	Thallium sulfate	[Marlatt, 1930]

[a]References in brackets designate field evaluations.

and Muto, 1965; Sugawara et al., 1976; Tominaga and Sugawara, 1981; Tsuji, 1966) have been reported to be highly attractive to American and German cockroaches. However, these findings have not been substantiated (Reierson and Rust, 1977; Wileyto and Boush, 1983). American cockroaches recognize food odors by a complex pattern of perception utilizing unspecific odors and numerous receptor cell types (Sass, 1978). Two possibilities exist: "monotonous" odors do not attract cockroaches (Wileyto and Boush, 1983), or cockroaches cannot orient themselves toward food odors beyond a few centimeters (Ebeling and Reierson, 1974).

Considerably greater progress has been made in empirically identifying suitable foods and compounds to incorporate into baits. German cockroaches prefer food with carbohydrates such as powdered sucrose (Bare, 1945; Lofgren and Burden, 1958; Miesch and Howell, 1967; Tsuji, 1965), L-arabinose (Tsuji, 1965) maltose (Tsuji, 1965, 1966), and dehydrated potatoes (Tsuji, 1965), and alcohols such as mannitol (Tsuji, 1965), sorbitol (Tsuji, 1965), and oelyl alcohol (Tsuji, 1966). Saturated fatty acids such as *n*-caproic, *n*-capyrlic, *n*-capric, and *n*-dodecanoic acid increased feeding activity in American and smokybrown cockroaches. Methyl tetradecanoate (methyl myristate) increased adult female German cockroach bait consumption (Lofgren and Burden, 1958) and also served as a feeding stimulant for several *Periplaneta* spp. (Tsuji, 1966).

In feeding tests, silverfish preferred baits containing wheat flour, beef extracts (Wakeland and Waters, 1931) and combinations of wheat flour, salt, and sugar (Mallis, 1941; Wakeland and Waters, 1931). Addition of water made dry baits more palatable (Mallis, 1941).

In general, ants are extremely omnivorous as different species prefer feeding on one or more of the following: honey-dew excreted from insects, sugars, meats, fats, and dead insects (Smith, 1965). However, very little laboratory or field research has been conducted on feeding preferences. Ant baits generally contain combinations of at least sugar, honey, corn syrup, peanut butter, mealworms, or liver. Comparative feeding studies with Argentine ants indicate that liquid baits containing about 25% sucrose are preferred (L. K. Gaston, pers. comm.).

Numerous toxicants and bait bases have been evaluated, but many finished bait formulations have been unacceptable because the toxicant was repellent or the bait was unpalatable. Stringer, Lofgren, and Bartlett (1964) have concluded that ant bait toxicants must exhibit delayed action over at least 10- to 100-fold dosage range; not be repellent to ants when combined with bait; be readily transferred between ants; and kill the recipient. Avoidance of baits containing repellent concentrations of toxicants is an important factor influencing their efficacy in control programs against a variety of insects. In choice tests firebrats were repelled by paris green (Wakefield and Waters, 1931) and sodium fluoride (Richardson and Seiferle, 1940; Wakefield and Waters, 1931). Even though baits with higher concentrations of sodium fluoride provided faster kill than arsenic trioxide baits (Snipes, Hutchins, and Adams, 1936), they were more repellent (Mallis, 1944). Baits containing sodium arsenite, potassium tartrate, sodium fluoride, sodium fluosilicate, and white arsenic were also repellent to silverfish (Mallis, 1941). Similarly, baits containing sodium fluoride repelled German cockroaches (Bare, 1945). Some other active ingredients reported to be repellent are propoxur (Cochran, 1982*a;* Reierson and Rust, 1984; Tabaru, Ono, and Tsuji, 1974), fenitrothion (Lofgren and Burden, 1958), iodonfenphos (Barson, 1982), lindane (Tabaru, Ono, and Tsuji, 1974), and diazinon (Tabaru, Ono, and Tsuji, 1974). Baits containing amidinohydrazone, chlordecone, boric acid, and borax have been shown not to repel cockroaches (Bare, 1945; Barson, 1982; Lofgren and Burden, 1958; Reierson and Rust, 1984).

In general, there is an inverse relationship between the rate of acute insecticidal activity and the repellency of baits. A similar relationship exists between the activity of residual spray or dust deposits and the repellency of insecticides against German cockroaches (Ebeling, Reierson, and Wagner, 1967). To date the most effective baits contain slow-acting compounds. For example, the LT-50s for a 50% boric acid bait against German and American cockroaches are 3.3 and 3.5 days, respectively (Barson, 1982). Chlordecone baits provide LT-50s for three cockroach species in 2 to 5 days (Hawkes, 1974). Baits containing 300 ppm avermectin and 1% amidinohydrazone require 138 to 161 hours, respectively, to provide 50% kill of German cockroaches (Ballard and Gold, 1983).

PROBLEMS CONFRONTING HOUSEHOLD PEST MANAGEMENT

Education of the pest control industry and the public is one of the major problems confronting urban integrated pest management (Byrne et al., 1984; Frankie and Levenson, 1978; Levenson and Frankie, 1981; National Research Council, 1980). Pest identification and knowledge of the habits, biology, and behavior of pests in the urban environment are essential if alternative pest management practices incorporating trapping, baiting, biological control, and insect growth regulators are to be utilized. The need for information about the chemistry, toxicology, and environmental impact of the chemicals used in the urban environment will increase in importance in the next decade. Surveys indicate that the public relies on friends and professional pest control operators for most of their information concerning pests and pest control strategies (Byrne et al., 1984; Frankie and Levenson, 1978; Levenson and Frankie, 1981; Wood et al., 1981). Additional education and training should be mandatory for those making pest control recommendations and applying pesticides. The pest control industry, universities, and public agencies must take a leadership role to improve information transfer from research laboratories to the public and private sector. If urban integrated pest management is to succeed, increased educational programs will be an important factor.

Economics is one of the biggest problems facing the pest control industry and urban pest management programs. Often the lack of control can be attributed to the lack of sanitation, structural modification, or cooperation with the homeowner or client. Frequently, to secure business, agencies and companies will submit minimal bids and provide inadequate service. To properly incorporate integrated pest management into existing programs additional labor, effort, and materials will be necessary. Household pest management may be cost effective in a long-term program (Slater et al., 1979), but it will not necessarily be inexpensive.

Physiological insecticide resistance is one of the major biological problems contributing to decreased efficacy and performance of many conventional insecticides and pest management strategies used for years by the industry. The existence of multiple resistance to five or more categories of insecticide has been documented in *Culex* mosquitoes, confused flour beetle, rice and granary weevils (Metcalf, 1983), and in the German cockroach. Other groups that may be experiencing a similar development of resistance include fleas, lice, and ticks. The existence and historical development of insecticide resistance in German cockroaches has been documented (Barson and McCheyne, 1978; Batth, 1977; Cochran, 1982*a*, 1982*b*; Cornwell, 1976). Cross-resistance to compounds that have had limited or no use for cockroach control has also been documented (Barson and McCheyne, 1978; Nelson and Wood, 1982; Rust and Reierson, 1978). When compiled with the inherent behavioral repellency of certain insecticides, the impact of physiological resistance on control can be pronounced (Rust and Reierson, 1978).

Introduction of exotic pests and expansion of an existing pest's distribution will continue to be a problem. In recent years the repeated introduction of Khapra beetle and various species of fruit fly and the spread of Formosan termite, smokybrown cockroach, German yellowjacket, Pharaoh ant, and imported fire ants poses a serious problem to the urban community. Continued surveillance, education, and research are necessary to implement effective pest management programs.

OPPORTUNITIES AVAILABLE IN HOUSEHOLD PEST MANAGEMENT

In the last several years, there has been an increasing acceptance and use of the so-called third generation insecticides, including insect growth regulators (IGRs) and chitin synthesis inhibitors. IGRs mimic the action of the naturally occurring juvenile hormones to interfere with certain normal physiological processes and prevent immature insects from completing development into functionally reproductive adults. Effects such as abnormal molting, twisted wings, darkening of the cuticle, and loss of mating behavior and fecundity in surviving female adultoids has been reported in several cockroach species (Das and Gupta, 1977; Riddiford, Ajamii, and Boake, 1975). The chitin synthesis inhibitors such as alsystin and diflubenzuron interfere with the molting and developmental processes and show some ovicidal activity. In addition, oothecae of female German cockroaches that food on alsystin failed to hatch (Weaver, Begley, and Kondo, 1984). Alsystin and diflubenzuron are active against stored product pests (Mian and Mulla, 1982*a*) and flies of public health importance (Ali and Lord, 1980; Johnson and Mulla, 1982; Mian and Mulla, 1982*b*).

Currently two IGRs, methoprene and hydroprene, are registered and being used in structural pest control against Pharaoh ants, cat fleas, and German cockroaches. Hydroprene prevented normal metamorphosis in 97% of the German cockroaches continuously confined to as little as 5 ppm deposits (Riddiford, Ajamii, and Boake, 1975). At 7 months, the resulting adultoids did not reproduce. In the same tests, methoprene was considerably less active and higher concentrations were required. Nymphal German cockroaches exposed to food treated with experimental IGRs died (Cruikshank and Palmere, 1971; Das and Gupta, 1977) or metamorphosed into sterile adult females with hypertrophied oocytes and accessory glands (Das and Gupta, 1977). Even though some preliminary reports of field testing with hydroprene have been released (Bennett, Runstrom, and Bertholf, 1984), there is not enough data at this time to determine its efficacy. Preliminary laboratory data with another IGR, fenoxycarb, also look promising. It remains to be seen whether a pest management program against cockroaches incorporating such slow acting compounds can be developed and integrated with existing programs.

Laboratory results show that oriental rat and cat flea larvae are extremely sensitive to methoprene (Chamberlain, 1979; Rust and Reierson, 1983) and fenoxycarb (Chamberlain, 1983; Rust and Reierson, 1983). As with conventional insecticides, the availability of the IGR was affected by the type of substrate and formulation (Rust and Reierson, 1983). Little information is available concerning the performance of these IGRs in homes and apartments infested with fleas. Data from recent tests in homes infested with cat fleas indicate that 0.05–0.15% concentrations of methoprene and fenoxycarb aerosols applied thoroughly to carpets and floors inhibited adult development for up to 60 days and provided excellent control indoors. The utility of various adulticides such as synergized pyrethrins, chlorpyrifos, DDVP, and propoxur added to the IGRs have never been adequately examined. Likewise, the efficacy of IGRs as a preventive treatment applied before the flea season begins has not been thoroughly explored, and additional field studies are necessary.

Another potential use for IGRs in urban pest management programs is the control of social insects such as ants and termites. Baits incorporating methoprene have been used against Pharaoh ants (Edwards and Clarke, 1978; Newton, 1980; Wilson and Booth, 1981) but required up to 21 weeks to eliminate the queens (Edwards and Clarke, 1978; Newton, 1980). Tests with subterranean termites belonging to the genus *Reticultitermes* indicate that IGRs produce several effects, including induction of presoldiers and soldiers, loss of hindgut symbionts, and inhibition of feeding (Haverty and Howard, 1979; Howard, 1983; Jones, 1984). Definitive studies on nonlaboratory populations have not been conducted, and the potential of IGRs for termite control remains in doubt.

Another exciting area for future research is the use of pheromones to reduce the repellency of and increase the efficacy of insecticides and for inspecting and monitoring pest populations. When aggregation pheromones secreted in the feces of German cockroaches were mixed with insecticides, the combination provided increased efficacy and control of German cockroaches (Glaser, 1980; Rust and Reierson, 1977*a*, 1977*b*). Similarly, when Periplanone B, a major component of female sex pheromone of the American cockroach, was mixed with a toxicant, the combination produced increased kill of American cockroaches (Bell et al., 1984). Other cockroach species and insects such as ants, stored-product beetles, and termites have various aggregation, trail, and sex pheromones. The use of these pheromones as adjuvants for toxicants may have broader applicability than to just cockroaches. Some sex pheromones of the stored product insects are currently being used to attract and detect active infestations. Periplanone B may also have some utility in increasing the effectiveness of traps for various species of *Periplaneta* (Bell et al., 1984). Trail-following compounds produced by subterranean termites (*Reticulitermes* spp.) may be useful in inspecting and monitoring structures.

Release of sterile insects and genetic control strtegies have had limited success in insect control. However, research with genetically induced sterile male German cockroaches offers some promise. Sterile male German cockroaches released on ships infested with German cockroaches effectively competed for females with wild males and were capable of reducing populations in limited areas. Limited movement of sterile adult males and mated females prevented population reductions over the entire ship (Ross, Keil, and Cochran, 1981). There may be some application of this technique to other cockroaches and household insects.

Pyrethroids such as bioresmethrin, cismethrin, resmethrin, and tetramethrin have been utilized as thermal fogs and sprays against American and German cockroaches (Chadwick and Shaw, 1974; Chadwick, Martin, and Marin, 1977; Elliott, Janes, and Potter, 1978). However, the photostable and persistent pyrethroids such as cyfluthrin, cypermethrin, decamethrin, fenvalerate, and permethrin look extremely promising against German cockroaches, some species of flies, and fabric pests. Cypermethrin and permethrin provided good to excellent control of German cockroaches in apartments (Bennett and Runstrom, 1979; Rust and Reierson, 1981*a*). Results with fenvalerate have not been as promising, possibly because of decreased performance on porous substrates (Grayson, 1980; Rust and Reierson, 1978), at temperatures above 26°C and repellency (Rust and Reierson, 1978). Permethrin has been shown to be a durable mothproofer and fabric protectant and an active direct contact spray against fabric pests (Bry et al., 1979). Even though pyrethroids are extremely toxic to German cockroaches and fabric pests, they are not extremely active against cat fleas and will probably have limited use against this important household pest.

Several new groups of compounds, avermectins, amidinohydrazones and formamidines with insecticidal and acaricidal activity seem extremely promising for use in pest management strategies. The formamidines, such as chlordimeform and amitraz, possess some very unusual properties. In addition to insecticidial activity, sublethal doses of chlordimeform produce an anorectic effect in American cockroaches (Beeman and Matsumura, 1978) and locomotor stimulation and reduced fecundity in ticks. The exact mode of action of formamidines is unknown, but they may act as octopamine mimics and affect biogenic amine receptors (Giles and Rothwell, 1983). The mode of action of the amidinohydrazones is also unknown, but there is some speculation that they interfere with cellular metabolic processes. The avermectins, originally developed as antihelminthics, are believed to mimic neurotransmitters. A promising feature of these compounds is their novel modes of action and potential for use against organophosphate and carbamate resistant insects. Amidinohydrazone is already being used for ant and cockroach control, and the future of such novel insecticides looks promising.

Research in resistance management has progressed steadily against flies and mosquitoes. Several strategies, including resistance management by moderation or saturation with contrasting low or high dosages of insecticides and by multiple attack with mixtures or alternation of chemicals (Georghiou, 1983), have been proposed. Development of formulations such as encapsulation has increased the residual activity and availability of diazinon (Sakurai et al., 1982) and has provided improved control of resistant German cockroaches (Rust and Reierson, 1978). Additional research with other encapsulated insecticides may also help counteract physiological resistance.

Some limited tests have been conducted with alternative strategies, but considerable research is still necessary to determine which strategies will overcome resistance and extend the use of existing insecticides.

In summary, the concept of urban integrated pest management is just beginning to flourish. If nothing else, a review of urban pest control has taught us that there are no simple solutions. The number of urban research programs at universities has doubled in the last five years and the development of pest management procedures and technology will rapidly expand in the next decade. Interest and support by basic insecticide manufacturers and the pest control industry have also been responsible for making increased research activity possible. Urban pest management research for household insect pests has a bright and promising future.

REFERENCES

Adams, J. A. 1933. Biological notes upon the firebrat *Thermobia domestica* Packard. *J. New York Entomol. Soc.* **41:**557-562.

Ahmed, S. M. 1976. Comparative efficacies of various insecticides in a new bait formulation against *Periplaneta americana. Int. Pest Control* **18:**4-6.

Ali, A., and J. Lord. 1980. Experimental insect growth regulators against some nuisance Chironomid midges of central Florida. *J. Econ. Entomol.* **73:**243-249.

Baker, L. F., and N. D. Southam. 1977. Detection of *Blattella germanica* and *Blatta orientalis* by trapping. *Int. Pest Control* **19:**8-11.

Ballard, J. B., and R. E. Gold. 1982. The effect of selected baits on the efficacy of a sticky trap in the evaluation of German cockroach populations. *J. Kansas Entomol. Soc.* **55:**86-90.

Ballard, J. B., and R. E. Gold. 1983. Efficacy of bait formulation of avermectin B1 and Amdro upon male German cockroaches, 1982. *Insectic. Acaric. Tests* **8:**58.

Barak, A. V., M. Shinkle, and W. E. Burkholder. 1977. Using attractant traps to help detect and control cockroaches. *Pest Control* **45**(10):14-16, 18-20.

Bare, O. S. 1945. Boric acid as a stomach poison for the German cockroach. *J. Econ. Entomol.* **38:**407.

Barson, G. 1982. Laboratory evaluation of boric acid plus porridge oats and iodofenphos gel as toxic baits against the German cockroach, *Blattella germanica* (L.) (Dictyoptera: Blattellidae). *Bull. Entomol. Res.* **72:**229-237.

Barson, G., and M. Lole. 1981. Cockroach control using a simple baiting technique. *Int. Pest Control* **23:**140-142.

Barson, G., and N. G. McCheyne. 1978. Resistance of the German cockroach (*Blattella germanica*) to bendiocarb. *Ann. Appl. Biol.* **90:**147-153.

Batth, S. S. 1977. A survey of Canadian populations of the German cockroach for resistance to insecticides. *Can. Entomol.* **109:**49-52.

Beatson, S. H. 1968. Eradication of Pharaoh's ants and crickets using chlordecone baits. *Int. Pest Control* **10:**8-10.

Beeman, R. W., and F. Matsmumura. 1978. Anorectic effect of chlordimeform in the American cockroach. *J. Econ. Entomol.* **71:**859-861.

Bell, W. J., J. Fromm, R. Quisumbing, and A. F. Kydonieus. 1984. Attraction of American cockroaches (Orthopter: Blattidae) to traps containing Periplanone B and to insecticide-Periplanone B mixtures. *Environ. Entomol.* **13:**448-450.
Bennett, G. W. 1977. Cockroach manual with test yourself quiz. *Pest Control* **45:**24, 26, 28-30.
Bennett, G. W., and R. D. Lund. 1977. Evaluation of encapsulated pyrethrins (Sectrol) for German cockroach and cat flea control. *Pest Control* **45:**44, 46, 48-50.
Bennett, G. W., and S. H. Robertson. 1978. Field testing of Ficam W for flea control. *Pest Control* **46:**29-30.
Bennett, G. W., and E. S. Runstrum. 1979. New developments in pest control insecticides. *Pest Control* **47:**14-16, 18, 20.
Bennett, G. W., and E. S. Runstrom. 1980. Efficacy of new insecticide formulations in urban pest control. *Pest Control* **48:**19-22, 24.
Bennett, G. W., E. S. Runstrom, and J. A. Wieland. 1983. Pesticide use in homes. *Bull. Entomol. Soc. Am.* **29**(1):31-38.
Bennett, G. W., E. S. Runstrom, and J. Bertholf. 1984. Examining the where and why and how of cockroach control. *Pest Control* **52:**42-43, 46, 48, 50.
Bills, G. T. 1965. Kepone and diclorvos for controlling cockroaches and vinegar flies. *Int. Pest Control.* **7:**8-11.
Bledsoe, B., V. A. Fadak, and M. E. Bledsoe. 1982. Current therapy and new developments in indoor flea control. *J. Am. Anim. Hosp.* **18:**415-422.
Bry, R. E., R. E. Boatright, J. H. Lang, and R. A. Simonaitis. 1979. Spray applications of permethrin against fabric pests. *Pest Control* **47:**14-17.
Burden, G. S., and B. J. Smittle. 1975. *Blattella germanica, Periplaneta americana* and *Monomorium pharaonis:* Control with insecticidal baits in an animal building and insectaries. *J. Med. Entomol.* **12:**352-353.
Byrne, D. N., E. H. Carpenter, E. M. Thoms, and S. T. Cotty. 1984. Public attitudes toward urban arthropods. *Bull. Entomol. Soc. Am.* **30**(2):40-44.
Chadwick, P. R., and R. D. Shaw. 1974. Cockroach control in sewers in Singapore using bioresmethrin and piperonyl butoxide as a thermal fog. *Pestic. Sci.* **5:**691-701.
Chadwick, P. R., M. Martin, and J. Marin. 1977. Use of thermal fogs of bioresmethrin and cismethrin for control of *Periplaneta americana* (Insecta: Blattidae) in sewers. *J. Med. Entomol.* **13:**625-626.
Chamberlain, W. F. 1979. Methoprene and the flea. *Pest Control* **47:**22, 24, 26.
Chamberlain, W. F. 1983. Laboratory evaluation of IGR's and toxicants against the larvae of the oriental rat flea. *Insectic. Acaric. Tests* **8:**59.
Cheng, T. H., and F. L. Campbell. 1940. Toxicity of phosphorus to cockroaches. *J. Econ. Entomol.* **33:**193-199.
Cochran, D. G. 1982*a.* German cockroach resistance. *Pest Control* **50:**16, 18, 20.
Cochran, D. G. 1982*b. Cockroaches: Biology and Control,* WHO/Vector Biology and Control No. 82.856. World Health Organization, 47p.
Cole, A. C., Jr., 1932. The olfactory responses of the cockroach (*Blatta orientalis*) to the more important essential oils and a control measure formulated from the results. *J. Econ. Entomol.* **25:**902-905.
Coler, R. R., R. G. van Driesche, and J. S. Elkinton. 1984. Effect of an oothecal parasitiod, *Comperi merceti* (Compere) (Hymenoptera: Encyrtidae), in a popula-

tion of brownbanded cockroach (Orthoptera: Blattellidae). *Environ. Entomol.* **13:**603-606.

Cornwell, P. B. 1968. *The Cockroach,* vol. 1. Hutchinson, London. 400p.

Cornwell, P. B. 1976. *The Cockroach,* vol. 2. St. Martin's, New York, 557p.

Cowan, F. 1865. *Curious Facts in the History of Insects.* J. B. Lippencott, Philadelphia, 396p.

Cruickshank, P. A., and R. M. Palmere. 1971. Terpenoid amines as insect juvenile hormones. *Nature* **233:**488-489.

D'Agnese, J. J. 1984. German cockroaches and cruise ships don't dance together very well. *Pest Control* **52:**26, 28, 30.

Das, Y. T., and A. P. Gupta. 1977. Abnormalities in the development and reproduction of *Blattella germanica* (L.) (Dictyoptera: Blattellidae) treated with insect growth regulators with juvenile hormone activity. *Experientia* **33:**968-970.

Ebeling, W. 1960. Control of the tropical rat mite. *J. Econ. Entomol.* **53:**475-476.

Ebeling, W. 1971. Sorptive dusts for pest control. *Annu. Rev. Entomol.* **16:**123-159.

Ebeling, W. 1975. *Urban Entomology.* Division of Agricultural Science, University of California, Los Angeles.

Ebeling, W. 1978. Past, present, and future directions in the management of structure-infesting insects. In *Perspectives in Urban Entomology,* G. W. Frankie and C. S. Koehler (eds.). Academic Press, New York, pp. 221-247.

Ebeling, W., and D. A. Reierson. 1973. Should flushing agents be added to blatticides? *Pest Control* **41:**24, 46, 48, 50-51.

Ebeling, W., and D. A. Reierson. 1974. Bait trapping silverfish, cockroaches, and earwigs. *Pest Control.* **42:**24, 36-39.

Ebeling, W., and R. E. Wagner. 1959. Rapid desiccation of drywood termites with inert sorptive dusts and other substances. *J. Econ. Entomol.* **52:**190-207.

Ebeling, W., and R. E. Wagner. 1964. Built-in pest control. *Pest Control.* **32:**20-22, 24, 26, 28, 31-32.

Ebeling, W., D. A. Reierson, and R. E. Wagner. 1967. Influence of repellency on the efficacy of blatticides, 2, Laboratory experiments with German cockroaches. *J. Econ. Entomol.* **60:**1375-90.

Ebeling, W., D. A. Reierson, and R. E. Wagner. 1968*a.* The influence of repellency on the efficacy of blatticides, 3, Field experiments with German cockroaches with notes on three other species. *J. Econ. Entomol.* **61:**751-761.

Ebeling, W., D. A. Reierson, and R. E. Wagner. 1968*b.* Pyrethrin aerosol and boric acid: A combination treatment for cockroach control. *PCO News* **28:**12-14, 22-26, 29.

Ebeling, W., R. E. Wagner, and D. A. Reierson. 1965. Cockroach control with Dri-die and Drione. *PCO News* **25:**16-19.

Ebeling, W., R. E. Wagner, and D. A. Reierson. 1966. Influence of repellency on the efficacy of blatticides, 1, Learned modification of behavior of the German cockroach. *J. Econ. Entomol.* **59:**1374-88.

Ebeling, W., R. E. Wagner, and D. A. Reierson. 1977. Cockroach proofing at the time of construction. In *Proceedings, 1977 Seminar on Cockroach Control,* K. A. Westphal (ed.). New York State Public Health, New York, pp. 33-38.

Ebeling, W., D. A. Reierson, R. J. Pence, and M. S. Viray. 1975. Silica aerogel and boric acid against cockroaches: External and internal action. *Pestic. Biochem. Physiol.* **5:**81-89.

Eckert, J. E., and A. Mallis. 1937. *Ants and Their Control in California,* Agricultural Station Circular 342. University of California, 39p.

Edwards, J. P., and B. Clarke. 1978. Eradication of Pharaoh's ants with baits containing the insect juvenile hormone analogue methoprene. *Int. Pest Control* **20:**5-10.

Elliott, M., N. F. Janes, and C. Potter. 1978. The future of pyrethroids in insect control. *Ann. Rev. Entomol.* **23:**443-469.

Farmer, B. R., and W. H. Robinson. 1984. Is caulking beneficial for cockroach control? *Pest Control.* **52**(6):28, 30, 32.

Fleet, R. R., and G. W. Frankie. 1975. Behavioral and ecological characteristics of a Eulopid egg parasite of two species of domiciliary cockroaches. *Environ. Entomol.* **4:**282-284.

Fowler, H.G. 1983. Urban structural pests: Carpenter ants (Hymenoptera: Formicidae) displacing subterranean termites (Isoptera: Rhinotermitidae) in public concern. *Environ. Entomol.* **12:**997-1002.

Frankie, G. W., and H. Levenson, 1978. Insect problems and insecticide use: Public opinion, information,and behavior. In *Perspectives in Urban Entomology,* G. W. Frankie and C. S. Koehler (eds.). Academic Press, New York, pp. 359-399.

Frishman, A. M. 1982. Cockroaches. In *Handbook of Pest Control,* A. Mallis (ed.), 6th ed. Franzak and Foster, Cleveland, Ohio, pp. 101-153.

Fulton, B. B. 1923. Some experimental poison baits for European earwigs. *J. Econ. Entomol.* **16:**369-376.

Georghiou, G. P. 1983. Management of resistance in arthropods. In *Pest Resistance to Pesticides.* G. P. Georghiou, and T. Saito (eds.). Plenum, New York, pp. 769-792.

Giles, D. P., and D. N. Rothwell. 1983. The sub-lethal activity of amidines on insects and acarids. *Pestic. Sci.* **14:**303-312.

Glaser, A. E. 1980. Use of aggregation pheromones in the control of the German cockroach (*Blattella germanica*). *Int. Pest Control* **22:**7, 8, 21.

Grayson, J. M. D. 1980. Pydrin on different surfaces for control of resistant cockroaches. *Pest Control* **48:**19-21.

Green, S. G. 1982. Mites. In *Handbook of Pest Control,* A. Mallis (ed.). Franzak and Foster, Cleveland, pp. 739-775.

Griffiths, J. T., and O. E. Tauber. 1943. Evaluation of sodium fluoride as a stomach poison and as a contact insecticide against the roach, *Periplaneta americana. J. Econ. Entomol.* **34:**536-540.

Gupta, A. P., and Y. T. Das. 1976. Inner-city control of German roaches with resmethrin/boric acid. *Pest Control* **45:**43-44, 55.

Gupta, A. P., Y. T. Das, J. R. Trout, D. S. Adam, and G. J. Bordash. 1973. Effectiveness of spray -dust-bait combination. *Pest Control* **41:**46-51.

Hagmann, L. E. 1982. Ant Baits. *Pest Control* **50:**30, 32.

Hawkes, C. 1974. Experiments with chlordecone in the formulation of baits for cockroach control. *Int. Pest Control* **16:**12-17.

Haverty, M. I., and R. W. Howard. 1979. Effects of insect growth regulators on subterranean termites: Induction of differentiation, defaunation, and starvation. *Ann. Entomol. Soc. Am.* **72**(4):503-508.

Hopkins, G. H. E., and M. Rothschild. 1953. *An Illustrated Catalogue of the Rothschild Collection of Fleas in the British Museum (Natural History),* vol. 1, *Tungidae and Pulicidae.* British Museum, London.

Howard, R. W. 1983. Effects of methoprene on binary caste groups of *Reticultitermes flavipes* (Kollar) (Isoptera: Rhinotermitidae). *Environ. Entomol.* **12:**1059-1063.

Iida, Y., Y. Tominaga, and R. Sugawara. 1981. Attractiveness of methyl cyclohexyl n-alkanoates to the German cockroach. *Agric. Biol. Chem.* **45:**469-473.

Johnson, G. D., and M. S. Mulla. 1982. Suppression of nuisance aquatic midges with a urea insect growth regulator. *J. Econ. Entomol.* **75:**297-300.

Jones, S. C. 1984. Evaluation of two insect growth regulators for the bait-block method of subterranean termite (Isoptera: Rhinotermitidae) control. *J. Econ. Entomol.* **77:**1086-1091.

Kamble, S. T., R. E. Gold, and E. F. Vitzhum. 1978. *Nebraska Pesticide Survey: Structural,* Environmental Progress Report No. 1. University of Nebraska, Lincoln.

Kamble, S. T., R. E. Gold, and A. M. Parkhurst. 1982. *Nebraska Residential Pesticide Use Survey,* Environmental Progress Report No. 1. University of Nebraska, Lincoln.

Levenson, H., and G. W. Frankie. 1981. Pest control in the urban environment. In *Progress in Resource Management and Environmental Planning,* vol. 3, T. O'Riordan and R. K. Turner (eds.). Wiley, New York, pp. 251-272.

Linsley, E. G. 1946. Some ecological factors influencing the control of carpet beetles and clothes moths. *Pests* **14:**10, 12, 14, 16, 18.

Lofgren, C. S., and G. S. Burden. 1958. Tests with poison baits against cockroaches. *Florida Entomol.* **41:**103-110.

Mallis, A. 1941. Preliminary experiments on the silverfish *Ctenolepisma urbani* Slabaugh. *J. Econ. Entomol.* **34:**787-791.

Mallis, A. 1944. Concentrations of sodium fluoride-flour mixtures for silverfish control. *J. Econ. Entomol.* **37:**842.

Mallis, A. 1960. *Handbook of Pest Control.* MacNair Dorland, New York, 1,132p.

Mallis, A. 1982*a.* Clothes moths. In *Handbook of Pest Control,* A. Mallis (ed.), 6th ed. Franzak and Foster, Cleveland, Ohio, pp. 353-385.

Mallis, A. 1982*b.* Hide and carpet beetles. In *Handbook of Pest Control,* A. Mallis (ed.), 6th ed., Franzak and Foster, Cleveland, Ohio, pp. 387-423.

Mallis, A., and R. Carr. 1982. Silverfish. In *Handbook of Pest Control,* A. Mallis (ed.), 6th ed. Franzak and Foster, Cleveland, Ohio, pp. 79-92.

Marlatt, C. L. 1907. *House Ants,* Circular 34. U.S. Department of Agriculture, Washington, D.C., 4p.

Marlatt, C. L. 1930. *House Ants: Kinds and Methods of Control,* Farmers' Bulletin 740. U.S. Department of Agriculture, Washington, D.C., 14p.

Metcalf, R. L. 1983. Implications and prognosis of resistance to insecticides. In *Pest Resistance to Pesticides,* G. P. Georghiou and T. Saito (eds.). Plenum, New York, pp. 703-734.

Mian, L. S., and M. S. Mulla. 1982*a.* Biological activity of IGRs against four stored-product Coleopterans. *J. Econ. Entomol.* **75:**80-85.

Mian, L. S., and M. S. Mulla. 1982*b.* Biological and environmental dynamics of insect growth regulators (IGRs) as used against Diptera of public health importance. *Residue Rev.* **84:**27-112.

Miesch, M. D., Jr. 1964. Ecological and physiological mechanisms influencing food finding in *Blattaria*. Ph.D. diss. Oklahoma State University, Stillwater, 91p.

Miesch, M. D., Jr., and D. E. Howell. 1967. An evaluation of baits for cockroaches. *Pest Control* **35:**16, 18, 20.

Moore, R. C. 1972. Boric acid-silica dusts for control of German cockroaches. *J. Econ. Entomol.* **65:**458-461.

Moore, R. C. 1973. *Cockroach Proofing: Preventive Treatments for Control of Cockroaches in Urban Housing and Food Service Carts,* Bulletin 740. Connecticut Agricultural Experiment Station, 13p.

Moore, R. C. 1977. Cockroach proofing: Preventive treatments for control of cockroaches in urban housing and food service carts. In *Proceedings: 1977 Seminar on Cockroach Control,* K. A. Westphal (ed.). New York State Public Health, New York, p. 60-68.

Moore, W. S., and T. A. Granovsky. 1983. Laboratory comparisons of sticky traps to detect and control five species of cockroaches (Orthoptera: Blattidae and Blattellidae). *J. Econ. Entomol.* **76:**845-849.

National Research Council. 1980. *Urban Pest Management.* Report prepared by the Committee on Urban Pest Management, Environmental Studies Board, Commission on Natural Resources. National Academy Press, Washington, D.C., 272p.

Neely, J. M., and C. L. Mattingly. 1966. Control of American cockroaches (*Periplaneta americana*) with Kepone ant and cockroach paste. *Pest Control* **34:**16-17.

Nelson, J. O., and F. E. Wood, 1982. Multiple and cross-resistance in a field-collected strain of the German cockroach (Orthoptera: Blattellidae). *J. Econ. Entomol.* **75:**1052-1054.

Newell, W. 1909. Measures suggested against the Argentine ant as a household pest. *J. Econ. Entomol.* **2:**324-332.

Newton, J. 1980. Alternatives to chlordecone for Pharaoh's ant control. *Int. Pest Control* **22:**112-114, 135.

Nuhn, T. P., and C. G. Wright. 1979. An ecological survey of ants (Hymenoptera: Formicidae) in a landscaped suburban habitat. *Am. Midl. Nat.* **102:**353-362.

Osbrink, W. L. A., and M. K. Rust. 1985. Cat flea (Siphonaptera: Pulicidae): Factors influencing host-finding behavior in the laboratory. *Ann. Entomol. Soc. Am.* **78:** 29-34.

Osbrink, W. L. A., M. K. Rust, D. A. Reierson, and M. J. Lawton. 1983. Evaluation of insecticides for cat flea control inside urban dwellings, 1982. *Insectic. Acaric. Tests* **8:**85.

Owens, J. M. 1980. Some aspects of German cockroach population ecology in urban apartments. Ph.D. diss., Purdue University, West Lafayette, Ind., 116p.

Owens, J. M., and G. W. Bennett. 1983. Comparative study of German cockroach (Dictyoptera: Blattellidae) population sampling techniques. *Environ. Entomol.* **12:**1040-1046.

Pence, R. J. 1961. *Tribolium* and cockroach control with Kepone bait in fabric insect culture cabinets. *J. Econ. Entomol.* **54:**821-823.

Piper, G. L., and G. W. Frankie. 1978. Integrated management of urban cockroach populations. In *Perspectives in Urban Entomology,* C. S. Koehler and G. W. Frankie (eds.). Academic Press, New York, pp. 249-266.

Piper, G. L., R. R. Fleet, G. W. Frankie, and R. E. Frisbie. 1975. *Controlling Cockroaches without Synthetic Organic Insecticides,* Leaflet 1373. Texas Agricultural Experiment Station and Extension Service.

Piper, G. L., G. W. Frankie, and J. Loehr. 1978. Incidence of cockroach egg parasites in urban environments in Texas and Louisiana. *Environ. Entomol.* **7:**289-293.

Piquett, P. G. 1948. Benzene hexachloride and cornstarch as a roach-control combination. *J. Econ. Entomol.* **41:**326-327.

Quattrochi, L. P. 1968. Let's talk about the advantages of bait. *Pest Control* **36:**8-10, 12, 14.

Rachesky, S. 1978. When are people concerned about pests? *Pest Control* **46:**18-21.

Rehn, J. A. G. 1945. Man's uninvited fellow traveler—the cockroach. *Sci. Monthly* **61:**265-276.

Reierson, D. A., and M. K. Rust. 1977. Trapping, flushing, counting German roaches. *Pest Control* **45:**40, 42-44.

Reierson, D. A., and M. K. Rust. 1984. Insecticidal baits and repellency in relation to control of the German cockroach, *Blattella germanica* (L.). *Pest Manage.* **2:**26-32.

Reierson, D. A., M. K. Rust, A. M. Van Dyke, and A. G. Appel. 1983. Control of German cockroaches with amidinohydrazone bait, 1982. *Insectic. Acaric. Tests* **8:**54.

Richardson, C. H., and E. J. Seiferle. 1940. Barium compounds as poisons in firebrat baits. *J. Econ. Entomol.* **33:**857-861.

Riddiford, L. M., A. M. Ajamii, and C. Boake. 1975. Effectiveness of insect growth regulators in the control of populations of the German cockroach. *J. Econ. Entomol.* **68:**46-48.

Ross, M. H. 1981. Trapping experiments with German cockroach, *Blattella germanica* (L.) (Dictypoptera: Blattellidae), showing differential effects from the type of trap and the environmental resources. *Proc. Entomol. Soc. Wash.* **83:**160-163.

Ross, M. H., C. B. Keil, and D. G. Cochran. 1981. The release of sterile males into natural populations of the German cockroach *Blattella germanica. Entomol. Exp. Appl.* **30:**241-253.

Runstrom, E. R., and G. W. Bennett. 1984. Efficacy of hydroprene on field populations of German cockroaches, 1982. *Insectic. Acaric. Tests* **9:**409.

Rust, M. K., and D. A. Reierson. 1977*a*. Using pheromone extract to reduce repellency of blatticides. *J. Econ. Entomol.* **70:**34-38.

Rust, M. K., and D. A. Reierson. 1977*b*. Increasing blatticidal efficacy with aggregation pheromone. *J. Econ. Entomol.* **70:**693-696.

Rust, M. K., and D. A. Reierson. 1978. Comparison of the laboratory and field efficacy of insecticides used for German cockroach control. *J. Econ. Entomol.* **71:**704-708.

Rust, M. K., and D. A. Reierson. 1981*a*. Control of German cockroaches in low-income apartments, 1980. *Insectic. Acaric. Tests* **6:**187.

Rust, M. K., and D. A. Reierson. 1981*b*. Attraction and performance of insecticidal baits for German cockroach control. *Int. Pet Control* **23:**106-109.

Rust, M. K., D. A. Reierson. 1983. Laboratory evaluation of IGR's against larval cat fleas, 1982. *Insectic. Acaric. Tests* **8:**60.

Sakurai, M., M. Kurotaki, S. Asaka, T. Umino, and T. Ikeshoji. 1982. Characteristic effects of the microencapsulated diazinon against the German cockroach, *Blattella germanica* L. *Jap. J. Sanit. Zool.* **33:**301-307.

Sass, H. 1978. Olfactory receptors on the antenna of *Periplantea* responses constellations that encode food odors. *J. Comp. Physiol.* **A128:**227-233.

Seiferle, E. J., J. A. Adams, C. M. Nagel, and W. C. Ho. 1938. Effectiveness of fluorine compounds as food poisons for the firebrat. *J. Econ. Entomol.* **31:**55-60.

Sheltar, D. J., and V. E. Walter. 1982. Ants. In *Handbook of Pest Control,* A. Mallis (ed.), 6th ed. Franzak and Foster, Cleveland, Ohio, pp. 425-487.

Skovmand, O., and N. D. Christensen. 1980. Fleas. *Danish Pest Inf. Lab. Annu. Rep. 1979,* pp. 61-62.

Slater, A. J., L. McIntosh, R. B. Coleman, and M. Hurlbert. 1979. German cockroach management in student housing. *J. Environ. Health* **42:**21-24.

Slater, A. J., M. J. Hurlbert, and V. R. Lewis. 1980. Biological control of brownbanded cockroaches. *Calif. Agric.* **34:**(8-9):16-18.

Smith, M. R. 1965. *House-infesting Ants of the Eastern United States: Their Recognition, Biology, and Economic Importance,* Technical Bulletin 1326. U.S. Department of Agriculture, Washington, D.C., 105p.

Snipes, B. T., R. E. Hutchins, and J. A. Adams. 1936. Effectiveness of sodium fluoride, arsenic trioxide and thiodiphenylamine as food poisons for the firebrat. *J. Econ. Entomol.* **29:**421-427.

Stringer, C. E., Jr., C. S. Lofgren, and F. J. Bartlett. 1964. Imported fire ant toxic bait studies: Evaluation of toxicants. *J. Econ. Entomol.* **57:**941-945.

Strong, R. G., and G. T. Okumura. 1958. Insects and mites associated with stored foods and seeds in California. *Dept. Agric. Bull.* **47:**233-249.

Sugawara, R., S. Kurihara, and T. Muto. 1965. Attraction of the German cockroach to cyclohexyl alkanoates and n-alkyl cyclohexaneacetates. *J. Insect Physiol.* **21:**957-964.

Sugawara, R., Y. Tominaga, M. Kobayashi, and T. Muto. 1976. Effects of side chain modification on the activity of propyl cyclohexaneacetate as an attractant to the German cockroach. *J. Insect Physiol.* **22:**785-790.

Tabaru, Y., S. Ono, and H. Tsuji. 1974. Laboratory evaluation of several insecticides as feeding inhibitors against the German cockroach, *Blattella germanica* (L.). *Jap. J. Sanit. Zool.* **25:**147-152.

Tarshis, I. B. 1961. Laboratory and field studies with sorptive dusts for the control of arthropods affecting man and animal. *Exp. Parasitol.* **11:**10-33.

Tarshis, I. B. 1964*a*. A sorptive dust for the control of the northern fowl mite. *Ornitonysous sylvarium,* infesting dwellings. *J. Econ. Entomol.* **57:**110-111.

Tarshis, I. B. 1964*b*. The use of the silica aerogel insecticides, Dri-die 67 and Drione, in new and existing structures for the prevention and control of cockroaches. *Lab. Anim. Care* **1:**167-184.

Tarshis, I. B., and M. R. Dunn. 1959. Control of brown dog tick. *Calif. Agric.* **13:**11.

Thompson, P. H. 1983. Fire ants continue their southern attack. *Pest Control* **51:**48-49, 52, 54.

Tominaga, Y., R. Sugawara. 1981. Effects of deuteration on the attractiveness of propyl cyclohexaneacetate to the German cockroach, *Blattella germanica* L. (Orthoptera: Blattidae). *Appl. Entomol. Zool.* **16:**279-287.

Truman, L. C., G. W. Bennett, and W. L. Butts. 1976. *Scientific Guide to Pest Control Operations,* 3rd ed. Purdue University/Harvest, Cleveland, Ohio, 276p.

Tsuji, H. 1965. Studies on the behaviour pattern of feeding of three species of cockroaches, *Blattella germanica* (L.), *Periplaneta americana* L., and *P. fuliginosa* S., with special reference to their responses to some constituents of rice bran and some carbohydrates. *Jap. J. Sanit. Zool.* **16:**255-262.

Tsuji, H. 1966. Attractive and feeding stimulative effect of some fatty acids and related compounds on three species of cockroaches. *Jap. J. Sanit. Zool.* **17:**89-97.

Tsuji, H., and S. Ono. 1969. Laboratory evaluation of several bait factors against the German cockroach, *Blattella germanica* (L.). *Jap. J. Sanit. Zool.* **20:**240-247.

Tsuji, H., and S. Ono. 1970. Wide application of baits against field populations of the German cockroach, *Blattella germanica* (L.). *Jap. J. Sanit. Zool.* **21:**36-40.

Tyler, P. S. 1964. Kepone bait for the control of resistant German cockroaches. *Int. Pest Control* **6:**10-11, 13.

Wagner, R. E. 1983. Effects of Amdro fire ant insecticide mouind treatments on southern California ants, 1982. *Insectic. Acaric. Tests* **8:**257.

Wakeland, C., and H. Waters. 1931. *Controlling the Firebrat in Buildings by Means of Poisoned Bait,* Bulletin 185. Agricultural Experiment Station, University of Idaho, 14p.

Weaver, J. E., J. W. Begley, and V. A. Kondo. 1984. Laboratory evaluation of alsystin against the German cockroach (Orthoptera: Blattellidae): Effects on immature stages and adult sterility. *J. Econ. Entomol.* **77:**313-317.

Weesner, F. M. 1965. *The Termites of the United States: A Handbook.* National Pest Control Association, Elizabeth N.J.

Wileyto, E. P., and G. M. Boush. 1983. Attraction of the German cockroach, *Blattella germanica* (Orthoptera: Blattellidae), to some volatile food components. *J. Econ. Entomol.* **76:**752-756.

Williams, D. F., and C. S. Lofgren. 1983. Effectiveness of EL-468, and AC 217,300 for control of imported fire ants, 1980. *Insectic. and Acaric. Tests* **8:**258.

Williams, D. F., C. S. Lofgren, W. A. Banks, C. E. Stringer, and J. K. Plumley. 1980. Laboratory studies with nine amidinohydrazones: A promising new class of bait toxicants for control of red imported fire ants. *J. Econ. Entomol.* **73:**798-802.

Wilson, G. R., and M. J. Booth. 1981. Pharaoh ant control with IGR in hospitals. *Pest Control* **49:**14-16, 74.

Wood, F. E., W. H. Robinson, S. K. Kraft, and P. A. Zungoli, 1981. Survey of attitudes and knowledge of public housing residents towards cockroaches. *Bull. Entomol. Soc. Am.* **27:**9-13.

Wright, C. G. 1965. Identification and occurrence of cockroaches in dwellings and business establishments in North Carolina. *J. Econ. Entomol.* **58:**1032-1033.

Wright, C. G., and R. C. Hillman. 1973. German cockroaches: Efficacy of chlorpyrifos spray and dust and boric acid powder. *J Econ. Entomol.* **66:**1075-1076.

Wright, C. G., and D. Stout. 1978. For better Pharaoh ant control. *Pest Control* **46:**26, 28, 32.

Wright, C. G., H. C. McDaniel, H. E. Johnson, and C. E. Smith. 1973. American cockroach feeding in sewer access shafts on paraffin baits containing propoxur or Kepone plus a mold inhibitor. *J. Econ. Entomol.* **66:**1277-1278.

15

Public Health Pest Management

Michael J. Sinsko
Indiana State Board of Health

The control of arthropods affecting people has been of concern to mankind since the beginning of the human race. Historical accounts of battles with pestiferous insects and the devastating effects of vector-borne disease outbreaks begin with the earliest written records. These have been summarized by a number of authors (Harwood and James, 1979). The historical record continues, however, with each new addition to our knowledge of the basic biology and ecology of medically important arthropods and the complex disease transmission cycles of which many are a part.

The challenge presented by the need for management of arthropods over populated areas is great. It requires of the control specialist every bit of biological and sociological skill at his disposal. It also increases with the growing size and population density of our urban areas. Public health officials have always been concerned about the effects of toxicants used for the control of arthropods in populated areas. They have pioneered the development of safer chemicals and alternative control techniques (National Academy of Sciences, 1976). Much of the expertise used in the development of alternatives such as sanitation and source management was developed by the U.S. military before and following World War II (Coates and Hoff, 1955). The simultaneous use of a variety of methods (chemical, physical, and biological) for the control of disease vectors therefore represents the application of the principles of integrated pest management (IPM) for years before the phrase was coined.

Arthropods of medical importance comprise a large and diverse group. Control of some members of this group is best handled on a personal or a household basis. Others, which occur over large geographic areas, threatening entire communities, may often be managed more reasonably by public health or other governmental authorities. The scope of this chapter will be limited to those groups that are best handled by community-wide management programs.

TICKS

Harwood and James (1979) have eloquently listed the reasons for the high-vector potential of ticks. These reasons included the fact that they are persistent bloodsuckers, are slow feeders, have a wide range of hosts, are long-lived, exhibit transovarial transmission of pathogens, are relatively free from predation and environmental stress, and have extremely high capacities for reproduction. It is not surprising therefore that they have been incriminated as vectors of two major diseases that have been reported from both rural and urban areas of the United States.

The first of these, Rocky Mountain Spotted Fever (RMSF) is a rickettsial disease that was originally described from the Rocky Mountain states. In recent years, however, the distribution of the majority of human cases has ranged from the midwest to the southeastern seaboard (Burgdorfer, 1975; Hattwick, O'Brien, and Hanson, 1976). The primary vectors in these areas are *Dermacentor variabilis,* the American dog tick, and *Amblyomma americanum,* the lone star tick.

Traditional control methods for ticks have employed the use of acaracides to tick-infested areas, habitat modification, control of animal hosts, and educational campaigns. Studies recently indicate, however, that spraying with acaracides may be less effective in the control of adult *A. americanum* on cattle than habitat modification and animal control (Meyer, Lancaster, and Simco, 1982). G. A. Mount (1983), however, found that applications of acaracides to wooded habitats resulted in significant reductions of *A. americanum* larvae.

This approach may work in areas that have been identified as a known harborage for vector tick populations on a limited basis. Educational programs still serve as the one main method to reduce disease prevalence statewide (Loving et al., 1978). Generally, these programs inform citizens in high-risk areas of the value of rapid and proper tick removal, prevention of attachment through the proper use of clothing and tick repellents, and the need to remember tick attachment episodes in the event of the onset of illness (Burgdorfer, Adkins, and Priester, 1975; Centers for Disease Control, 1978*a*). Information should also be provided to the owners of pets, since dogs have been found to harbor both the rickettsiae and the ticks, which can be brought into the home (Keenan et al., 1977; Magnarelli et al., 1982).

A recent case involving the application of management techniques for tick control in an urban area has been reported by S. W. Gordon (1983). An urban area close to downtown Columbus, Ohio, had a long history of RMSF. Thirty-eight cases with four fatalities had been reported within 17 years. The Columbus City Department of Health initiated a program of weed control, quarantine of dogs, public education, tick surveillance, and the use of acaracides for tick control to combat the problem.

The second major tick-borne disease that has been reported from both rural and urban settings is Lyme disease. This disease, based on numbers of human cases reported in the last few years (Centers for Disease Control, 1984; Minnesota Department of Health, 1984) shows the potential for becoming the most prevalent vector-borne disease in the United States. The agent of the disease, a spirochete, *Borrelia burgdorferi* has only recently been described (Burgdorfer et al., 1982). It is vectored by the bear tick, *Ixodes dammini* (Steere and Malawista, 1979), the distribution of which appears to be spreading by as of yet unexplained mechanisms. *I. dammini* has also been identified as the vector of human babesiosis on Nantucket Island (Spielman et al., 1979). In addition to the bear tick, however, *Amblyomma americanum*, the lone star tick, may also be involved as a vector in the southern part of the United States (Schulze et al., 1984).

With the recent identification of Lyme disease as a significant threat to public health, many research institutions and health agencies are working on the description of the natural history of its cycle. At this time, much still needs to be done in the identification of the occurrence, spread, natural history, and control of this disease.

In addition to RMSF and Lyme disease, H. Hoogstraal (1981) has identified several other tick-borne diseases that have been diagnosed in the United States. His account relates changes in the epidemiology of these diseases to the evolution of human societies from sylvan rural to suburban and urban settings. Since this evolution remains constant, the identification of other major tick-borne diseases is possible.

MOSQUITOES

Mosquitoes are, as a group, the most notorious carriers of known disabling and fatal diseases. They are the main vectors of four species of human malaria, two or more species of Filarioidea, and 80 arboviruses which are known at the very least to elicit a human antibody response in nature (Mattingly, 1973). The diversity of pathogens transmitted by mosquitoes is only surpassed by the large numbers of genera and species of Culicidae that have been implicated as vectors. The vectors of human disease, however, represent only a fraction of the species of mosquitoes in the world. Many of these have been the objects of control because of the tremendous impact they have as nuisance biters. Indeed, vast sections of Florida were uninhabitable prior to the establishment of organized mosquito management programs (Gaiser, 1980).

Historically, efforts to control mosquitoes revolved around malaria and the association people made between its prevalence and impounded water. Epidemics that occurred in the 1870s, long before the implication of mosquitoes as vectors, were viewed with respect to surface water retention. Often, control involved the removal of these impoundments (U.S. Public Health Service and Tennessee Valley Authority, 1947).

The years following the confirmation of mosquitoes as vectors of malaria, filiariasis, and yellow fever at the turn of the nineteenth century were filled with efforts at control. The use of kerosene as a larvacide was described by Howard in 1882 (Harwood and James, 1979). Removal of breeding sites continued to be the most appropriate method to control mosquitoes. Eventually, a few other chemicals such as paris green were utilized in areas where impounded water could not be removed.

The availability of DDT and other chlorinated hydrocarbons following World War II revolutionized mosquito control. The eradication of diseases such as malaria, which had plagued the human race for centuries, was thought to be possible. By the 1960s, however, the limitations to the use of chemicals became apparent (World Health Organization Expert Committee on Malaria, 1979). The vast array of research that has been conducted on mosquitoes and mosquito-borne disease since that time has in many respects been a response to the perceived need to broaden control strategies.

As a general rule, the proper control of mosquitoes the last few decades has taken on a universal pattern. Target mosquito populations are identified to species in a particular geographic area, and larval breeding sites are located and mapped. These sites are then either removed or modified to provide an inhospitable environment for the larvae. If these physical alterations cannot be accomplished immediately, biological or chemical control agents are used as a temporary expedient to reduce the larval population. Adult control (adulticiding), often employing an insecticide in a thermal fog or as a ultra low volume (ULV) application is used to reduce adult populations as a supplement to larval control. Mosquito population densities are then monitored, often with the use of light traps or oviposition traps, to determine the effectiveness of control. This methodology is described well in manuals such as the excellent publication by the Centers for Disease Control on the control of mosquitoes of public health importance (Pratt, Darsie, and Littig, 1976). These general concepts are then altered by a number of management programs to fit the constraints provided by different mosquitoes, habitats, and disease cycles in various parts of the country (Beams and Challet, 1984; Breeland and Mulrennan, 1983; Sutherland, 1985).

Current trends in many management programs involve the use of specific control techniques that are designed on the basis of research on the bionomics of target mosquito species. As researchers piece together the intricate elements of complex disease transmission cycles, such as the mosquito-borne encephalitides, the rationale for management of vector populations only to the point of reduction of pathogen transmission becomes apparent.

A prime example of this approach is outlined by J. E. Parry (1983). He describes the use of research conducted on the natural history of *Aedes triseriatus* and the LaCrosse encephalitis virus cycle to determine a management strategy to control this disease in LaCrosse, Wisconsin. This strategy involved the filling of all treeholes and removal of artificial containers in endemic areas to eliminate breeding sites. Oviposition traps were used to remove *A. triseriatus* eggs. Educational programs to inform the public of the disease and its natural cycle were conducted. Adulticiding with ULV equipment was used to supplement other management strategies when monitoring indicated a problem.

The results of this effort show a decline of human LaCrosse Encephalitis cases between 1978 and 1982. This is an excellent example of a rational program, designed on the basis of current research, to accomplish a specific management goal without concurrent damage to the environment or nontarget organisms.

Similar publications are available to summarize research and outline specific control strategies for other target mosquitoes and transmission cycles. These include St. Louis encephalitis (Centers for Disease Control, 1976; Monath and Reeves, 1980), dengue (Centers for Disease Control, 1979*a*), *Aedes aegypti* (Centers for Disease Control, 1979*b*), and Western equine encephalitis (Centers for Disease Control, 1978*b*).

Current advances in mosquito control research involve the continued search for specific biological control agents. For many years *Gambusia affinis affinis* had been used as the sole biological control agent by a number of programs. This species is the mosquito fish, a highly efficient predator on aquatic immature stages. Problems involving adaptability and lack of compatibility with nontarget aquatic life have hastened the quest for other natural enemies. Reviews of the biological control of mosquitoes by H. C. Chapman (1974, 1985) excellently summarize the work that has been done to date with predators, competitors, pathogens, and the use of genetic manipulation on mosquitoes. The results of these efforts in practical terms are relatively sparse (Laird, 1985). The only biological control agents that are readily available to management programs on a commercial basis are the mosquito fish and *Bacillus thuringiensis israelensis*. Several formulations of the latter are available from three different manufacturers. The potential for a related microbial agent, *Bacillus sphaericus* also appears promising.

Advances have also been made in physical control through environmental management and manipulation (World Health Organization, 1982). The engineering technology required for this approach is advancing with the experience gained by many water management projects throughout the world.

The use of chemical agents has in recent years come under greater public and scientific scrutiny. The efficiency of time-honored application techniques such as adulticiding with ground-level ULV equipment has similarly been seriously questioned. Several studies have shown that the level of control seen through the use of this method against a variety of target species has been less than had been desired (Leiser, Beier, and Craig, 1982; Parsons, 1982; Strickman, 1979). For every publication questioning the value of this technique, however, several others appear confirming its efficacy against a variety of mosquito species using a number of chemical agents in varied geographical locations. The answer to these questions may lie not only in the ability to effectively monitor the effects of adulticiding on natural populations but in a greater understanding of the limitations of ULV application under various environmental and atmospheric conditions. In the meantime many management programs will continue to adulticide, accepting the technique for its value while keeping in perspective its relative efficiency as opposed to larval control (Sutherland, 1983).

The chemical industry, under pressure from a concerned public, has been working to produce safer agents and innovative formulations. Some examples of these are monomolecular surface films and methoprene.

The development of monomolecular surface films is a step in the direction of effective products with minimal toxic effects that can be used in environmentally sensitive areas. These films are drowning agents that affect mosquito larvae, egg rafts, and ovipositing and emerging adults by eliminating surface tension on the breeding site (Levy et al., 1982). Products such as methoprene, the synthetic insect growth regulator, or the various *Bacillus thuringiensis israelensis* formulations can similarly be used in environmentally sensitive areas such as coastal salt marshes.

The selection of an appropriate product, however, still has to be made with the constraints of water quality, target species, and environmental characteristics in mind. As safe as these products may be under certain conditions, such as high levels of organic pollution or dense aquatic vegetative cover, some physiological toxins may be more effective yet safe when properly used.

Finally, modern corporate management techniques are being utilized by sophisticated mosquito control agencies. The size and complexities of many control programs dictates the use of concepts such as the systems approach

(Walker, 1980) and computerization of survey and monitoring data (Russo and McCain, 1979). The increased efficiency derived from methods such as these coupled with a constant awareness of advances in mosquito research and the flexibility to try new approaches plus a willingness to vary management schemes to accommodate these changes can only serve to increase the effectiveness of these programs.

YELLOWJACKETS

Yellowjacket problems have intensified in many urban areas of the country during the last several years. This may, to some degree, be associated with the westward movement of *Vespula germanica*, the German yellowjacket. This species has been introduced several different times into the northeastern part of the country but did not become firmly enough established to migrate westward until 1968 (MacDonald, Akre, and Keyel, 1980).

The problems that have been reported from urban areas due to the German yellowjacket in this country and other parts of the world (Davis, 1978) are a direct result of the affinity this species shows toward nesting in human structures and eating our foods. The number of stings, therefore, increases because of the close association of scavenging workers with people.

MacDonald's (1976) evaluation of control strategies, unfortunately, do not provide much hope for the area-wide reduction of this species. While possible directions for future research, such as the use of pheromones, may provide an eventual effective control strategy, control efforts on the part of public health authorities will have to involve education. The dissemination of information to citizens regarding the safe elimination of individual colonies can be extremely valuable in the reduction of the risk to homeowners in the immediate vicinity of the colony. Control information for both ground and structural nesting colonies is provided in the scholarly publication by R. D. Akre and associates (1980). Information is also provided regarding the steps that may be taken to reduce the attractiveness of a limited area to scavenging yellowjacket workers.

RED IMPORTED FIRE ANTS

Red imported fire ants (RIFA) have created significant health problems in localized areas in the southern part of the United States. Studies conducted in Georgia by Adams and Lofgren document the impact of stings on residents (Adams and Lofgren, 1981, 1982).

The importance of two species, *Solenopsis invicta* and *S. richteri* resulted in the spread of *S. invicta* to a large area extending from the Carolinas to Texas (Lofgren, Banks, and Glancey, 1975). The use of pesticides for control over wide areas using a baited formulation of mirex produced a controversy regarding safety. This product was later banned by the U.S. Environmental Protection Agency in 1971.

The bibliography by Wojcik and Lofgren (1982) is a helpful introduction to the voluminous literature recently published on the biology and control of the fire ant. Little can be found in these articles, however, that promises immediate relief from the fire ant problem. A review of biological control efforts (Jouvenaz, Lofgren, and Banks, 1981) is a similar listing of research on various biological control agents that have been screened with no positive results.

In summary, RIFA poses another potential problem for health officials for which there is no rational approach other than the use of commercially available baits for the control of individual colonies.

HEAD LICE

Human head louse cases within the last several years have reached proportions unheard of since World War II. The exact magnitude of the problem is not known, since pediculosis is not a reportable disease. In other words, physicians are not required to report cases to public health authorities (Graham, 1979). The amount of concern reflected by school officials, news media, and health departments, however, indicates that this problem has reached epidemic proportions throughout the country.

Head lice comprise a group of insects that would normally be handled solely by the person affected through the care of a physician. The mass hysteria that has besieged communities, though, dictates that public health departments play a role in the coordination of the management of the problem on a community level. This role is appropriate, since efforts to gain control over the situation require cooperation and understanding by all community members and organizations.

Infestations of head lice are often initially identified in elementary schools. School nurses periodically conduct screening programs to assess the prevalence of pediculosis at various grade levels. Studies conducted by the Centers for Disease Control have highlighted several important points about the dynamics of *Pediculosis capitis* in the modern school system (Juranek, 1977):

1. Prevalence is higher in whites than blacks. The reasons for this are unknown.
2. Hair length does not appear to be a factor, although females had a tendency to be infested at a higher rate than males.
3. Children at younger grade levels had a higher prevalence than older children.

4. Groups of children who shared lockers at schools had higher prevalence rates than those with assigned wall hooks. This suggests coats and hats as a mechanism for transmission.
5. At least one other household member was found to be infested in 59% of the cases in which a child was found positive.

These data coupled with the basic biology of the head louse provide important factors in determining the actions that should be taken to reduce an outbreak of head lice in a community. Of utmost importance are the facts that schools can play a large role in reducing transmission in their buildings by providing adequate opportunity for the separation of students' hats and coats. Screening programs might be more sensitive to infestation within the school system by concentrating more on the lower grades. Finally, it is important that when a child is found to be infested, all other family members are examined for the presence of lice. The degree of association between infested school children and other family infestations illustrates the possibility of the maintenance of a reservoir in the home if only the child is treated.

Screening programs as the basic tool for identifying the existence and scope of pediculosis have had occasional problems with misdiagnosis. One episode involved a school district that had misdiagnosed an outbreak of head lice on the basis of pseudonits which were mistaken for head louse nits (Gemrich et al., 1974). Pseudonits, or hair casts, the origins of which are uncertain (Kohn, 1977), can easily be mistaken for head louse nits if screening is done by persons unfamiliar with nits without the aid of suitable optical equipment and microscopic examination.

Once an outbreak is identified, the proper management of the problem involves several procedures:

1. Treatment of individuals infested with one of several pediculicides that are available as either prescription or over-the-counter products.
2. Inspection of evidence of treatment before the children are readmitted to school.
3. Notification of parents within the school system to be on the alert for head lice.
4. Education of parents and the public about pediculosis and how it can best be managed.
5. Separation of childrens' outer winter garments by the school system.
6. Examination of family members of infested individuals and treatment of any found to be positive.
7. Periodic rescreenings to determine effectiveness of the control strategy.
8. All patients retreated with pediculicide 10 days after the initial treatment.

It is important that at no place in this strategy is the application of pesticides to the environment indicated. This is an important consideration to keep in mind through the hysteria that often follows the identification of a pediculosis problem in a school. Frequently, immediate demands are made by the public to spray or fumigate schools and school buses. The rationale behind recommendations not to spray is the fact that head lice are obligatory ectoparasites with a narrow range of temperature tolerance. They do not survive well in any stage off the head of the host (Keh, 1979).

Misinformation regarding the basic biology of head lice often fuels the irrationality that accompanies outbreaks. Public health authorities can play a significant role by disseminating factual information on the basic life cycle of the head louse and reasons behind management strategies. Citizens who believe that lice can jump or fly or that the spraying of pesticides will solve the problem need to become the targets of rigorous health education campaigns before this community-level problem can be solved. Many informational publications have been developed by health agencies (Keh, 1979; Slonka and McKinley, 1975). Similar brochures are frequently available from any public health agency in the country.

PROBLEMS AND PROSPECTS FOR THE FUTURE

The management of arthropods of public health importance has developed greatly in the last several years. Our knowledge of the basic biology of many of our vectors has been increasing at a phenomenal rate. Disease transmission cycles that have been a mystery are becoming better understood. Problems such as malaria and filth flies in urban areas in this country have been controlled. The fact remains, however, that many of these problems still plague urban areas in other parts of the world.

The emphasis during the last 20 years on the search for biological control agents has yielded discouragingly few positive results. The failures of control strategies such as genetic manipulation of mosquito populations because of the inability of treated or laboratory-reared mosquitoes to compete with wild populations emphasizes one fact. We still do not know enough about the basic field biology of our target organisms.

We are quickly realizing that the interactions of wild populations are so complex that not only do we have difficulty measuring various ecological parameters but we have failed to identify all the parameters in need of measurement. The resulting dilemma faced by researchers is the difficulty of measuring the effects of a perturbation on a system if all the interactions within the system are not understood. Hence, controversies such as that over the efficacy of ULV application for mosquito control rage on because we are unable to even accurately measure the effects of the treatment on the wild, free-flying target population.

Basically, we need to go back to the study of the natural history of our disease vectors, transmission cycles, and other medically important arthropods that have been able to withstand our best control efforts. Perhaps when several missing pieces in our array of knowledge of these organisms are found, we will have a chance to spot the weaknesses in our red imported fire ants and German yellowjackets, and the mechanism of spread of *Ixodes dammini* will be understood.

The trend of the chemical industry to develop innovative products that are biologically compatible with nontarget organisms is gratifying. The concern of the public in recent years over the unknown chronic effects of exposure to toxic products dictates that this trend should continue.

Finally, public health officials have often been frustrated with the fact that the control of many of our problems should be relatively simple. This is particularly true in the case of vector mosquitoes, which breed in artificial containers or other man-made situations. Since we are dealing with a pest system in which the affected organism is an intelligent being, the possibilities for control in some situations should be boundless. We are, however, failing greatly in our efforts to solicit the help of our fellow citizens. Health education has often been said to be the greatest challenge in public health and also the key to a great number of solutions. It is time we consider the human factor as one of the elements of integrated pest management.

REFERENCES

Adams, C. T., and C. S. Lofgren. 1981. Red imported fire ants (Hymenoptera: Formicidae): Frequency of sting attacks on residents of Sumter County, *Georgia J. Med. Entomol.* **5:**378-382.

Adams, C. T., and C. S. Lofgren. 1982. Incidence of stings or bites of the red imported fire ant (Hymenoptera: Formicidae) on other arthropods among patients at Ft. Stewart, Georgia, USA. *J. Med. Entomol.* **19:**366-370.

Akre, R. D., A. Greene, J. F. MacDonald, P. J. Landolt, and H. G. Davis. 1980. *Yellowjackets of America North of Mexico,* Agriculture Handbook No. 552. U.S. Department of Agriculture, Washington, D.C., 102p.

Beams, F. B., and G. L. Challet. 1984. California's Orange County offers well-rounded vector control program. *Pest Control* **52:**18-24.

Breeland, S. G., and J. A. Mulrennan, Jr. 1983. Florida mosquito control: State's program is not a matter of chance. *Pest Control* **51**(2): 16-24.

Burgdorfer, W. 1975. A review of Rocky Mountain spotted fever (tick-borne typhus), its agent, and its tick vectors in the United States. *J. Med. Entomol.* **12:**269-278.

Burgdorfer, W., T. R. Adkins, Jr., and L. Priester. 1975. Rocky Mountain spotted fever (tick-borne typhus) in South Carolina: An educational program and tick/rickettsia survey in 1973 and 1975. *Am. J. Trop. Med. Hyg.* **24:**866-872.

Burgdorfer, W., A. G. Barbour, S. F. Hayes, J. L. Benach, E. Grunwaldt, and J. P. Davis. 1982. Lyme disease: A tick-borne spirochetosis? *Science.* **216:**1317-1319.

Centers for Disease Control. 1976. *Vector Control Topics No. 1: Control of St. Louis Encephalitis.* U.S. Department of Health, Education and Welfare, Atlanta, Georgia, 35p.

Centers for Disease Control. 1978*a. Ticks of Public Health Importance and Their Control,* Publication CDC 78-842. U.S. Department of Health, Education, and Welfare, Atlanta, Georgia, 37p.

Centers for Disease Control. 1978*b. Vector Topics No. 3: Control of Western Equine Encephalitis.* U.S. Department of Health, Education, and Welfare, Atlanta, Georgia, 35p.

Centers for Disease Control. 1979*a. Vector Topics No. 2: Control of Dengue.* U.S. Department of Health, Education and Welfare, Atlanta, Georgia, 39p.

Centers for Disease Control. 1979*b. Vector Topics No. 4: Biology and Control of* Aedes aegypti. U.S. Department of Health, Education, and Welfare, Atlanta, Georgia, 68p.

Centers for Disease Control. 1984. Update: Lyme disease, United States. *Morbidity and Mortality Weekly Rep.* **33:**268-270.

Chapman, H. C. 1974. Biological control of mosquito larvae. *Annu. Rev. Entomol.* **19:**33-59.

Chapman, H. C., ed. 1985. *Biological Control of Mosquitoes,* Bulletin No. 6. American Mosquito Control Association, Fresno, Calif., 218p.

Coates, J. B., Jr., and E. C. Hoff. 1955. *Preventive Medicine in World War II,* vol. 2, *Environmental Hygiene.* Department of the Army, Washington, D.C. 404p.

Davis, H. G. 1978. Yellowjacket wasps in urban environments. In *Perspectives in Urban Entomology,* G. W. Frankie and C. S. Koehler (eds.). Academic Press, New York, pp.163-185.

Gaiser, D. 1980. The importance of mosquito control to tourism in Florida. *Proc. Florida Anti-Mosq. Assoc.* **51:**7-8.

Gemrich, E. G., J. G. Brady, B. L. Lee, and P. H. Parham. 1974. Outbreak of head lice in Michigan misdiagnosed. *Am. J. Public Health* **64:**805-806.

Gordon, S. W. 1983. Current status of Rocky Mountain spotted fever and the tick testing program in Ohio. In *Proceedings, 7th Annual Meeting,* R. Russo (ed.). Indiana Vector Control Association, pp. 36-39.

Graham, B., ed. 1979. Pediculosis: A heady problem for many public health officials. *Alabama's Health* **12:**4-5.

Harwood, R. F., and M. T. James. 1979. *Entomology in Human and Animal Health,* 7th ed. Macmillan, New York. 548p.

Hattwick, M. A. W., R. J. O'Brien, and B. F. Hanson. 1976. Rocky Mountain spotted fever: Epidemiology of an increasing problem. *Ann. Intern. Med.* **84:**732-739.

Hoogstraal, H. 1981. Changing patterns of tickborne diseases in modern society. *Annu. Rev. Entomol.* **26:**75-99.

Jouvenaz, D. P., C. S. Lofgren, and W. A. Banks. 1981. Biological control of imported fire ants: A review of current knowledge. *Bull. Entomol. Soc. Am.* **27:**203-208.

Juranek, D. D. 1977. Epidemiological investigations of pediculosis capitis in school children. In *Scabies and Pediculosis,* H. I. Maibach and L. S. Parish (eds.). J. B. Lippincott, Philadelphia, pp. 168-173.

Keenan, K. P., W. C. Buhles, Jr., D. L. Huxsoll, R. G. Williams, P. K. Hildebrandt, J. M. Campbell, and E. H. Hall. 1977. Pathogenesis of infection with *Rickettsia rickettsii* in the dog: A disease model for Rocky Mountain spotted fever. *J. Infect. Dis.* **135:**911-917.

Keh, B. 1979. Answers to some questions frequently asked about pediculosis. *Calif. Vector Views.* **26:**51-62.

Kohn, S. R. 1977. Hair casts or pseudonits. *J. Am. Med. Assoc.* **238:**2058-2059.

Laird, M. 1985. Conclusion. In *Biological Control of Mosquitoes,* Bulletin No. 6, H. C. Chapman (ed.). American Mosquito Control Association, Fresno, Calif., pp. 216-218.

Leiser, L. B., J. C. Beier, and G. B. Craig. 1982. The efficacy of malathion ULV spraying for urban *Culex* control in South Bend, Indiana. *Mosq. News* **42:**617-618.

Levy, R., J. J. Chizzonite, W. D. Garrett, and T. W. Miller, Jr. 1982. Efficacy of organic surface film isostearyl alcohol containing two oxyethylene groups for control of *Culex* and *Psorophora* mosquitoes: Laboratory and field studies. *Mosq. News.* **42:**1-11.

Lofgren, C. S., W. A. Banks, and B. M. Glancey. 1975. Biology and control of imported fire ants. *Annu. Rev. Entomol.* **57:**695-698.

Loving, S. M., A. B. Smith, A. F. Di Salvo, and W. Burgdorfer. 1978. Distribution and prevalence of spotted fever group rickettsiae in ticks from South Carolina, with an epidemiological survey of persons by infected ticks. *Am. J. Trop. Med. Hyg.* **27:**1255-1260.

MacDonald, J. F., R. D. Akre, and R. E Keyel. 1980. The German yellowjacket (*Vespula germanica*) problem in the United States (Hymenoptera: Vespidae). *Bull. Entomol. Soc. Am.* **26:**436-442.

MacDonald, J. F., R. D. Akre, and R. W. Matthews. 1976. Evaluation of yellowjacket abatement in the United States. *Bull Entomol. Soc. Am.* **22:**397-401.

Magnarelli, L. A., J. F. Anderson, R. N. Phillip, and W. Burgdorfer. 1982. Antibodies to spotted fever-group rickettsiae in dogs and prevalence of infected ticks in southern Connecticut. *Am. J. Vet. Res.* **143:**656-659.

Mattingly, P. F. 1973. Culicidae (Mosquitoes). In *Insects and Other Arthropods of Medical Importance,* K. G. V. Smith (ed.). British Museum, Natural History, London, pp. 37-107.

Meyer, J. A., J. L. Lancaster, Jr., and J. S. Simco. 1982. Comparison of habitat modification, animal control, and standard spraying for control of the lone star tick. *J. Econ. Entomol.* **75:**524-529.

Minnesota Department of Health. 1984. Lyme disease: An emerging public health problem. *Dis. Control Newsletter.* **11:**21-24.

Monath, T. P. and W. C. Reeves, eds. 1980. *St. Louis Encephalitis.* American Public Health Association, Washington, D.C. 680p.

Mount, G. A. 1983. Area control of larvae of the lone star tick (Acari: Ixodidae) with acaricides. *J. Econ. Entomol.* **76:**113-116.

National Academy of Sciences. 1976. *Pest Control: An Assessment of Present and*

Alternative Technologies, vol. 5, *Pest Control and Public Health.* National Academy Press, Washington, D.C., 282p.

Parry, J. E. 1983. Control of *Aedes triseriatus* in LaCrosse, Wisconsin. In *California Serogroup Viruses,* C. H. Thompson and W. H. Thompson (eds.). Alan R. Liss, New York, pp. 355-363.

Parsons, R. E. 1982. Effect of ground ultra low volume insecticide applications on natural mosquito populations. *J. Florida Anti-Mosq. Assoc.* **53:**31-35.

Pratt, H. D., R. F. Darsie, Jr., and K. S. Littig. 1976. *Mosquitoes of Public Health Importance and Their Control.* U.S. Department of Health, Education and Welfare, Centers for Disease Control, Atlanta, Georgia, 68p.

Russo, R. J., and T. L. McCain. 1979. The use of computerized information retrieval in mosquito control. *Mosq. News.* **39:**333-338.

Schulze, T. L., G. S. Bowen, E. M. Bosler, M. F. Lakat, W. E. Parkin, R. Altman, B. G. Ormistan, and J. K. Shisler. 1984. *Amblyomma americanum:* A potential vector of Lyme disease in New Jersey. *Science* **224:**601-603.

Slonka, G. F., and T. W. McKinley. 1975. *Controlling Head Lice.* U.S. Department of Health, Education and Welfare, Centers for Disease Control, Atlanta, Georgia, 16p.

Spielman, A., C. M. Clifford, J. Piesman, and M. D. Corwin. 1979. Human babesiosis on Nantucket Island, USA: Description of the vector, *Ixodes (Ixodes) dammini,* N. sp. (Acarina: Ixodidae). *J. Med. Entomol.* **15:**218-234.

Steere, A. C., and S. E. Malawista. 1979. Cases of Lyme disease in the United States: Locations correlated with the distribution of *Ixodes dammini. Ann. Intern. Med.* **91:**730-733.

Strickman, D. 1979. Malathion as ground applied ULV evaluated against natural populations of *Culex pipiens* and *Cx. restuans. Mosq. News.* **39:**64-67.

Sutherland, D. J. 1983. The value and efficacy of adulticiding in New Jersey. *J. Florida Anti-Mosq. Assoc.* **54:**41-43.

Sutherland, D. J. 1985. New Jersey battles mosquitoes from several different fronts. *Pest Control* **53:**30-36.

U.S. Public Health Service and Tennessee Valley Authority. 1947. *Malaria Control on Impounded Water.* Washington, D.C., 422p.

Walker, G. M., Jr. 1980. A systems approach to program control: A method for program development. *Misc. Publ. Entomol. Soc. Am.* **11:**50-61.

World Health Organization. 1982. *Manual on Environmental Management for Mosquito Control,* WHO Offset Publication No. 66. Geneva, 283p.

World Health Organization Expert Committee on Malaria. 1979. *Seventeenth Report,* WHO Technical Report Series No. 640. Geneva, 71p.

Wojcik, D. P., and C. S. Lofgren. 1982. Bibliography of imported fire ants and their control: First supplement. *Bull. Entomol. Soc. Am.* **28:**269-276.

Index

About the Editors

GARY W. BENNETT is a professor in the Department of Entomology at Purdue University. He is Extension Coordinator for the department, heads a research program on German cockroach ecology and management, and is responsible for the undergraduate teaching program in urban and industrial pest control. Other responsibilities include chairing the annual Purdue University Pest Control Conference and directing Purdue's correspondence course in pest control technology. He is the author of two books—*Scientific Guide to Pest Control Operations* and *Practical Insect Pest Management: Pests of Man's Household and Health*. He received the Ph.D. from North Carolina State University.

JOHN M. OWENS is an entomology group leader in the Corporate Research Division of S. C. Johnson & Son, Inc., Racine, Wisconsin. He directs worldwide entomology testing on Johnson Wax's Raid™ household crawling insect control products. He received the B.S. in natural resources from the University of Rhode Island and the M.S. and Ph.D. in urban and industrial entomology from Purdue University. He is a member of the Entomological Society of America, the American Registry of Professional Entomologists, Pi Chi Omega (National Pest Control Fraternity) and Thomas Say Entomological Society. He has authored or coauthored numerous papers on entomology.